BIOS INSTANT NOTES

Microbiology

FOURTH EDITION

BIOS INSTANT NOTES

Microbiology

FOURTH EDITION

Simon Baker
Department of Biological and Medical Sciences
Oxford Brookes University

Caroline Griffiths
Department of Biological and Medical Sciences
Oxford Brookes University

Jane Nicklin
Department of Biological Sciences
School of Science
Birkbeck College, University of London

GS Garland Science
Taylor & Francis Group
NEW YORK AND LONDON

Garland Science
Vice President: Denise Schanck
Editor: Elizabeth Owen
Editorial Assistant: Louise Dawnay
Production Editor: Ioana Moldovan
Copyeditor: Sally Livitt
Typesetting and illustrations: Phoenix Photosetting, Chatham, Kent
Proofreader: Jo Clayton
Printed by: MPG Books Limited

ISBN 978-0-4156-0770-4

Library of Congress Cataloging-in-Publication Data
Baker, Simon, 1966-
    Microbiology / Simon Baker, Jane Nicklin, Caroline Griffiths. — 4th ed.
        p. ; cm. — (BIOS instant notes)
    Rev. ed. of: Microbiology / S. Baker ... [et al.]. 3rd ed. 2007.
    Includes bibliographical references and index.
    ISBN 978-0-415-60770-4 (pbk.)
    1. Microbiology—Textbooks.   I. Nicklin, J. (Jane) II. Griffiths, Caroline.
    III. Microbiology. IV. Title. V. Series: BIOS instant notes.
    [DNLM: 1. Microbiology—Outlines. 2. Microbiological Processes—Outlines.
    QW 18.2]
    QR41.2.B34 2011
    579—dc22   2011003161

Published by Garland Science, Taylor & Francis Group, LLC, an informa business,
270 Madison Avenue, New York NY 10016, USA, and 2 Park Square, Milton Park, Abingdon,
OX14 4RN, UK.

15 14 13 12 11 10 9 8 7 6 5 4 3 2 1

GS Garland Science
Taylor & Francis Group

Visit our web site at http://www.garlandscience.com

Preface

The Fourth Edition of *Instant Notes in Microbiology* builds on the changes made in the third edition, and this edition integrates more fully with other books in the series. We have chosen to omit sections on the biochemistry of central metabolic pathways and general DNA metabolism (which can be found in much more detail in other *Instant Notes* texts). Notes on the biochemistry of the Bacteria and Archaea have been expanded to compensate, but for the moment we have chosen to avoid detailed discussions on the next generations of molecular techniques. While high throughput genome sequencing, micro-arrays, and other nucleic acid techniques have a huge impact on microbiology, we felt that these are general methods that would be better discussed in other texts. We also decided to remove the section on Bacterial infections, since again this is better dealt with in more detail in a companion *Instant Notes* text. However, we have still referred to human pathogenesis in all sections, as micro-biology cannot be divorced from medicine entirely. This is especially true of the virology section, which has been updated to reflect changes in nomenclature and practice in the field.

Once again, the authors have chosen to use the word Bacteria to mean those members of the Kingdom Bacteria, and not in its older usage to denote non-eukaryotic microbes. For this we have used 'prokaryotes' to include both the Bacteria and the Archaea. Our prediction in the third edition that microbial taxonomy would once again be overhauled by the publication of this edition was incorrect with respect to the prokaryotes, but instead we have made major changes to the sections on the eukaryotes to reflect the current thinking in tax-onomy. We have also undertaken a major revision of the classification of the fungi reflecting the great advances made in the subject area. It seems likely that this branch of the tree of life will receive further attention over the next few years.

The authors would like to thank the members of the Department of Biologi-cal and Medical Sciences at Oxford Brookes University and the Department of Biological Sciences at Birkbeck College for help in the preparation of this book. As ever, we would also like to thank our friends and families for their continu-ing support, as well as those who gave us valuable feedback on how we could improve the previous edition.

Contents

A1 The microbial world

Key Note

Microorganisms are found in all three major kingdoms of life: the Bacteria, the Archaea, and the Eukarya. The presence of a nucleus defines the eukaryotes, while both the Bacteria and Archaea can be defined as prokaryotes. Apart from the nucleus, there are many physiological and biochemical properties distinguishing the prokaryotes from the eukaryotes.

What are microbes?

Microbes are a diverse group of organisms that can be divided into the viruses, unicellular groups (Archaea, Eubacteria, protista, some fungi, and some chlorophyta), and a small number of organisms with a simple multicellular structure (the larger fungi and chlorophyta). These larger microorganisms are characterized by having a filamentous, sheet-like or parenchymous thallus that does not display true tissue differentiation. Most microbes cannot be seen without the aid of a microscope.

Microbiology

Microbiology is defined as the study of microorganisms. The discipline now includes their molecular biology and functional ecology as well as the traditional studies of structure and physiology. The discipline began in the late 17th century with Leeuwenhoek's discovery of bacteria using simple microscopy of mixed natural cultures. Through the 1850s and 60s, Louis Pasteur's simple experiments using sterilized beef broth finally refuted the long-held theory of spontaneous generation as an origin for microbes, and microbiology moved into mainstream science.

The early days were characterized by studying environments like soil and sediments, natural fermentations and infections, and it was not until Robert Koch developed techniques for pure culture in the late 19th century that the science moved to a reductionist phase, where microbes were isolated and characterized in the laboratory.

Through the 20th century microbiologists focused on the discovery and characterization of many different microorganisms, including a new kingdom of microorganisms (the Archaea), new eubacterial pathogens including *Legionella* and MRSA (methicillin-resistant *Staphylococcus aureus*), and the complex of pathogens associated with HIV (human immunodeficiency virus) including the fungal pathogen *Pneumocystis*. The discovery of the unique communities found in extreme environments with their temperature-tolerant DNA (deoxyribonucleic acid) polymerase enzymes has further opened up the new field of molecular biology.

The rapid advances of techniques in molecular biology have allowed microbiology to return to the natural environments. Techniques such as DGGE (denaturing gradient gel electrophoresis), SCCP (single-stranded conformation polymorphism), DNA chips, and *in situ* hybridization now give us the tools to study microbial community ecology at the molecular level. Microbiology has returned to its roots!

Bacteria, Archaea, and Eukaryotes

The microbial world has three main cell lineages within it, all of which are thought to have evolved from a single progenitor (Figure 1). The lineages are formally known as domains and were established from the DNA sequence of genes common to all organisms (Section B3). The three domains are the Bacteria (previously called the Eubacteria), the Archaea (previously called the Archaeabacteria), and the Eukarya. The defining property of the Eukarya compared to the Archaea and Bacteria is the presence of a nucleus. It is convenient to group the annucleate lineages (the Bacteria and Archaea) together as the prokaryotes. The prokaryotes are, with a very few exceptions (Section C6), all microorganisms, but the Eukarya include not only microbial fungi, chlorophyta, and protists (Sections H and I) but also macroorganisms such as higher plants and animals.

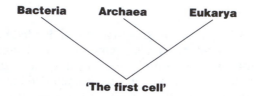

Figure 1. The three cell lineages evolved from a common ancestor.

Prokaryotic cell structure is characterized by the absence of a nucleus, but it also lacks energy-generating organelles such as mitochondria and chloroplasts. Instead, prokaryotes generate energy by cytoplasmic substrate-level phosphorylation and oxidative phosphorylation across their cell membranes (Section E). Apart from these major differences, there are a multitude of distinctive biochemical and physiological properties, the most important of which are listed in Table 1. The differences that exist between the Bacteria and the Archaea (Section C6) are discussed elsewhere in more detail.

Subjects covered in this volume

In this edition of the book, we have decided not to discuss the pathogenicity of bacteria and have focused on aspects of systematics, physiology, and biochemistry. Bacterial pathogens are the best studied among microorganisms due to their importance to humankind, and a detailed description of the biology of disease can be found in *Instant Notes in Medical Microbiology*.

Systematics

Bacterial systematics allows the microbiologist to name, classify, and identify Bacteria and Archaea in a rational way. The importance of molecular biology to microbiology is emphasized by the prominence of 16S rRNA (ribosomal ribonucleic acid) sequencing in microbial phylogeny.

General microbiology

Microbiology as a science has had a long and varied history, but we are only just beginning to appreciate the full ecological, biochemical, and genetic diversity of microbes.

Table 1. Some of the major differences between the prokarya and the eukarya

Prokaryotes	Eukaryotes
Organization of genetic material and replication	
DNA free in the cytoplasm	DNA contained in a membrane-bound structure
Chromosome – frequently haploid, single, and circular	More than one chromosome – frequently diploid and linear
DNA complexed with histone-like proteins	DNA complexed with histones
May contain extrachromosomal DNA as part of the genome	Organelles may have separate chromosomes, but rarely find free DNA in the cytoplasm
Cell division by binary fission or budding	Cells divide by mitosis
Transfer of genetic information can occur via conjugation, transduction, and transformation	Transfer of genetic information can only occur during sexual reproduction
Cellular organization	
Reinforcement of the cytoplasmic membrane with hopanoids (not Archaea)	Reinforcement of the cytoplasmic membrane with sterols
Cell wall made up of peptidoglycan and lipopolysaccharides or teichoic acids. Variety of cell wall constructions in Archaea	Often lack a rigid cell wall, but cell reinforced via a cytoskeleton of microtubules. Where cell wall is present it is made up of a thin layer of chitin or cellulose
Energy generation across the cell membrane	Energy generation across the membrane of mitochondria and chloroplasts
Internal membranes for specialized biochemical pathways (e.g. ammonia oxidation, photosynthesis)	Internal membranes in most cells (endoplasmic reticulum, Golgi apparatus, etc.)
Flagella made up of a single protein (flagellin)	Multi-protein flagella made up of a 9 + 2 arrangement of fibrils
Ribosomes are small (70S)	Ribosomes are larger (80S)

Sophisticated methods have been developed to measure the growth of microbes in the laboratory. Our understanding of microorganisms has improved so that we can now appreciate the fine structure of the prokaryotic cell, rather than just considering it as a bag of enzymes. The understanding of microbial cell division and movement has also led to important breakthroughs in eukaryotic biology. Although the microbiology of human disease is well studied, we are beginning to find that microbes play an essential global role in the biogeochemical cycling of the elements.

Microbial growth

The way in which most prokaryotic cultures divide in batch and continuous culture can be modeled mathematically to reveal the limitations imposed by laboratory conditions. From these models it can be shown that the design of any growth vessel should primarily optimize the oxygen requirements of the culture growing in it.

Molecular biology

Microbiology has always been intrinsic to advances in genetics, DNA metabolism, and *in vitro* and *in vivo* genetic manipulation across the whole of biology. The principles of DNA replication, transcription to mRNA and translation to protein have all been characterized in *Escherichia coli* in the first instance. Coupled to our detailed knowledge of the control of transcription and the mechanisms of DNA transfer between cells, it is now possible to use bacteria as powerful tools in recombinant DNA technology.

Eukaryotic microbes

A number of distinct groups of eukaryotic microbes are considered using the classification based on the Tree of Life website. The general cell biology and cell division of eukaryotic microbes is then described, followed by a more detailed consideration in separate sections of the structure, physiology, and reproduction of the Fungi, and the photosynthetic and the nonphotosynthetic protista. The beneficial and detrimental effects of each group of microbes on its environment are also examined, with a more extensive review of the taxonomy and pathogenicity of protistan parasites.

Viruses

Viruses are a hugely diverse group of sub-microbial agents that are active only when parasitizing a living cell (animal, plant, bacterium, etc.). We have selected medically relevant viruses to illustrate fundamental concepts within virology, beginning with virus structures and their molecular components. These are important taxonomically but also, as later sections illustrate, in the search for preventative vaccines and curative antiviral treatments. We review the common stages of the virus replication (propagation) cycle, highlighting some of the major variations, and describe the range of infectious outcomes at the level of the whole organism. We have included a short introduction to less well understood sub-microbial agents such as viroids, found within plant systems, and prions, which cause neurological diseases in mammals.

B1 Prokaryotic systematics

Key Notes

Classification and taxonomy	Classification is a method of organizing information. Microorganisms can be classified on their growth properties (e.g. chemolithotroph, denitrifier), but are formally classified using the Linnaean system. The full classification of a microorganism is its taxonomy. Microorganisms can be uniquely identified solely by the use of their genus and species names.
Identification of prokaryotes	Identification is the placing of new isolates into the taxonomic framework, normally to the level of genus and species. However, the definition of species is still less clear in the prokaryotes than it is in the higher eukaryotes.
Phylogeny of prokaryotes	The evolutionary relationship of a microorganism among and between taxa is its phylogeny. The reliance on DNA sequences to elucidate these relationships has led to the emergence of phylogenetics. The taxonomy of Bacteria and Archaea for the most part reflects their phylogeny.
Related topics	(B2) Identification of Bacteria (C2) Prokaryotic diversity (B3) Inference of phylogeny (C11) Prokaryotes and their from rRNA gene sequence environment

Classification and taxonomy

With the advent of molecular methods, the distinction between identification, classification, and evolutionary relationships has become blurred. In bacteriology, a **classification** is simply a method of organizing information. This organization may have an underlying meaning, or no meaning at all. We could choose to classify bacteria on the color their colonies have when grown on agar plates, a classification that would give prominence to the few yellow and red colored bacteria, while the majority would be classified in a group of white to cream colony forms. In early microbiology, organisms were classified according to shape, with the bacillus shape (now also called a rod) forming the largest group, and cocci, filaments, and so on smaller ones. It is important to stress that classifications were arbitrary; however, they do still have a use today. Microbiologists classify organisms according to their growth properties (anaerobe, chemolithotroph, methylotroph, etc., Sections C2 and D1).

The primary means of classification in microbiology, in common with the rest of biology, is the Linnaean system. This is a hierarchical system, with major divisions sequentially separated down to the lowest level species (Figure 1).

A full classification of *Escherichia coli* is Prokaryota (domain), Bacteria (kingdom), Proteobacteria (phylum), λ-proteobacteria (class), Enterobacteriales (order), Enterobacteriaceae (family), *Escherichia* (genus) and *coli* (species). Each of these levels is described as a **taxon** (plural **taxa**). This system of classification allows biologists a unique identification

Domain
Kingdom
Phylum
Class
Order
Family
Genus
Species

Figure 1. The full Linnaean system of classification.

across all the domains and kingdoms solely by using the appropriate genus and species. The full name of *E. coli* is its **taxonomy**, and the Linnaean system is a **taxonomic classification.**

In this text the correct domain and kingdom names are used to denote both the Archaea and Bacteria (prokaryotes). Slightly older systems give the kingdom names as Eubacteria (true bacteria) and Archaeabacteria, which in many ways is slightly more descriptive.

Identification of prokaryotes

With a taxonomy for prokaryotes in place, the microbiologist can now begin to place organisms within this framework, using the taxonomy to describe and identify species. When first purified, an organism is described as an **isolate** and is generally given a number that helps to distinguish it from others in the laboratory. Isolates may be genetically identical, but have been taken from their natural environments at separate times or geographical locations.

Normally it is uncontroversial to place a new bacterium in an appropriate genus, so this isolate may become, for example, *Paracoccus* **strain** NCIMB 8944. The **species concept** is more difficult to apply to Bacteria and Archaea than in the animal kingdom because of the enormous genetic diversity, so the subspecies definition of **biovar** is sometimes applied to organisms of the same species with slightly different properties; for example the plague organism, *Yersinia pestis*, is divided into biovars, including Orientalis and Mediaevalis. The differences between strains of the same species are sometimes defined by subspecies, for example the plant pathogen *Pectobacterium carotovorum* has subspecies including atroseptica and carotovorum. In this case the subspecies defines members of the same species that cause different plant pathologies. In microbiological research, the problem of species definition means that it is always best to leave the original isolate code in place when writing about an organism, so that its history in various laboratories can be traced.

Although the definition of the individual Bacterial and Archaeal taxa down to genus level has gained consensus among microbiologists, there has yet to be agreement on what constitutes the fundamental unit of biological diversity, the species. Species definition in one genus (e.g. sharing less than a certain percentage of DNA homology) does not necessarily hold true in another, a problem particularly true in taxa dominated by human pathogens (Section B2). If we cannot properly and consistently define a species, then identification down to species level becomes problematic. The methods used for prokaryotic identification, that is the way in which we assign a classification to our new isolate, are explored further in Section B2.

Phylogeny of prokaryotes

Phylogeny is a description of the evolutionary relationships among and between taxa. In microbiology there is a heavy reliance on DNA sequence rather than morphology (as seen in the plants and animals) as data for phylogeny and thus is normally referred to as **phylogenetics**. The classification of Bacteria and Archaea for the most part reflects their phylogeny as we currently understand it. There is, of course, always room for debate in any system and a lively discourse continues over the exact phylogenetic or taxonomic position of many species or genera.

B2 Identification of Bacteria

<table>
<tr><td colspan="2">Key Notes</td></tr>
<tr><td>Identification of Bacteria</td><td>Until recently, identification in microbiology laboratories was deduced from the biochemistry of the new isolate. However, sequencing data are used more frequently as a primary attribute when characterizing a new isolate, with physiology and biochemistry added in the later stages of formalizing the classification, if at all.</td></tr>
<tr><td>Identification from growth characteristics</td><td>The growth on media can be used to aid identification of isolates, in a selective or differential (diagnostic) manner. A numerical taxonomy can be built up by scoring the ability of the organism to grow on a range of sugars and its possession of key enzymes. The numerical taxonomy can then be used to identify the organism, either by consulting commercial libraries or by reference to Bergey's Manual of Systematic Bacteriology.</td></tr>
<tr><td>Other methods of identification</td><td>Microorganisms can be identified by the fatty acids that they produce under defined conditions (FAMEs – fatty acid methyl ester analysis), or by examination of their 16S rRNA sequence.</td></tr>
<tr><td>Identification of pathogens</td><td>Medical laboratories tend to rely more on biochemical classification from pure isolated culture rather than the molecular techniques used in nonclinical research. As PCR (polymerase chain reaction) becomes cheaper, more medical laboratories are beginning to use molecular techniques. The identification of pathogenic bacteria relies more on their disease-causing properties than their phylogeny, which leads to some anomalies in Bacterial classification.</td></tr>
<tr><td>Related topics</td><td>(B1) Prokaryotic systematics   (B3) Inference of phylogeny from rRNA gene sequence</td></tr>
</table>

Identification of Bacteria

Although molecular methods (Section B3) are becoming increasingly dominant in the identification and classification of bacteria, most identification work in clinical laboratories throughout the world is done using cheaper growth and biochemical methods. Furthermore, a full identification of any bacterium to publishable standard should include a polyphasic approach, i.e. including the characteristics of the strain determined molecularly, biochemically, and from growth studies. If bacteria can be grown from a clinical or environmental sample, the first step in identification or classification is growth studies, followed by an analysis of the enzymes that may be present in the strain, and lastly a molecular analysis of the genome.

Identification from growth characteristics

An unknown bacterial isolate may be subcultured on many different sorts of solid and liquid media to aid its identification. Broadly these media fall into two types: **selective media** allow the growth of one type of bacterium while inhibiting that of others; **differential** or **diagnostic media** usually contain some sort of visual indicator, a change in which is linked to a unique biochemical property of a group of microorganisms (Section C3). If the strain can be purified, then a Gram stain and an examination of morphology might be performed. Once an overall picture of the organism's growth has been obtained, the ability to use sugars (and if acid is produced during their use) and an assessment of the possession of certain enzymes is performed in detail. These properties can be combined and scored against the known properties of other organisms to form a basic **numerical taxonomy**. Kits can be obtained commercially that semi-automate this process, but are limited to certain groups of Prokaryotes, particularly the enterics. With most environmentally isolated, nonpathogenic Bacteria and Archaea, guidance must be sought from the standard reference text for identification, *Bergey's Manual of Systematic Bacteriology*.

Other methods of identification

If an organism can be grown under the same conditions as many other reference strains, other methods of identification are also available to the microbiologist. The lipids of a pure culture can be extracted, esterified, and then quantified by GC (gas chromatography). This FAMEs method then requires the GC trace to be compared by computer against other organisms grown in exactly the same way on the same medium. This is a rapid and inexpensive procedure, but interpretation can be difficult and is unsuitable for microorganisms that grow under unusual conditions. The FAME profile of any organism alters depending on the medium used for growth. An adaptation of this method is phospholipid-linked fatty acid analysis (PLFA), which is a more specific and sensitive technique.

As DNA sequencing becomes easier and cheaper, it has also become standard practice to complement the results from biochemical and physiological tests with the results of 16S rRNA gene sequencing (Section B3). Although this practice blurs the distinction between identification and phylogeny further, the results can be obtained quickly and easily, but may be misleading if only the sequencing data are considered.

Identification of pathogens

Medical microbiology has, to some extent, fallen behind the rest of microbiology in its approach to the identification of disease-causing organisms. Due to the costs involved, routine identification of pathogens is still carried out by the classic bacteriological methods. This would normally entail the isolation or enrichment of bacteria from a clinical specimen using broth or agar, procurement of a pure culture from the primary culture, identification of the bacterium by microscopy, growth characterisation, and perhaps PCR. The reliance on the culture of pathogens means that some common pathogens have been overlooked. For example, for many years the existence of *Campylobacter* as the most frequent causative agent of food poisoning was not known because of the difficulty in cultivating this genus in the laboratory. As the cost of PCR continues to fall, medical microbiology will embrace such concepts as **viable but nonculturable** (meaning they can be detected but cannot be cultured in the laboratory), which are currently accepted in disciplines such as environmental microbiology, with a concurrent increase in the reliability of diagnosis.

Pathogen identification and naming are driven by patient symptoms rather than the overall properties of the microorganism. For example, the enteric genera are all very closely related on a genetic level, but cause a variety of human diseases (Table 1). It is now becoming clearer from examination at the molecular level that the pathogenic members of the Gram-positive genus *Bacillus* (Table 2) may even be the same species, but with different sets of plasmids.

Table 1. Disease caused by species of the enteric bacteria

Genus	Disease
Escherichia	Enteropathogenic diarrhea
Shigella	Shigellosis
Salmonella	Typhoid fever, gastroenteritis
Vibrio	Cholera, gastroenteritis
Klebsiella	Pneumonia
Yersinia	Plague

Table 2. Disease caused by closely related species of *Bacillus*

Species	Disease
B. subtilis	Nonpathogenic
B. anthracis	Anthrax
B. cereus	Gastroenteritis
B. thuringiensis	Insect pathogen

B3 Inference of phylogeny from rRNA gene sequence

Key Notes

Bacterial phylogeny

The recent advent of molecular phylogenetic methods has made phylogeny more accessible to the laboratory microbiologist.

The molecular clock concept

The changes in DNA or protein sequence over long periods of time can be used to measure overall evolutionary change from a common ancestor. The evolutionary chronometer chosen should be universally distributed, functionally homologous, and possess sequence conservation. To be able to distinguish between rapid and slow periods of change, the molecule chosen should also have regions of conservation and hypervariability.

Ribosomal RNA (rRNA)

Cytochrome *c* has been suggested as a suitable chronometer, but 16S rRNA sequence has gained widespread acceptance. The size is convenient for most sequencing protocols, but some doubts remain as to the validity of some phylogenies. About 1.4 million Bacterial rRNA sequences have been entered in to the Ribosomal Database Project.

Acquisition of 16S rRNA gene sequence

16S rRNA genes from many different organisms are amplified using universal primer sets, although currently no primer set has been found capable of amplifying every single known species. In mixed populations, primer sets can sometimes generate false results due to the formation of chimeric PCR products from more than one template.

16S rRNA gene bioinformatics

Once a 16S rRNA PCR product has been amplified and sequenced, it must be placed in the context of its phylogenetic relationships with other sequences. It is first aligned against similar sequences and then clipped so that all the data in the alignment are the same length. Phylogenetic trees can then be constructed by neighbor joining or maximum parsimony methods. These methods can still generate many different trees from the same dataset, so bootstrapping is used to assign confidence levels to the existence of each branch of the tree.

Related topics

(B1) Prokaryotic systematics   (B2) Identification of Bacteria

Bacterial phylogeny

The relatedness of Bacteria to one another is discussed elsewhere in this book (Section C6) in the context of diversity. Most of the major phyla (Gram-positive, *Cyanobacteria*, and so on) have been deduced from classic taxonomic methods and have been in place for several decades. The more recent advent of **molecular phylogenetic methods** has made phylogeny more accessible to the laboratory microbiologist.

The molecular clock concept

A microbiologist should be able to place any organism in the context of its relationship to other organisms and its evolution from a common ancestor. To be able to do this an evolutionary clock must be identified, which reflects small changes in the organism over time. Cellular macromolecules, such as proteins and nucleotides, have the potential to act as **evolutionary chronometers**, but to be ideal they must meet the following criteria:

- **Universally distributed** – i.e. present in all known (and presumably yet to be discovered) organisms.

- **Functionally homologous** – i.e. the molecule must perform the same action in all organisms. Molecules with different functions could be expected to become too diverse to show any relevant sequence similarity.

- **Possess sequence conservation** – i.e. an ideal chronometer should have regions of sequence that are highly **conserved** and thus expected to change only very slowly over long periods of time coupled with other regions with moderate variability or **hypervariability** to illuminate more recent changes.

Many macromolecular chronometers have been proposed, including cytochrome *c* (Section E2), ATPase (Section E2), RecA (Section F9), and 16S/18S rRNA. Most of the protein chronometers have failed to satisfy the universality requirement, particularly when examining the extremophilic Bacteria and Archaea that lack a conventional electron transport chain. To date, the most widely used chronometer is 16S rRNA of Bacteria and Archaea along with its 18S rRNA equivalent in Eukarya.

Ribosomal RNA (rRNA)

The small, medium, and large rRNA molecules are all ideal chronometers. They have been found to perform the same function in all known organisms and have regions of conservation as well as hypervariability. The interaction of RNA with ribosomal proteins means that rRNA buried deep within the protein structure is less likely to change, as any change must also be reflected in the protein sequence. Changes in either ribosomal protein or rRNA in these regions that lead to an unstable ribosome are lethal to the organism and do not persist in subsequent generations. However, many parts of the rRNA molecules do not have any direct interaction with the ribosomal proteins, and so can accumulate mutations much more easily, i.e. are hypervariable.

Of the three rRNA molecules available in prokaryotes (5S, 16S, and 23S, see Section F5), 16S rRNA provides the ideal balance between information content (5S is too short) and ease of sequencing (23S is too long). The attractiveness of the molecule is shown by the number of entries in the **Ribosomal Database Project**, standing at 1 483 016 Bacterial sequences for release 10 (v23) in December 2010. There are a few drawbacks to the use of the molecule, primarily that many Bacteria have more than one copy of the 16S rRNA gene on their genome, frequently with a different sequence. This can cause confusion and dispute, depending on which sequence is used, emphasizing the need for a poly-phasic approach.

Acquisition of 16S rRNA gene sequence

DNA fragments containing all or part of the 16s rRNA gene are generally obtained by PCR. The primers are designed to anneal to the conserved regions within the gene and sometimes this enables the use of one primer set to amplify 16S from many phyloge-netically diverse bacteria (**a universal primer set**). However, no one set of primers can amplify all the genes from all the Bacteria and all the Archaea, and many primer sets have been designed that are phylum- or group-specific. The use of combinations of these sets means that most known microorganisms can be amplified from any environmental, pure culture or mixed culture.

Although acquisition of sequence by PCR is quick, there are limitations imposed by the technique itself. PCR can generate **chimeras**, PCR products that are composed of the 5′ end of one species' gene coupled to the 3′ end of another. Although computer programs exist to eliminate these false sequences from the final results, it is sometimes difficult to detect them when dealing with rare or undiscovered organisms. The existence of the division *Korarchaeota* in the kingdom Archaea was in doubt for precisely this reason. Ten years of research into how to cultivate the organism confirmed that it did indeed form a deeply branched division of the Archaea.

16S rRNA gene bioinformatics

Once a 16S PCR product has been amplified and sequenced, it must be placed in the context of its phylogenetic relationships with other sequences. A preliminary idea of the close relatives can be gained by the use of an **alignment** to one or more known sequences. A program such as BLAST (National Center for Biotechnology Information, US National Library of Medicine) can do this, although it is extremely limited in detail and will only give information relating to one other sequence at a time. To gain a true idea of phylog-eny, the 16S rRNA gene sequence should be compared to as many other sequences as possible simultaneously. Only 10 years ago this was impossible to achieve on anything but a supercomputer; however, recent advances in computing power now mean that most personal computers can carry out some or all of this process. Many web-based pro-grams also allow free access to the more powerful computers that may be needed.

To begin with, the newly acquired sequence must be aligned with all or some of the sequences obtained previously. As there is some variation in length of 16S rRNA genes, gaps must be inserted to achieve a perfect alignment, though this can be done by pro-grams such as CLUSTAL (European Bioinformatics Institute (EBI)). The aligned sequences are then **clipped** so that the 5′ and 3′ ends are equivalent bases and the alignment is sent to a program capable of generating **phylogenetic trees**.

An ideal representation of phylogeny would be multidimensional, but given the con-straints of our three-dimensional universe in general, and the scientific predilection for presentation in two-dimensional form on paper in particular, the 'tree' is a good compro-mise. Two main algorithms are used: **neighbor joining** and **maximum parsimony**. Neigh-bor joining is an evolutionary distance method, based on a matrix of differences in the dataset. The resulting tree has branches of lengths proportional to evolutionary distance, statistically corrected for back mutation. Maximum parsimony is a more difficult concept to grasp, in that the resulting tree has branches whose length is proportional to the mini-mum amount of sequence change necessary to enable the creation of a new branch.

For both methods, it is possible to generate trees differing in details, such as the number of branches, from one dataset. A process known as **bootstrapping** is applied to get an idea of the sum of all the possible trees, and this gives a confidence value for the presence of each branch. In addition, neighbor joining and parsimonious trees generated from the same dataset can give quite different results, and to date neither method is considered to be more 'right' than the other. Thus any tree should be considered as the best possible result with the data available, and should not necessarily over-rule any other information.

C1 Discovery and history

Key Notes

The history of bacteriology	Robert Hooke (1660) and his contemporary Antonie van Leeuwenhoek are considered to be the first microbiologists, but the theory of spontaneous generation did not allow for the existence of microorganisms to be placed in their true context. However, Pasteur's swan-necked flask experiments (1861) showed that food-spoilage organisms were microscopic and airborne. Cohn (1875) founded the science of bacteriology, and was followed by other late Victorian scientists, such as Koch, Beijerinck, and Winogradsky. Bacteria became the model system for biochemistry throughout the 20th and early 21st centuries, culminating in the sequencing of the *Haemophilus influenzae* genome by Venter and colleagues in 1995.
Major subgroups	The Bacteria were first subdivided by use of the Gram stain, but now are separated into many phyla. The Bacteria are a separate kingdom from the Archaea, and each of their phyla contains species with broad nutritional, physiological, and biochemical properties.
Related topics	(B1) Prokaryotic systematics (C2) Prokaryotic diversity (C6) The major prokaryotic groups

---

The history of bacteriology

It was not until the development of the first microscopes (by the Janssen brothers around 1590) that microbes were observed as minute structures on surfaces. Robert Hooke began showing the fruiting structures of molds around the courts of Europe and published the first survey of microbes (*Micrographia*) in 1660. The first person to observe prokaryotes microscopically was Antonie van Leeuwenhoek in 1676. He published his observations on these 'animicules' to the Royal Society of London. However, the theory of spontaneous generation stopped much further investigation, since it claimed that the intervention of divine power led to the spontaneous creation of molds and other spoilage organisms (including mice) should food be left unattended. The belief that living organisms could arise from otherwise inert materials began around the time of Aristotle (384–322 BC) and eventually this theory was disproved (but not entirely discarded) by the experiments of Pasteur in 1861 using swan-necked flasks. These allowed the preservation of beef broth for long periods, with spoilage only initiated once an airborne, invisible contaminant was reintroduced into the broth.

Ferdinand Cohn is credited with founding the science of bacteriology, proposing a morphological classification for bacteria and using the term '*Bacillus*' for the first time in 1875. This was soon followed by the seminal work by Koch over the period between 1876 and 1884, with his system for firmly establishing the link between bacteria and disease (Koch's postulates). Martinus Beijerinck developed the technique of enrichment culture, establishing the first pure culture of *Rhizobium* in 1889, a year before Winogradsky demonstrated the link between oxygen and nitrification in bacteria. Beijerinck also went on to found the science of virology, while working for a company producing yeast in the Netherlands, establishing the strong link between microbiology and biotechnology. By the turn of the 20th century, the first microbiology journal had been published (the American Society for Microbiology's *Journal of Bacteriology*, still in publication today), and many of the organisms we study now were named, although not necessarily classified as they are today. Bacteria became the model systems for biochemistry throughout the 20th and early 21st centuries, culminating in the sequence of the genome of the first free-living organism (*Haemophilus influenzae*) being decoded by Venter and colleagues in 1995.

Major subgroups

The work of the early bacteriologists (Cohn, Koch, and Beijerinck) was focused on identification and classification of bacteria (Section B1). However, this work was based around the morphology (gross shape) of the organism, and since most bacteria are rod-shaped when viewed under the microscope (Section C2) and form white or cream colonies on agar plates, the success of this approach is limited. The development of the Gram stain in 1884 by the Danish physician Hans Christian Gram, allowed the separation of the prokaryotes into two classes, which was later found to be based around cell wall structure. Gram-positive bacteria retained the purple dye (crystal violet) while the Gram-negative bacterial cell wall allowed the stain to be washed away. The Gram-positive Bacteria have remained a valid taxonomic group (Section C6), but the Gram-negative Bacteria have been proved to be more phenotypically and genetically diverse, and for a while included members of the Archaea.

Although we can now separate the Bacteria from the Archaea and further subdivide the Bacteria on the basis of phylogeny, many physiological properties are broadly spread throughout the kingdom. Bacteria with the capability of causing a human pathology are not restricted to any single subgroup, nor are the abilities to denitrify, grow anaerobically or photosynthesize. The majority of Bacterial species have unique properties, which may be very different from members of the same genus, but superficially similar to those that could be considered to be phylogenetically very distinct.

C2 Prokaryotic diversity

Key Notes

Morphological diversity

One of the most common shapes for Bacteria and Archaea is the rod or bacillus. Shape is dictated to some extent by the method of cell division, most commonly binary fission, although cocci, vibroid, spiral, filamentous, and even star-shaped prokaryotes have been isolated. The smallest bacteria are spheres only 0.3 μm in diameter, while the largest is 0.75 by 0.25 mm. Large Bacteria are composed mostly of gas vesicles and seem to have little more cytoplasmic content than *E. coli*.

Habitat diversity

Prokaryotes can grow at extremes of temperature, pH, oxygen, and radiation dosage, and can be named after their ability to grow under certain conditions (e.g. heat preferential bacteria are thermophiles). Prokaryotes can grow at 4°C and below, while the highest recorded growth temperature is 96°C (Bacteria) or 110°C (Archaea). Organisms growing at the limits of life are known as extremophiles.

Related topics

(C11) Prokaryotes and their environment

Morphological diversity

The model organism used in bacteriology, *Escherichia coli*, is a rod-shaped organism, about 3 μm in length and 1 μm in diameter. Many very different Bacteria have this rod-like morphology, sometimes called a bacillus. Bacilli can be found in all the taxonomic groups of the Bacteria, as well as in the Archaea. This means that, unlike the classification of higher organisms, shape is not a reliable characteristic when classifying prokaryotes (Table 1), even though it is one of the few visible differences between cells when using a light microscope. However, the shape of some microorganisms has had an influence on the naming of some prokaryotes. The Gram-positive Bacterial genus *Bacillus* is, of course, made up of rod-shaped species, while another Gram-positive genus, *Streptococcus*, is made up of species of spherical bacteria (or cocci) about 1 μm in diameter.

Bacteria and Archaea are considered to be independent single-celled microorganisms dividing by a process of binary fission (Section C9) and so both mother and daughter cells are the same size after division. Most bacteria are of simple, regular shape (spherical, cylindrical, etc.) but a few exceptions, such as the star-shaped bacterium *Stella humosa*, exist. Some bacteria, such as the Actinomycete *Streptomyces coelicolor*, grow as multinucleate filaments in which the concept of a discrete cell becomes more esoteric.

Many prokaryotes form smaller cells when stressed, and these are thought to be resting stages. The smallest vegetative prokaryotes are the marine ultramicro bacteria, such as *Sphingopyxis alaskensis*, whose diameter is less than 0.3 μm and has an estimated cellular volume of less than 1 μm^3. The mycoplasmas and chlamydia are the smallest bacteria found colonizing humans – *Mycoplasma genitalium* is only 0.2 by 0.3 μm. Although these

Table 1. Morphology and classification of selected prokaryotes

Shape (morphology)	Singular	Plural	Examples	Classification
Spherical or ovoid	Coccus	Cocci	Streptococcus pneumoniae	Bacteria – Gram-positive – Lactobacillales
			Deinococcus radiodurans	Bacteria – Gram-variable – Deinococcus/Thermus
			Neisseria meningitidis	Bacteria – Gram-negative – β-proteobacteria
			Desulfurococcus fermentans	Archaea – Crenarchaeota – Thermoprotei
Cylindrical	Rod	Rods	Escherichia coli	Bacteria – Gram-negative – γ-proteobacteria
			Bacillus subtilis	Bacteria – Gram-positive – Bacilli
			Methanobacterium oryzae	Archaea – Euryarchaeota – Methanobacteria
3D comma	Vibrio	Vibrio	Vibrio cholerae	Bacteria – Gram-negative – γ-proteobacteria
Curved rod	Spirillum	Spirilla	Spirillum pleomorphum	Bacteria – Gram-negative – β-proteobacteria
Filamentous			Streptomyces coelicolor	Bacteria – Gram-positive – Actinobacteria
Star			Stella humosa	Bacteria – Gram-negative – α-proteobacteria
Flat square or box-like			'Haloquadratum walsbyi'	Archaea – Euryarchaeota – Halobacteria

seem tiny compared with eukaryotic cells (2–200 μm) the theoretical size limit of life was calculated by the US National Academy of Sciences Space Study Board to be a spherical cell of 0.17 μm. This minimum size assumes the presence of a minimum genomic complement plus the ribosomes and other subcellular components required for existence. Although 'nanobacteria' of about this size have been observed in nature, it is now thought that these are produced by crystallization, despite the presence of membranes bounding the particles.

The largest known prokaryote is the sulfur bacterium *Thiomargarita namibiensis*, which is 750 μm in length and has a diameter between 100 and 250 μm. Close relatives of this bacterium can form filaments several centimeters long. These and Bacteria such as *Epulopiscium fishelsoni* (a surgeonfish symbiont, 600 μm long and 75 μm at its widest) can just be seen with the naked eye. Under the light microscope, organisms such as *Bacillus megaterium* (4 × 1.5 μm) and the cyanobacterium *Oscillatoria* (50 × 8 μm) seem unusually large. These large bacteria have little more cytoplasm than a single *E. coli*, and most of the cellular volume is made up of vacuoles, since diffusion is the main method of movement of all prokaryotic substrates and metabolites. All the cytoplasm of a large cell must be close to a cellular membrane, so the vacuole structure allows the organism to grow to larger sizes.

Habitat diversity

Prokaryotes can grow at extremes of temperature, pH, oxygen, and radiation dosage. This has allowed them to colonize all parts of the earth that may provide the cells with substrates for growth. The way in which microorganisms grow has given us a means of describing them, according to their ability to use certain compounds for growth (-trophy) or their tolerance of physiochemical conditions (-phily). Thus, an organism that can use many organic compounds to grow can be described as a heterotroph, and if it can tolerate atmospheric oxygen concentrations it will be called an aerophile. The range of descriptions is given in Table 2.

Bacteria and Archaea have been found growing in snow, while the upper temperature limit of bacterial life appears to be around 96°C (seen in *Aquifex aeolicus*). However, the Archaea hold the record for the highest growth temperature, with *Pyrolobus fumarii* growing at above 113°C. Such high temperatures are not found on terrestrial earth, but are restricted to deep-sea hydrothermal vents. Broadly speaking, the Archaea have more representative species growing at very high or very low temperature and pH, and are commonly held to be predominantly **extremophiles**.

Table 2. Commonly used description of microorganisms

Growth requirement	Description	Adjective	Example
Carbon from organic compounds	Heterotroph	Heterotrophic	*Escherichia coli*
Carbon from carbon dioxide	Autotroph	Autotrophic	*Paracoccus denitrificans*
Carbon from C1 compounds	Methylotroph	Methylotrophic	*Methylobacterium extorquens*
Energy from light	Phototroph	Phototrophic	*Rhodobacter sphaeroides*
Energy from inorganic compounds	Lithotroph	Lithotrophic	*Acidithiobacillus ferrooxidans*
Energy from organic compounds	Heterotroph	Heterotrophic	*Escherichia coli*
Molecular nitrogen	Diazotroph	Diazotrophic	*Rhizobium leguminosarum*
Atmospheric oxygen	Aerophile	Aerobic	*Pseudomonas aeruginosa*
Reduced levels of oxygen	Microaerophile	Microaerophilic	*Campylobacter jejuni*
Absence of oxygen	Anaerobe	Anaerobic	*Clostridium tetani*
High salt	Halophile	Halophilic	*Halobacterium salinarum*
High alkalinity	Alkaliphile	Alkaliphilic	*Bacillus pseudofirmus*
High acidity	Acidophile	Acidophilic	*Lactobacillus acidophilus*
Low temperature (< 0 to 12°C)	Psychrophile	Psychrophilic	*Psychromonas profunda*
'Normal' temperatures (10 to 47°C)	Mesophile	Mesophilic	*Escherichia coli*
Moderately high temperatures (40 to 68°C)	Thermophile	Thermophilic	*Geobacillus (Bacillus) stearothermophilus*
High temperatures (68 to 98°C)	Hyperthermophile	Hyperthermophilic	*Thermus aquaticus*
Very high temperatures (90 to above 110°C)	Extreme hyperthermophile	Hyperthermophilic	*Pyrodictium occultum*

The terms can be combined to describe more than one physiological property, e.g. the Archaea *Acidianus breyerli* can be described as a hyperthermophilic chemolithoautotroph. The term obligate can be used to denote those organisms that are restricted to one mode of growth, e.g. *Methanococcus capsulatus* can only grow on methane so is described as an obligate methanotroph.

C3 Culture of bacteria in the laboratory

Key Notes

Growth media

Bacteria can be grown on surfaces, or in aqueous suspension on solid, or in liquid media. Solid agar media are normally held in Petri dishes, and inoculated by streaking or plating with a loop or spreader. Inoculants may be mixed (composed of many species) or pure (composed of only one species). The process of inoculation while maintaining culture purity is called aseptic technique.

Defined media contain known amounts of simple chemical compounds, also known as minimal or synthetic media. Many organisms will not grow on such media without the addition of vitamins and trace elements, while auxotrophs require the addition of amino acids as well. Universal growth media are complex, containing compounds whose exact chemical formula is variable or uncharacterized. By changing the composition of a medium, they can be selective for a group of microorganisms, or even diagnostic for a genus or species.

Storage and revival of microorganisms

Bacteria and Archaea can be stored in a 50% glycerol solution for many years, and can also withstand freeze-drying in the lyophilization process. Short-term storage is accomplished by streaking out onto solid media. Cells are revived from storage by growth in a complex medium to reduce stress.

Sterilization

Pasteurization will kill many pathogens, but does not kill all Bacteria. Autoclaving is used to sterilize media and other apparatus used in microbiology, but the harsh temperatures mean that the equipment must be made of glass, steel, or polypropylene. Heat-sensitive media or apparatus can be tyndallized, UV (ultraviolet)-treated, gamma irradiated or filter sterilized if liquid.

Related topics

(C5) Looking at microbes
(D1) Measurement of microbial growth

(D2) Batch culture in the laboratory

Growth media

The microbiologist has many choices to make if wishing to grow microorganisms in the laboratory. The biomass introduced into the growth medium is known as the **inoculant**, and one of the first questions that need to be addressed is whether to **inoculate** onto **solid** or into **liquid** media. The medium will be chosen to reflect the origin of the inoculant, which might be a **mixed culture** of many different species of microorganism, or a **pure culture** of only one species. Solid media are normally held in circular sterile plastic containers with lids (**Petri dishes**), the solidity being provided by **agar**. The culture is **streaked** or **plated** onto the surface of the medium using a sterile wire or plastic **loop**, or a sterile glass **spreader**, respectively.

Microorganisms have an enormous metabolic diversity (Sections C2, C5 and E4), so require a medium made up of components to suit. Some organisms can synthesize all their cellular components from simple carbon and energy sources, such as glucose, and so will grow on a **minimal** (**synthetic/defined**) medium. Such a medium will contain sources of nitrogen, phosphate, sulfur, calcium, magnesium, potassium, and iron as inorganic salts, and may be supplemented with **trace elements**, such as zinc, manganese, boron, cobalt, copper, nickel, chromium, and molybdenum. These elements are required in minute quantities by the organism, and are used as prosthetic groups in some enzymes (e.g. alcohol dehydrogenase contains zinc). A truly minimal medium will not contain any **vitamins**, but organisms will often grow more quickly if provided with riboflavin, thiamine, nicotinic acid, pyridoxine HCl, calcium pantothenate, biotin, folic acid, and vitamin B_{12}. Again, these are provided in minute quantities, enough to be used during enzyme synthesis but not in sufficient amounts to act as a carbon and/or energy source.

When growing some **auxotrophs** (Section F3) or bacteria with a requirement for amino acids, it may be necessary to supplement the minimal medium with some or all of the 20 possible acids. This is common in pathogens, where a protected lifestyle in an animal host has led to the loss of one or all the enzymes involved in amino acid synthesis pathways. Identifying amino acid requirements is often time-consuming and costly, in which case a **complex medium** can be used.

Complex media are defined in the sense that absolute quantities of buffering ions are added to a solution, as well as known amounts of plant, animal, or yeast extracts. The common complex medium Lauria Bertoni broth (LB) contains 5 g l^{-1} NaCl, but carbon, energy, trace elements, and other growth factors are provided by 5 g l^{-1} 'yeast extract' and 10 g l^{-1} '**Tryptone**'. Tryptone is casein (milk solids) digested with pancreatic enzymes, so the exact composition in terms of molar concentrations of amino acids, short peptides, and so on is unknown and will vary between manufacturers. Similarly, **yeast extract**, a hydrolysate of baker's yeast (*Saccharomyces cerevisiae*) is of undefined composition. Pathogenic bacteria associated with bacteremia are frequently grown on a complex medium containing whole or partially hydrolyzed blood, which not only provides the essential growth factors but can also give an indication of the presence of hemolytic organisms by clear haloes in the blood red medium around colonies.

Liquid media can be placed in a variety of containers appropriate to the oxygen requirements of the organism to be grown. Facultative anaerobes and anaerobes can be grown in bottles, with gentle shaking to mix the culture, while the more commonly used aerophiles are generally grown in batch culture (Section D2) in **Erlenmeyer flasks**. These flasks are adapted chemistry apparatus, conical flasks between 5 ml and 5 l in volume. They are filled to 10% of the total volume so that the liquid medium provides sufficient surface area for oxygen transfer to the culture (Section D2). The inoculation of liquid and solid media and the transfer of cultures from one container to another without the ingress of contaminating organisms have become known as **aseptic technique**.

Media have been developed over the last 100 years in both composition and utility. **Selective** or **differential** media are of a composition that only allows the growth of one type or group of organisms. For example, a minimal medium containing methanol as a carbon and energy source will select for methylotrophs, those organisms able to use reduced C1 compounds. The medium can be enhanced further to be **diagnostic**. For example, solid **Baird Parker** medium will allow the growth of only a handful of genera, including *Micrococcus* and *Staphylococcus*, but only *Staphylococcus aureus* will grow as gray-black shiny colonies with a narrow white entire margin surrounded by a zone of clearing 2–5 mm. This 'egg yolk' colony form is used as a first indication of the presence of potential pathogens before more detailed tests are carried out.

Most microbiological media are adapted to the study of aerobes, but the true anaerobes, particularly those that are damaged by exposure to oxygen (some *Clostridia* and members of the Archaea), need special culture conditions. The total exclusion of oxygen is difficult, but a series of methods named after their inventor (the **Hungate** techniques) achieve this.

Storage and revival of microorganisms

The Bacteria and Archaea are remarkably resistant to extreme conditions and many can survive freezing or desiccation without ill effect, even in their vegetative states. This means that many, but by no means all, prokaryotes can be stored as pelleted biomass for decades. In the laboratory, long-term storage of biomass frozen at −70°C in a 50% glycerol solution is often used, although this can sometimes cause cell death. Freezing in dimethyl sulfoxide (DMSO) can be an alternative, but prokaryotes rarely require cryopreservation in liquid nitrogen as eukaryotic cells can do. Larger laboratories may purchase a freeze-drying apparatus to **lyophilize** cultures. This is by far the most efficient method of long-term storage. Short-term laboratory storage (for up to a week) is normally done by streaking the biomass out onto solid media held in a suitable container (Petri dish or 30 ml bottle). After storage, the cells are in a starved state and must be revived with a complex medium so as little stress as possible is placed on them.

Sterilization

Once a microbiological experiment has been completed, the live organisms should be safely destroyed. Similarly, before an experiment starts all living cells present should be inactivated so that only the inoculum desired is present. This poses problems for the microbiologist, particularly one working with pathogens, thermophiles, or sporulating bacteria. The process of pasteurization (71.7°C for 15 seconds) will kill many common human bacterial pathogens without affecting the medium a great deal and so is therefore used in food preparation. Moist heat in the form of steam or boiling will kill most vegetative cells as well as some viruses, but thermophiles and endospores will survive. **Autoclaving** contaminated equipment (121°C for 15 minutes, 15 psi (100 kPa) above atmospheric pressure) kills all cells as well as endospores, but is not suitable for use with polycarbonate (a common material for making plastic containers) and will cause many carbohydrate medium components to caramelize. Several methods (outlined below) have been used to deal with heat-sensitive components:

- **Tyndallization** – repetitive heating to 90–100°C for 10 minutes, followed by cooling for 1–2 days. Allows endospores to germinate in the medium, which are then killed by the heating.

- **Ultraviolet radiation** – kills living cells but does not penetrate opaque containers or large volumes of solution well.

- **γ radiation** – kills living cells but causes brittleness in polycarbonate and polypropylene.

- **Filtration of media** – 0.22 μm filters can be suitable for aqueous solutions of heat-labile chemical constituents, but it is difficult to filter large (> 500 ml) quantities effectively while maintaining sterility.

C4 Enumeration of microorganisms

Key Notes

Obtaining a pure culture

The pure culture of a single species of prokaryote is central to the interpretation of many microbiological experiments. This is normally done by streaking on an agar plate to obtain single colonies, or diluting with sterile media until it is possible to grow a culture from a single cell.

Counting prokaryotes

The number of bacteria per milliliter of sample is important for industrial, food, and medical standards, as well as in microbiological research. This can be expressed as a total count of all cells living and dead, or a viable count of those cells that can be expected to grow. Total counts can be estimated with a hemocytometer or via indirect methods such as quantitative PCR (qPCR), but viable counts present more of a problem.

Classically, viable counting is done by serial dilution, plating, and estimation of colony forming units (cfu) or the most probable number (MPN) method. These have been enhanced by biochemical and molecular techniques such as quantitative reverse-transcription PCR (qRT-PCR) and estimation of ATP (adenosine 5'-triphosphate). One of the most accurate ways of determining the ratio of living to dead cells is via the use of dyes and fluorescence-activated cell sorting (FACS).

Related topics

(C5) Looking at microbes

Obtaining a pure culture

The concept of the pure culture is central to classical bacteriology, and is a central tenet of Koch's postulates (see *Instant Notes in Medical Microbiology*). If an organism can grow on agar, it can be streaked to obtain single colonies, each of which should have arisen as a result of a single prokaryotic cell (Figure 1). This works by means of the **dilution** effect of each round of streaking and sterilization. The first inoculum onto the plate might transport millions of bacteria to the agar but, each time the loop is dragged across the plate, these cells are removed further from their neighbors. Coupled with sterilization of the loop, fewer than 10 cells may be present in the last line of streaking before incubation.

Many prokaryotic cells will not grow on agar plates. If this is found to be the case then a similar process of dilution is carried out in liquid broth to obtain a pure culture. One milliliter of the primary liquid culture is taken with a sterile pipette and added to 9 ml of fresh sterile medium (a dilution of 10, written as 10^{-1}). The 10^{-1} dilution is mixed well,

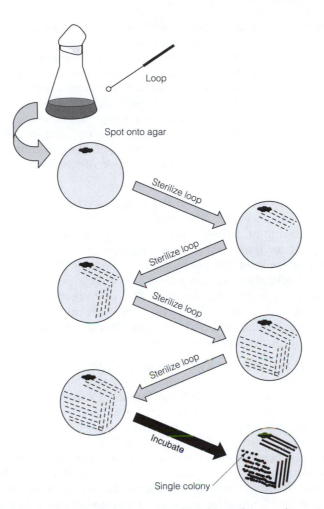

Figure 1. Diagram of the method of obtaining single colonies by streaking. Dashed lines show the path of the loop on the surface of the agar.

then 1 ml of that is removed and again added to 9 ml of fresh sterile medium (a dilution of 100, written as 10^{-2}). This process is repeated a further 10 or 12 times, and then the dilutions are incubated. The lower dilutions will show no growth, while the others will be turbid. The lowest dilution that shows growth is likely to have arisen from inoculation with less than 10 cells, so repetition of this process will eventually lead to a pure culture. This approach can be further enhanced by making 10-fold replicates of each dilution. This should mean that the tube with growth in, where less than three or four of the other replicates are growing, must have arisen from a single cell. This concept can be extended into a means of actually estimating the numbers in the original culture (most probable number (MPN) method, see below).

Counting prokaryotes

In microbiology, the number of cells per milliliter of sample is often important. We can be fairly sure of finding most species of prokaryote in a sample, provided that sample is large enough and was taken from a habitat that allows growth of that organism. It is likely that we might find *Salmonella* on eggs, but the important question we need to answer is 'Are the eggs safe to eat?' To be safe, there must be less than the intoxicating dose of *Salmonella* in an amount of egg that is likely to be consumed raw. The number of cells thought to cause *Salmonella* food poisoning is around 40, so we could reasonably expect safe eggs to carry less than, for example, 40 cells per dozen eggs. Such limits exist in industrial standards for most foods, so a good estimation of bacterial numbers is crucial to the food industry. In medicine, the presence of only one or two *Staphylococcus aureus* per 10 cm^2 of human skin could be considered quite normal, but 10^4 cells per mm^2 might reveal the underlying cause of a serious skin condition. In environmental microbiology the relative numbers of organisms per ml of a sample of river water might indicate the dominant species.

The numbers of prokaryotes in a sample can be expressed in two ways: the **total count** or the **viable count**. The former estimates the number of cells, alive or dead, the latter only those capable of growing under the conditions tested. Total counts are made by diluting the sample in a known amount of buffer and then counting the number of cells in each well of a **hemocytometer**. The hemocytometer is a specialized microscope slide and cover slip in which a grid of known size is displayed while viewing under the microscope. The count of cells per grid can then be multiplied up to reveal the number of cells per ml in the original sample. Flow cytometry is a method, similar in concept, in which the number of particles in a small sample is electronically counted by passing a laser shining across a capillary approximately one cell wide.

More recently these methods have been complemented by **quantitative PCR (qPCR)**. This method allows the counting of the number of copies of individual genes. For example, if the copy number per ml of the 16S rRNA gene is estimated, this can give an idea of the bacterial numbers. Although there are drawbacks to this method (we cannot be sure that all Bacteria have only one copy of this, or any other gene, per genome) we can simultaneously estimate the relative numbers of many different prokaryotes and eukaryotes by carrying out parallel experiments on the same sample using specific primer sets. Whether cells are alive or dormant can be related to the presence of ATP. Thus, an estimate can be made of the overall activity of a sample by measuring ATP (chemically). Similarly, only live cells can transcribe DNA, so an estimation of viability can be made by measuring mRNA (messenger RNA) concentrations from ribosomal genes, using **reverse-transcription qPCR (RT-qPCR)**. This presupposes that only active cells would accumulate these transcripts.

The most commonly used and informative method for enumeration of prokaryotes is the viable count. Classically, viable counts are made by **serial dilution** (Figure 2). This agar plate-based method gives a result in **colony forming units** (cfu) ml^{-1}. This is not equivalent to the true viable count, as the numbers only reflect those species that are capable of forming visible colonies under the conditions of medium and incubation chosen for the experiment. The number of cells in the original sample is estimated by back-calculating the number of dilutions made from the plate that has the highest number of easily discernible colonies. Normally this figure is less than 200 colonies per plate, but varies according to colony size. The equivalent method for prokaryotes not capable of growth on agar is the **most probable number (MPN)** technique, in which the pattern of growth in replicates of liquid cultures at various dilutions is used to deduce the number in the original sample.

One of the most accurate ways of counting any microscopic particle is by use of **FACS** (fluorescence-activated cell sorting). Fluorescent dyes can be obtained which differentially stain living and dead cells. The sample is introduced into a narrow capillary of only one

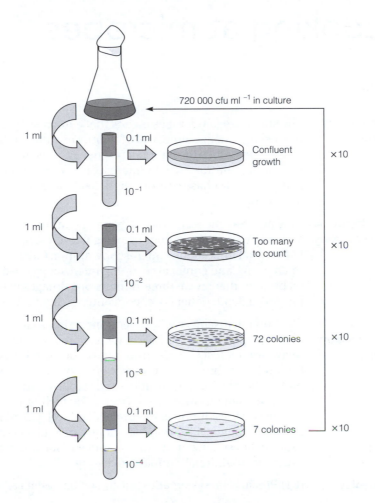

720 000 cfu ml $^{-1}$ in culture

1 ml | 0.1 ml → Confluent growth | ×10

10^{-1}

1 ml | 0.1 ml → Too many to count | ×10

10^{-2}

1 ml | 0.1 ml → 72 colonies | ×10

10^{-3}

1 ml | 0.1 ml → 7 colonies | ×10

10^{-4}

Figure 2. Diagram showing serial dilutions for the estimation of viable count. Test tubes contain 9 ml before addition of bacterial culture.

cell width in diameter, and passes by a laser. The laser excites the fluorescent dye and the excitation passes to a detector. The detector is linked to a gate just downstream of the laser, which will switch to move the cell into a receptacle. Nonfluorescent cells do not cause the gate to open and pass into a second vessel. The number of times the excitation gate opens can be counted and related to the flow rate past the laser to give an exact number of cells in the sample.

C5 Looking at microbes

Key Notes

Light microscopy – old school!

The small size of microbes means that a microscope is needed to visualize almost all of them. For some of the smallest microbes the light microscope is at the limit of its resolution. Contrast in light microscopy can be enhanced using stains or phase contrast, dark field, or fluorescence microscopy.

New developments in light microscopy

New microscopic techniques are greatly improving our ability to visualize microbes using light microscopy. Differential interference microscopy, atomic force microscopy, and confocal scanning microscopy are all techniques that create three-dimensional images with improved depth of field over conventional light microscopy.

Transmission and scanning electron microscopy

Electron microscopy utilizes electrons rather than photons to image specimens and has a much greater resolution than light microscopy. Electron microscopy can be used for the study of subcellular structures but requires extensive pretreatment of tissues (fixation and stabilization) to stabilize them for the electron beam. The technique has now improved to resolution at the molecular level, particularly when the chemical fixation techniques are replaced by cryo fixation and stabilization, which reduce tissue artifacts caused by traditional sample processing.

Related topics

(B1) Prokaryotic systematics   (C8) The bacterial cell wall

Light microscopy – old school!

Of the organisms classed as microbes, only the larger members of the fungi and chlorophyta are visible easily with the naked eye. A few members of the microbes are between 200 and 500 µm in size and can be seen using a hand lens, but by far the largest group of microbes have cell sizes of between 1 and 10 µm. This last group require substantial magnification before they can be viewed by the human eye and therefore different types of microscope have to be used to see these microbes.

The first microscopes were based on simple, single lenses, which provided sufficient magnification to see yeasts, protists, and larger Bacteria. The early microscopists Robert Hooke (1665) and Antonie van Leeuwenhoek (1684) both observed and drew microbes using these simple single-lens microscopes. Present day light microscopes are compound microscopes, they have an objective lens (up to 100-fold magnification) and an eyepiece lens (usually 10-fold magnification) (Figure 1), which together give a total magnification of 1000×. At maximum magnification, optical oil is used between the sample and the lens to optimize light collection and improve resolution (Figure 2). The resolution limit of the light microscope is about 0.2 µm (i.e. you can differentiate two small black spots 0.2 µm apart), limited by the physics of glass lenses. This level of resolution and magnification allows us to see most of the prokaryotes using a light microscope, particularly if stains are used to maximize contrast between the background and the specimen.

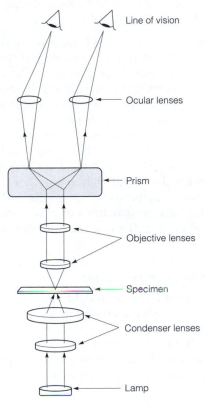

Figure 1. The light microscope.

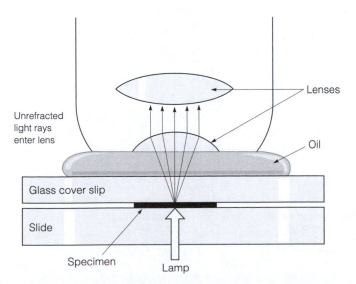

Figure 2. Effect of oil on the path of light through an objective lens.

A number of stains can be used to visualize microbial cells for light microscopy, including crystal violet and safranin, which are both used in the Gram stain (Section C1). Bacterial spores can be stained using malachite green, and fungal structures can be stained using cotton blue in lactophenol. All stains should be treated with great care as, by their very nature, they are toxic to cells, including yours!

Other microscopic techniques can be used to maximize contrast between specimen and background including phase contrast, fluorescence, and dark field microscopy. Phase contrast microscopy takes advantage of the change in phase of light that occurs when light passes through a cell. This change in phase alters the refractive index of the sample relative to the background and when this difference is amplified using a phase plate in the microscope the resulting image has enhanced contrast (Figure 3). Dark field microscopy uses side illumination of the specimen. Only scattered light from the specimen is seen through the microscope (Figure 4) and the object appears light on a dark background. Fluorescence microscopy takes advantage of the fact that some molecules will emit light (fluoresce) when irradiated with light of another wavelength. Some microbes contain naturally fluorescent compounds (chlorophylls and other pigments). Fluorescent dyes can be used to tag specific structures or processes. Light from the ultraviolet spectrum is commonly used to excite fluorescence and this requires a separate source of illumination.

New developments in light microscopy
Digital microscopes are a recent advance in light microscopy. Instead of objective and eyepiece lenses, a CCD camera produces a digital image of a specimen which is produced on a screen and images can be viewed, annotated, and stored on the computer. Other recent improvements in light microscopy include differential interference microscopy, atomic force microscopy, and confocal scanning microscopy. All these techniques can create three-dimensional images with improved depth of field over those of conventional light microscopy.

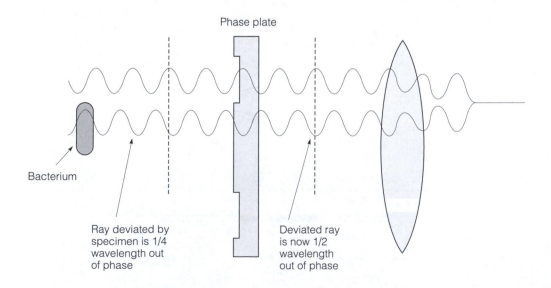

Figure 3. Phase contrast microscopy.

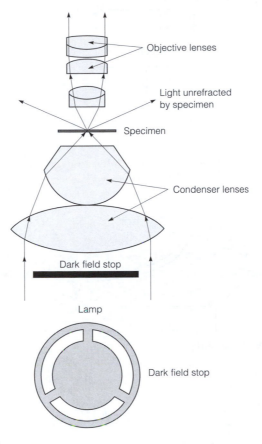

Figure 4. Dark field microscopy.

Differential interference contrast microscopy uses a polarized beam of light that is split into two and both light beams are passed through the specimen. The beams are then reunited in the objective lens. The image they produce is created from an interference effect caused by small changes in phase caused by passage through the specimen. Cell organelles viewed with DIC (differential interference contrast) microscopy have a three-dimensional quality.

In atomic force microscopy, a living, hydrated specimen is scanned using a microscopic stylus, so small that it records minute repulsive forces that exist between itself and the specimen. The stylus records changes in topography as it scans across the specimen (Figure 5). Data are then processed by computing to create detailed three-dimensional images. No chemical fixatives or coatings need be used with this technique and therefore the artifacts seen in SEM (scanning electron microscopy) (see below) are avoided.

Confocal scanning laser microscopy (CSLM) uses a laser light source and computing to create three-dimensional digital images of thick specimens. The precision of the laser beam, focused through a pinhole, insures that only a single plane of a specimen is illuminated at one time (Figure 6). By adjusting focus, different layers of a specimen can be viewed and complex images can be created from digital data. Fluorescent staining and artificial colors linked to depth or density differences in the specimen can be used to enhance the image. CSLM is particularly useful in the study of microbial ecology of biofilms and soils.

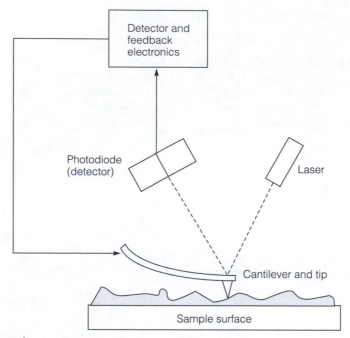

Figure 5. Atomic force microscopy.

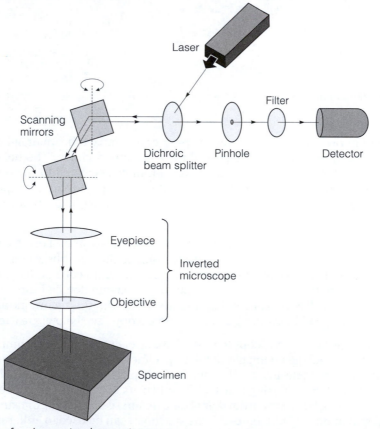

Figure 6. Confocal scanning laser microscopy.

Transmission and scanning electron microscopy

Transmission electron microscopy utilizes electrons rather than photons to image specimens. This technique was originally developed from X-ray crystallography and it took some years to develop the technique for biological specimens. In the electron microscope electron beams are focused by a series of electromagnetic lenses (Figure 7). The microscope is kept under high vacuum to insure unimpeded travel of electrons to and through the specimen. The electrons pass through the specimen on to a phosphor screen for visualization (we cannot see electrons with our eyes). Biological specimens are inherently unstable under vacuum, and therefore have to be fixed (usually glutaraldehyde and paraformaldehyde, followed by osmium tetroxide), dehydrated, and stabilized in resins before they can be placed in a vacuum. The depth of specimen that electrons can pass through is very small, which means that the biological specimen has to be sliced into thin sections, usually with a diamond knife, to produce 1 micron sections. These sections are electron transparent and specimen contrast has to be enhanced using metal stains, for example, lead citrate and uranyl acetate. This technique was used extensively for the study of subcellular structures, and now has moved on to resolution at the molecular level. When the chemical fixation techniques described above are replaced by cryo fixation and stabilization, tissue artifacts caused by traditional sample processing are reduced. The electron microscope can also be used to study virus and organelle structure using the technique of negative staining. Contrast is brought to the specimen by surrounding it with phosphotungstic acid.

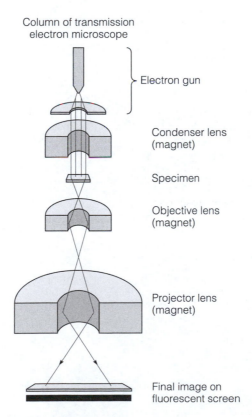

Figure 7. Transmission electron microscope.

External structures of microbes can be viewed at high magnification using the technique of scanning electron microscopy. The specimen is stabilized by either chemical or cryo fixation, covered by a layer of conductive material (carbon or gold/platinum) and the specimen is scanned with an electron beam. Electrons are either reflected or passed through the specimen and a digital image is built from the composite data, creating three-dimensional structures (Figure 8). This technique has been further developed for examination of specimens at low vacuum and ambient conditions.

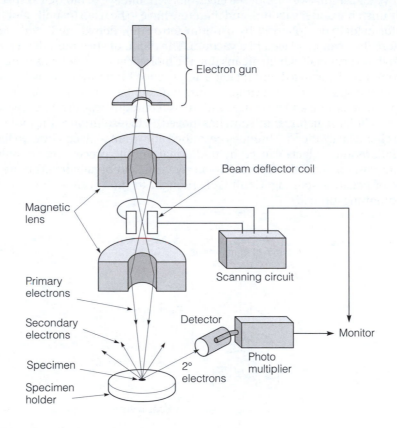

Figure 8. Scanning electron microscope.

C6 The major prokaryotic groups

Key Notes

The Prokaryotes

The Prokaryotes can be divided into two kingdoms, the Bacteria and the Archaea. This classification was first proposed because of the differences in 16S rRNA sequence.

The *Proteobacteria*

This phylum contains the largest number of known species, and is further subdivided into alpha (α), beta (β), gamma (γ), delta (δ), and epsilon (ε). The enteric group of the γ-*Proteobacteria* includes *Escherichia coli* and many well-known pathogens. Other well-known Bacteria in this phylum include *Pseudomonas* (γ-*Proteobacteria*) and *Campylobacter* (ε-proteobacteria).

The Gram-positive bacteria

The two main groups in this phylum are called the low GC and high GC Gram-positives (or Actinomycetes), a phylogenetically valid classification based on the %GC content of their genomes. Low GC Gram-positive organisms include *Bacillus*, *Clostridium*, and *Lactobacillus/Streptococcus*, while high GC organisms include *Mycobacterium* and *Streptomyces*.

Cyanobacteria

The *Cyanobacteria* are a uniformly phototrophic phylum, but differ fundamentally from apparently similar photosynthetic Proteobacteria, such as *Rhodococcus*. Examples include *Synechococcus*, *Anabaena*, and *Prochloron*. The phylum includes many species exhibiting differentiation of vegetative cells to form gas vesicles or heterocysts.

Planctomycetes

The *Planctomycetes*, such as *Brocadia anammoxidans*, are unusual in having budding rather than binary fission and genomic material bounded by a membrane. These are the only known prokaryotes to have a true nuclear organelle.

Spirochetes

Although there are many nonpathogenic representatives, this phylum includes the pathogens *Treponema pallidum* (syphilis) and *Borrelia burgdorferi* (Lyme disease). They are highly motile and helical in shape, with an unusual form of motility using endoflagella.

Deinococcus/ Thermus

The two representative genera of this phylum both resist extreme conditions: *Deinococcus radiodurans* can resist high doses of radiation while *Thermus aquaticus* grows at temperatures up to 80°C.

Aquifex and hyperthermophilic phyla

Several phyla near to the root of the 'tree of life' contain only a few species, most of which are thermophiles. This property coupled to the primitive nature of these bacteria supports

	the notion that life began at higher temperatures than we experience today.
Other phyla	The great diversity of the bacteria is reflected in many other phyla not discussed here. These include the numerically abundant members of the *Flavobacteria* as well as *Verrucomicrobia, Cytophaga*, green sulfur bacteria, *Chloroflexus,* and *Chlamydia*.
The kingdom Archaea	Thermophiles, halophiles, and other extremophilic Archaea are well known, but this kingdom also includes many mesophiles. It is becoming apparent that there is a similar or even greater physiological and biochemical diversity in the Archaea compared with the Bacteria.
The phylum *Crenarchaeota*	Most crenarchaeotes cultured in the laboratory are extremophiles capable of growth above 80°C. The best known examples are *Sulfolobus solfataricus* and *Pyrodictium abyssi*, the latter holding the current record for biological growth at high temperature (110°C). The morphology of *Pyrodictium* spp. is unusual, with disk-shaped cells interconnected by hollow tubes of unknown function (cannulae).
The phylum *Euryarchaeota*	This phylum of the Archaea includes both mesophiles and extremophiles, most notable among which are the methanogens and *Pyrococcus furiosus*.
The phylum *Korarchaeota*	Several phyla in both the Archaea and the Bacteria have been proposed on the basis of the existence of environmental 16S rRNA sequences. The *Korarchaeota* have been identified in this way, despite claims that the signature sequences were artifacts. A member of the phylum has now been cultured and its genome sequenced.
Related topics	(B1) Prokaryotic systematics (C9) Cell division (B2) Identification of Bacteria (C10) Bacterial flagella and (B3) Inference of phylogeny movement from rRNA gene sequence (C11) Prokaryotes and their (C7) Composition of a typical environment prokaryotic cell

The prokaryotes

The prokaryotes consist of many thousands of known species, to which some order has been applied with the advent of 16S rRNA sequencing. The resulting phylogenetic tree (Figure 1) has been used as an indication of how life started. To view the two kingdoms in the prokaryotes as primitive is an oversimplification, as both the Bacteria and the Archaea are extremely well adapted to their environments. A brief survey of the Bacteria and Archaea cannot hope to encompass their diversity, and may even give a misguided view that the Bacteria are predominantly pathogenic or that Archaea are predominantly thermophilic. However, the best studied Bacteria are pathogenic (despite the numerous examples of beneficial species) and the best studied Archaea are thermophiles.

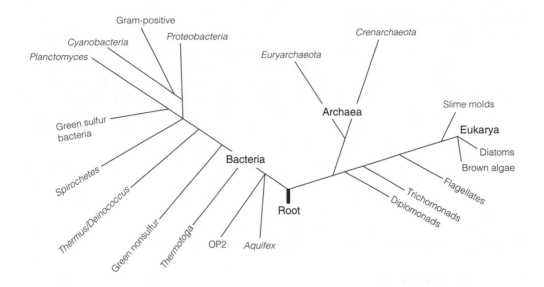

Figure 1. Phylogenetic tree showing the relationship of the Bacteria, Eukarya, and Archaea.

The *Proteobacteria*

The largest phylum in the Bacteria is the *Proteobacteria*, which include most of the species commonly encountered in the microbiology laboratory. The *Proteobacteria* are further subdivided into alpha (α), beta (β), gamma (γ), delta (δ), and epsilon (ε) classes. A few well-known examples from each subdivision are shown in Table 1. Key members of the phylum are discussed further below.

Table 1. Some members of the proteobacteria

Subdivision	Example genus/species	Notes
α-Proteobacteria	*Rhodobacter sphaeroides*	Photosynthetic organism
	Paracoccus denitrificans	Denitrifying bacterium
	Rickettsia prowazekii	Causes typhus
β-Proteobacteria	*Bordetella pertussis*	Whooping cough
	Burkholderia cepacia	Causes onion bulb rot and is also an opportunistic human pathogen
	Neisseria meningitidis	Causes bacterial meningitis
γ-Proteobacteria	*Acetobacter*	Used in the manufacture of vinegar
	Escherichia coli	The model bacterium
	Pseudomonas aeruginosa	Opportunistic pathogen
δ-Proteobacteria	*Aeromonas*	Food-spoilage organism
	Bdellovibrio	Preys on other Gram-negative bacteria
	Myxococcus	Exhibits differentiation
ε-Proteobacteria	*Campylobacter jejuni*	Most common food-borne pathogen
	Wolinella succinogenes	Possible symbiont of cattle

Escherichia coli and the enterics

The rod-shaped facultative anaerobic *Escherichia coli* (γ-*Proteobacteria*) has become the model organism for microbiology and the standard vector for molecular biology. It is essential in human food digestion as it is the main producer of vitamin K from undigested food in the large intestine. Despite these obvious benefits, it is also the most common cause of urinary tract infections and can also cause pneumonia or even meningitis. In the mind of the general public, it is closely linked with outbreaks of diarrhea, normally caused by the strain O157:H7 (enteropathogenic *E. coli*). Most mammals have *Escherichia* species in their digestive systems but they are rarely dominant.

The genus *Escherichia* has a single species, and is grouped with other medically important bacteria known as the **enterics**. This is a phylogenetically distinct group (Enterobacteriaceae) which includes *Salmonella, Shigella, Proteus,* and *Enterobacter*. The significance of these organisms in disease (Table 2) has become a barrier in rationalizing their taxonomy. *Escherichia* and *Salmonella* share over 50% genomic identity and *Shigella* over 70%, which would normally place them all as species in the same genus.

The *Pseudomonas* group

The straight rods of *Pseudomonas* (γ-*Proteobacteria*) are often encountered in enrichment cultures, due to their nutritional versatility – some species can utilize more than 100 different organic compounds. The genus *Pseudomonas* also includes a significant number of plant pathogens. Despite their ubiquitous nature, they are still associated with some serious human diseases, mostly opportunistic infections of wounds and cystic fibrosis. The best studied in the group is *Pseudomonas aeruginosa*, characterized by, among other things, the ability to produce a fluorescent pigment on certain media. Collectively the group including *Pseudomonas, Cupriavidus* (formerly *Ralstonia*), *Xanthomonas, Comamonas, Burkholderia,* and *Brevundimonas* is known as the **pseudomonads**, although these genera do not share any distinct phylogeny within the *Proteobacteria*.

Campylobacter

Despite the media fascination with *Salmonella* and *E. coli* food poisoning, the most common food-borne infections are caused by the *Campylobacter* species. Part of the reason why this genus is overlooked is because its members are obligately microaerophilic and difficult to grow in laboratories more used to dealing with exclusively aerobic organisms. The closely related organism *Helicobacter pylori* has been linked, either as the causative

Table 2. Pathogenic activities of the enteric bacteria

Species	Disease
Escherichia coli	Enterotoxin-induced diarrhea
Salmonella typhimurium	Typhoid fever
Shigella dysenteriae	Dysentery
Klebsiella pneumoniae	Pneumonia
Yersinia pestis	Plague
Proteus vulgaris	Urinary tract infections
Serratia marcescens	Mammary gland inflammation
Vibrio cholerae	Cholera

agent or as an opportunist, to gastric ulcers. The rod/spirillar Gram-negative bacteria are classed as ε-*Proteobacteria*.

The Gram-positive bacteria

The Gram-positive bacteria can be separated into two subgroups based on the percentage of guanine and cytosine in their genomes (Table 3) compared with adenine and thymine. The **low GC** Gram-positives have much less than 50% G+C, while the **high GC** Gram-positives have far more than 50% G+C. The latter are sometimes referred to as the actinobacteria.

Bacillus

Bacillus subtilis (low GC) is one of the oldest named bacteria, first identified by Ferdinand Cohn in 1872. The genus is characterized by the formation of endospores (Section C9). The spores survive so efficiently that it is possible to recover *Bacillus subtilis* from most environments. Rod-shaped *Bacillus* is a physiologically diverse genus, rationalized by the recent removal of many thermotolerant strains to *Geobacillus*. Closely related species include *Bacillus anthracis*, the causative agent of anthrax and *Bacillus thuringiensis*, which produces commercially employed insect larvicides.

Table 3. Some members of the Gram-positive Bacteria

Subdivision	Example genus/species	Notes
Low GC	*Staphylococcus aureus*	Causes a variety of pus-forming diseases in humans
	Lactobacillus delbrueckii	Used in the production of yogurt
	Streptococcus pyogenes	Implicated in 'Strep throat'
	Listeria monocytogenes	Causes listeriosis in humans
	Bacillus subtilis	The Gram-positive equivalent of *E. coli* but is less frequently associated with disease
	Clostridium perfringens	Invasive organism causing gas gangrene
	Geobacillus stearothermophilus	Thermotolerant, growing from 30 to 70°C, has extremely heat-resistant endospores
	Mycoplasma genitalium	Causes nongonococcal urethritis
High GC	*Corynebacterium diphtheriae*	Pathogen, causing diphtheria
	Propionibacterium acnes	Members of this genus are prominent in the fermentation of many foods, including production of holes in Swiss cheese. This species may cause acne but is also resident on normal skin
	Mycobacterium tuberculosis	Causative agent of tuberculosis
	Streptomyces coelicolor	Produces antibiotics against Gram-negatives, gives the soil its smell (production of geosmins)
	Bifidobacterium lactis	Obligate anaerobe found in the intestine of breast-fed infant humans
	Rhodococcus rhodochrous	Metabolically diverse bacterium capable of metabolizing nitriles and cyanides

Clostridium

The members of this genus (low GC) are phylogenetically distinct but morphologically similar to *Bacillus*. They are sometimes referred to as **clostridia**, and lack any respiratory chain. This renders them obligate anaerobes, generating ATP solely by substrate-level phosphorylation (Section E). Although organisms such as *Clostridium tetani* (tetanus), *C. perfringens* (gas gangrene, food poisoning), and *C. botulinum* (botulism) cause medical problems, *Clostridium acetobutylicum* can be used to biologically synthesize butanol, ethanol, isopropanol, and other alcohols rather than using fossil fuel-based alternatives.

Lactobacillus/Streptococcus

These low GC organisms are significant in the human and animal food industry, being involved in the fermentation of cheese, yogurt, sauerkraut, olives, and silage. Their metabolism tends to produce lactic acid so, along with the genera *Pediococcus*, *Micrococcus*, and others, they are often grouped together as the **lactic acid bacteria**. *Lactobacillus* species are rarely associated with human disease, but *Streptococcus* species cause throat infections and dental caries.

Mycoplasma

The members of *Mycoplasma* (low GC) are remarkable in the Bacteria in that they do not have cell walls at any stage in their life cycle. They are considered to be among the smallest organisms capable of growth outside a host cell, and can have a genome of less than 600 000 bp. Although they are phylogenetically true Gram-positive organisms, they do not retain the crystal violet–iodine complex in the Gram stain (Section C8) through lack of a cell wall and so appear to be Gram-negative. The absence of a rigid wall also renders the cells **pleiomorphic**, where they can display different shapes under different physiological or environmental conditions, or even in the same culture, appearing as small cocci, swollen rods or branched filaments. Lastly the lack of cell wall renders them resistant to antibiotics interfering with cell wall synthesis, such as penicillin and vancomycin, so other antibiotics such as kanamycin must be used in combating the pathogenic species.

Streptomyces

The members of this high GC genus produce many antibiotics including streptomycin, spectinomycin, neomycin, tetracycline, erythromycin, and chloramphenicol. Despite an enormous research effort to isolate and characterize the genes and proteins associated with antibiotic production, the significant role of *Streptomyces* in soil ecology has yet to be fully explored. The cells form filaments that can appear to be much like the fungal mycelium and, as the colony matures, aerial filaments with **sporophores** are formed. When released these **conidia** are similar in size but distinct from the endospores of *Clostridia* and *Bacillus*.

Cyanobacteria

The *Cyanobacteria* are a uniformly phototrophic phylum, but differ fundamentally from apparently similar photosynthetic *Proteobacteria* such as *Rhodococcus*. They can be subdivided into five groups according to their cellular morphology (filamentous or unicellular, branching, nonbranching, etc.), and in this case the morphology has significance in their phylogeny. All produce complex membranes in which photosynthesis takes place. The cells are green-blue in color due to the presence of chlorophyll α and phycobilins on the cell membrane, giving the phylum its present name. The filamentous varieties can show cellular structural differentiation, with a few cells forming **gas vesicles** (to aid in floating) or **heterocysts** (sites of nitrogen fixation). The genera *Synechococcus, Anabaena*, and *Prochloron* are the best studied.

Planctomycetes

The *Planctomycetes* lack a peptidoglycan cell wall, but instead rely on a well developed S-layer for rigidity (Section C8). They are characterized by a stalk for attachment to surfaces and divide by **budding** rather than binary fission. Although there is little doubt that members of the phylum (such as *Gemmata*) are prokaryotes, their genomic material is bounded by a membrane. The Anammox organism *Brocadia anammoxidans* (used in waste-water treatment) also has a true organelle in the form of the anammoxosome, where the anaerobic oxidation of ammonia to molecular nitrogen is carried out.

Spirochetes

The *Spirochetes* stain Gram-negative, but form a separate phylum from *Escherichia coli* and close relatives. They are highly motile and helical in shape with an unusual form of motility using endoflagella (Section C10). The most well-known example of this phylum is *Treponema pallidum*, which causes syphilis; however, the majority are free-living or obligately symbiotic nonpathogenic organisms. *T. pallidum* cannot be grown as a free-living organism in the laboratory but must be grown with animal cell culture. The other notable pathogen from this phylum is *Borrelia burgdorferi*, which causes the tick-borne **Lyme disease**.

Deinococcus/Thermus

The phylum containing only the genera *Thermus* and *Deinococcus* includes industrially significant bacteria. *Thermus* species have the ability to grow at temperatures of up to 80°C, and *T. aquaticus* is the source of one of the most important proteins in molecular biology, *Taq* polymerase. The extremely radiation-resistant organism *Deinococcus radiodurans* can survive exposure to gamma rays better than *Bacillus* endospores, while still retaining a vegetative mode. The allied resistance to chemical mutagens shows that this is achieved with a very efficient DNA repair system. This phylum contains no known human or animal pathogens.

Aquifex and hyperthermophilic phyla

Many hyperthermophilic species of Bacteria have been isolated that are phylogenetically close to the hypothetical root of the tree of life (Figure 1). These isolates have been arranged into several distinct phyla containing only one or two genera such as *Aquifex*, *Thermotoga, Methanopyrus*, and *Pyrolobus*, all characterized by having optimal growth temperatures above 80°C. This property, coupled to the phylogenetic position of these Bacteria, supports the notion that life began at higher temperatures than we experience today.

Other phyla

The great diversity of the Bacteria is reflected in many other phyla not discussed here. These include the numerically abundant members of the *Flavobacteria* as well as *Verrucomicrobia, Cytophaga*, green sulfur bacteria, *Chloroflexus*, and *Chlamydia*.

The kingdom Archaea

Until recently the Archaea were regarded as a solely extremophilic species, growing in environmental niches unsuitable for other forms of life, as extreme thermophiles, psychrophiles, halophiles or natronophiles (i.e. require salt to grow). With the advent of rRNA sequencing, we are now seeing species within the *Euryarchaeota* in arboreal forest soils, in temperate lakes, indeed in most of the places we expect to see Bacteria. This should have been no great surprise since we have known about the methanogenic Archaea in our digestive systems for many decades. Despite having as great an ability to behave as chemoorganotrophs or chemolithotrophs as the Bacteria and possessing a greater diversity in nucleic acid replication, there are currently no known human Archaeal pathogens.

As has been emphasized in other sections, the Archaea have many similarities physiologically to both the Bacteria and the Eukarya (Table 4). However, our understanding of the ecology, physiology, and biochemistry of this kingdom does not begin to approach the wealth of biological knowledge we have for such organisms as *E. coli* and *Saccharomyces cerevisiae*.

The phylum *Crenarchaeota*

Those *Crenarchaeota* that have been cultured in the laboratory are extremophiles with growth optima in excess of 80°C, including *Desulfurococcus, Thermoproteus, Sulfolobus,* and *Pyrodictium*. Presumed psychrophilic *Crenarchaeotes* suspended in Antarctic

Table 4. Selected features of the Archaea, compared to the Eukarya and Bacteria

Feature	Bacteria	Archaea	Eukarya
Cell wall lipids	Isoprenoids ester bonded to glycerol	Isoprenoids ether bonded to glycerol	Isoprenoids ether bonded to glycerol
Light mediated ATP synthesis	Photosynthetic	Bacterioruberins catalyzed system (e.g. Bacteriorhodopsin)	Photosynthetic
Chromosome(s)	At least one circular, rarely linear	Circular	Linear
Nuclear membrane	Absent	Absent	Present
Histones	Absent	Present	Present
DNA replication	Unique	Similar to Eukarya	Similar to Archaea
mRNA	May be polycistronic	May be polycistronic	Monocistronic
RNA polymerase holoenzyme	$\alpha_2\beta\beta'\sigma$	Up to 13 subunits	More than 33 subunits
rRNA	5S, 16S, 23S	5S, 16S, 23S	5S, 18S, 23S
Transcription	Unique. Generally initiated by derepression	Simplified Eukarya-like mechanism	Generally initiated by activation
Translation initiation	Via Shine Dalgarno sequence	Sometimes via Shine Dalgarno sequence	Scanning mechanism

waters have also been identified via rRNA sequencing. The majority of the bacteriology and enzymology of this division has focused on the biotechnological applications of crenarchaeotal proteins, so our overall understanding of these organisms is relatively poor.

The best known example of the *Crenarchaeota* is *Sulfolobus solfataricus* (the first *Sulfolobus* species was *S. acidocaldarius*, discovered by Thomas Brock in 1970). *S. solfataricus* was isolated from a hot spring emerging from the volcanic area around Naples in Italy. Morphologically the cells are nondescript, irregularly shaped cocci, but this archaeon is both an extreme thermophile with a growth optimum of around 80°C and an acidophile, growing at pH 3. It has proved easy to work with in the laboratory, especially as few air-borne contaminants survive in the culture medium. Work on this organism pioneered the concept that the RNA polymerase of Archaea has more in common with eukaryotes (e.g. the use of basal transcription factors) than Bacteria.

The crenarchaeote *Pyrolobus fumarii* is another extremely thermophilic acidophile, and holds the current record for biological growth at high temperature. Its optimum growth temperature is between 97 and 105°C, but it can still divide at 113°C. An obligate thermophile, it will not grow below 80°C and can tolerate temperatures of up to 140°C. Like *Sulfolobus* it is an acidophile, growing at an optimum pH of 5.5, but is also a strict anaerobe with a requirement for sodium chloride (0.7–4.2% w/v). Such conditions cannot be found anywhere within the terrestrial aqueous ecosystems on earth, so it is unsurprising that this bug was first isolated from a **black smoker** in the deep oceans. Volcanic fissures under the sea release large amounts of sulfide, hydrogen, and metals at high temperature, giving a plume of black liquid across the sea bed. As *P. fumarii* uses hydrogen as an energy source and sulfur as an electron acceptor instead of oxygen, these are ideal conditions for growth. As well as being biochemically remarkable, the morphology of the Archaeon is unusual as well. The cells themselves are disk-shaped and are interconnected with up to 100 of their neighbors by hollow tubes, known as **cannulae**. The function of these fine tubes has not been fully explored, but it is thought that they extend into the Archaeal equivalent of the periplasm, but not into the cytoplasm itself.

The phylum *Euryarchaeota*

This phylum also includes extremophiles, but also numbers mesophiles including some of the **methanogens** among its members. The biochemistry of methanogenesis is detailed in Section E4. The methanogens include such genera as *Methanococcus*, *Methanobacterium*, and *Methanosarcina*. The phylum is physiologically diverse, with representatives in the extreme thermophiles (e.g. *Picrophilus* and the cell wall free *Thermoplasma*), the halophiles (*Halococcus*, *Natronococcus*), and many oligotrophic marine Archaea. The phylum may also include the organism with the smallest known Archaeal genome, *Nanoarchaeum equitans*. This intracellular parasite of *Ignicoccus*, an extreme thermophile isolated from an undersea hot water system off the coast of Iceland, has been the subject of some debate. Some taxonomists maintain that the organism should be classified in its own phylum '*Nanoarchaeota*.'

Pyrococcus furiosus is an extremely thermophilic euryarchaeote and was the source of the *Pfu* polymerase, used as an alternative to *Taq* polymerase in the PCR reaction. Although *Taq*, of bacterial origin, and *Pfu*, of Archaeal origin, can be substituted for one another in the *in vitro* technique of PCR, the two proteins illustrate a difference between the two kingdoms. *Pfu* is the main replicative DNA polymerase of *Pyrococcus furiosus*, but has little in common with the eubacterial equivalent, DNA polymerase III (Section F3). *Taq* is a DNA polymerase I-type protein, with a function in DNA repair (Section F7), but it is not yet known whether *Pfu* or other proteins perform the equivalent role in *P. furiosus*. *P. furiosus* is attractive for PCR as it has the quickest doubling time of the Archaea, replicating its genome in only 37 minutes. To do this it must be growing at between 70 and 103°C, and studies on this organism reveal that even at its maximum growth temperature, the DNA within the cell remains double-stranded without breaks, something that cannot be achieved if pure DNA is boiled in water. It is also almost as resistant to radiation as *Deinococcus*.

The methanogens are a diverse group of the *Euryarchaeota*, including methanogenic thermophiles, acidophiles, halophiles, and mesophiles. They are linked by the ability to produce methane from C1 compounds or acetate (Section E4) and an obligately anaerobic way of life. The model organism for the genus is *Methanocaldococcus jannaschii*, one of several methanogens to have had their genomes sequenced. Comparative genomics reveals that *M. jannaschii* has similar enzymes to the Bacteria in its central metabolic pathways (apart from the specialist ones involved in methane generation) but a greater similarity to the Eukaryotes in DNA replication, transcription, and translation. Notwithstanding, 50% of the genes of *M. jannaschii* have absolutely no counterpart in any organism so far sequenced. The organism was named after Holger Jannasch, a pioneer in the field of deep sea microbiology. The organism was isolated from a white smoker in the East Pacific, and consequently the growth requirements of *M. jannaschii* are strict: carbon dioxide, hydrogen, a few mineral salts, no oxygen, temperature of around 85°C, and a significant amount of pressure. This barophile can withstand up to 200 atmospheres of pressure.

The phylum *Korarchaeota*

This phylum is still under scrutiny, and contains only a single representative '*Korarchaeum cryptofilum*.' Environmental 16S rRNA gene sequences were detected in a single spring in Yellowstone Park (USA) and it was first suggested that this group was an artifact, perhaps formed by other environmental 16S rRNA gene sequences ligating during PCR. However, cultivation of '*Korarchaeum cryptofilum*' in mixed culture at 85°C eventually enabled the assembly of a complete genome sequence. Analysis of genes confirmed that the organism was physiologically and phylogenetically distinct from other known Archaea. The 16S signature has also been found in a number of geographically separated environments, but little is known about the functions of these filamentous organisms except that they are anaerobic heterotrophs.

C7 Composition of a typical prokaryotic cell

Key Notes

The prokaryotic cell

The model Bacterial cell is that of *Escherichia coli*, a rod-shaped cell about 2 μm long and 1 μm wide. With a very few exceptions, there are no discernible subcellular components in prokaryotic cells. At the biochemical and physiological level, Bacteria have some similarity to the Archaea, yet the Archaea have similarities to both the Bacteria and the Eukaryotes.

The cytoplasmic membrane

The Bacterial cell membrane is made up of a lipid bilayer with proteins buried within it. The bilayer is impermeable to most molecules except gases such as oxygen, nitrogen, and CO_2, while passage of larger molecules and ions is mediated by protein. The membrane does not confer rigidity to the cell but is reinforced with hopanoids. The Archaeal cell membrane has a similar arrangement but has a unique monolayer structure.

The cytoplasm

Subcellular structures in the aqueous cytoplasm include ribosomes and storage structures (inclusion bodies, carboxysomes, and magnetosomes). Activity-specific structures also exist, for example, thylakoid membranes or chlorosomes involved in photosynthesis. The examination of subcellular structure in prokaryotes is sometimes complicated by the process of electron microscopy and this has been implicated in the formation of artifacts such as the mesosome.

Genomic material

The nucleoid is the tightly packed chromosomal material of prokaryotes, while plasmids are thought to float free in the cytoplasm. The chromosomal complement of a prokaryotic cell is always haploid and plasmid copy number is often more than one.

The periplasm

The periplasm of Gram-negative Bacteria is of lower water content than the cytoplasm. It is packed with proteins and plays a major role in secretion, environmental sensing, and many other key pathways.

The cell wall

All prokaryotic cells have a strong cell wall (apart from *Mycoplasma* and *Thermoplasma*). Bacteria are covered in peptidoglycan. The Archaea have a far greater variety in the composition of the outer wall but have a Bacterial ortholog in pseudomurein.

Endospores	*Clostridium* and *Bacillus* are able to form spores within the cytoplasm in response to cell starvation. Endospores have the capacity to survive heat, radiation, and chemical disinfectants.	
External features	All prokaryotes are covered in some form of slime capsule (or glycocalyx). In addition they may have filaments extending from the cell wall such as flagella for movement, or much finer fimbriae for attachment.	
Related topics	(C2) Prokaryotic diversity	(C9) Cell division
	(C6) The major prokaryotic groups	(C10) Bacterial flagella and movement
	(C8) The bacterial cell wall	(F2) Genomes

The prokaryotic cell

The model Bacterium is *Escherichia coli*, normally found in the colon of humans and other mammals. It has a limited capacity for independent existence, is motile and rod-shaped, about 2 μm long and 1 μm wide. Its cell wall is Gram-negative in structure (for a comparison of Gram-negative and Gram-positive Bacteria, see Section C8). Morphologically *E. coli* resembles many other Bacteria and some members of the Archaea, although the fine structure of the cell wall, biochemistry, and molecular biology differ considerably. The single circular chromosome (Section F2) is attached to the cytoplasmic membrane (Figure 1) and this DNA is free in the cytoplasm. The word 'prokaryote' is derived from the Greek *pro* (before) and *karyote* (literally a nut's kernel, but biologically adapted to mean nucleus), so this prenucleate state defines the prokaryotes. There are generally no discernible subcellular components in Bacterial cells (compared with eukaryotic cells, Section H2), the exception being the presence of layered membrane bodies of unknown function in the methanotrophic and nitrifying Bacteria (Section C11). The main subcellular differences are that the following are **absent** from Bacterial and Archaeal cells:

- Nuclear membrane (except *Planctomycetes*)

- Chloroplasts and mitochondria – energy generation takes place across the cell membrane rather than in any specialized bodies

- Golgi apparatus or endoplasmic reticulum – free ribosomes in the cytoplasm translate mRNA

Bacteria and Archaea come in a variety of shapes and sizes (Section C2) and are classified according to their biochemical rather than morphological differences (Section B). The differences between the Bacteria and Archaea are broad, covering the machinery used for replication, transcription, translation, and central metabolism (Section F). Overall they are linked by a common cell structure and share many ecological niches, but in this respect are a product of convergent evolution.

The cytoplasmic membrane

In common with all biological membranes, the Bacterial membrane between the cytoplasm and the cell wall is composed of a **lipid bilayer** (Figure 2), made up of two tiers

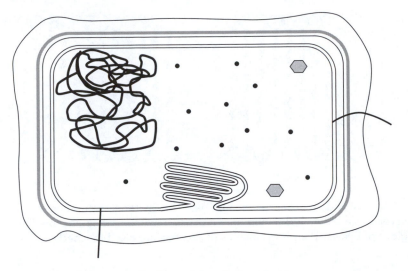

Figure 1. A generalized diagram of a Bacterial or Archaeal cell.

of phospholipids. These phospholipids are arranged so that their hydrophilic heads face into the cytoplasm or the outside world, while the hydrophobic tails are embedded within the membrane. Many proteins are partially buried within the membrane, or pass right through it. These proteins have many final functions but all mediate transport in one form or another – f_1f_0 ATPase regulates the transport of H^+, sugar transporters regulate the flow of carbohydrate, and so on. The lipid bilayer is impermeable to H^+, as without this property the organism would be unable to generate ATP (Section E2). The current model of the cell membrane allows passage of:

- Gases (O_2, CO_2, N_2) via **passive diffusion** directly through the membrane
- Water-soluble ions (Na^+, K^+, except where these substitute for H^+ in extremophiles) via small pores in the membrane
- Water itself up or down the **osmotic gradient**
- Small molecules via **facilitated transport** through protein channels but following an osmotic, chemical or potential gradient
- Molecules via **active transport**, normally at the expense of ATP

The cytoplasmic membrane does not give rigidity or shape to the cell as this is provided by the tougher cell wall. Bacterial cell membranes lack the sterols eukaryotes have but instead stabilize membrane structure with hopanoids.

The Archaeal cell membrane performs the same roles as the Bacterial equivalent, but it is unique in biology with a predominantly monolayer phospholipid structure (Figure 2).

The cytoplasm

Around 80% of the cytoplasm is water, the other fifth is protein, lipid, nucleic acid, carbohydrate, inorganic ions, and other low molecular weight compounds. The cytoplasm is the medium for enzyme activity and contains the ribosomes. Other subcellular structures that may be present are **inclusion bodies** (storage granules made of compounds such as the membrane-associated poly-β-hydroxybutyrate or cytoplasmic polyphosphate),

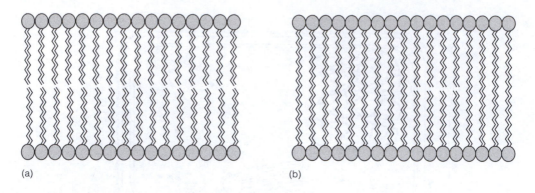

(a)   (b)

Figure 2. Bacterial and Archaeal cell membranes. The Bacterial membrane (a) is made up of a lipid bilayer, while the Archaeal membrane (b) has transmembrane phospholipids as well as some regions of bilayer. The phospholipids of Archaea are also branched.

carboxysomes (sites of CO_2 fixation) and **magnetosomes**. The magnetosome is found in magnetotactic bacteria such as *Magnetospirillum* and is thought to play some role in biomineralization.

Although Bacteria and Archaea lack true organelles, some species have developed membranous configurations to aid some biochemical processes. Cyanobacteria, such as *Anabaena*, carry out oxygenic photosynthesis at **thylakoid membranes**, which are lined with **phycobilisomes**. The green sulfur bacteria, such as *Chlorobium*, perform anoxygenic photosynthesis in **chlorosomes**. Unlike the thylakoid membranes, these are separate from the cell membrane. Another method of anoxygenic photosynthesis uses spherical or lamellar systems attached to the cell membrane (e.g. purple Bacteria such as *Rhodopseudomonas*).

The major drawback in the study of the fine structure of prokaryotes is the method used to view such small objects. Electron microscopy (EM) has provided us with many of the details given here and in other texts, but the harsh methods used to fix and develop the samples are prone to generation of artifacts. An example of this is the **mesosome**, a small vesicle seen in many EM preparations, often seen budding from the cell membrane. It is possible that these have one of the proposed roles in cell division, cell wall synthesis or chromosome resolution, but it seems more likely that mesosomes are created during the EM drying and fixation process.

Genomic material

The Bacterial chromosome(s) are often found in one area of the cell, normally close to the cytoplasmic membrane (Section F2). Chromosomes are tightly supercoiled and are associated with chromatin, proteins much like those found in eukaryotes but lacking basic histones. In contrast, it is thought that the chromatin of Archaea is very much like that of eukaryotic microorganisms. The tightly packed nuclear material is referred to as the **nucleoid** in prokaryotes and is complemented by plasmids floating free in the cytoplasm. The size of some plasmids approaches that of chromosomes and the distinction between the two is sometimes esoteric. However, the chromosomal complement of a prokaryotic cell is always **haploid** and plasmid copy number (Section F11) is often more than one.

The periplasm

Gram-negative Bacteria have a second phospholipid bilayer (the outer membrane) enclosing the cytoplasmic membrane. A **periplasm**, the space between the outer and cytoplasmic membranes, is thought to be aqueous but has much lower water content than the cytoplasm. An examination of the estimated total protein content suggests that the periplasm may even have a gel-like state. It is the site of several important reactions, taking a major role in the generation of ATP, environmental sensing, denitrification, and many other metabolic pathways that interact directly with the electron transport chain (Section E2). There has been continued debate about the existence of similar environments in Gram-positive Bacteria, with recent suggestions that a gel-like stratum exists outside the Gram-positive cell membrane, but is not bounded by another membrane.

The cell wall

Prokaryotic cells are strong compared with eukaryotic single cells. The basis of their strength comes from the cell wall, which is present in all prokaryotes apart from the Bacterium *Mycoplasma* and the Archaeal *Thermoplasma*. The Bacteria are covered in a semi-rigid compound called **peptidoglycan** (Section C8) in most cases, with the exception of the *Chlamydia*. In these Bacteria the cell wall rigidity is provided by cysteine-rich membrane proteins interlinked with disulfide bonds. The Archaea have a far greater variety in the composition of the outer wall, but have a Bacterial ortholog in **pseudomurein**.

Endospores

A small number of Bacterial genera, most significantly *Clostridium* and *Bacillus*, are able to form **spores** within the cytoplasm in response to cell starvation. These endospores have been shown to be able to survive desiccation for centuries and their formation is discussed in detail in Section C9. The ability of endospores to survive heat, radiation, and chemical disinfectants coupled to the pathogenic nature of some members of *Clostridium* and *Bacillus* have led to the efficient sterilization procedures (Section C3) developed for bacteriology and medicine.

External features

All prokaryotes secrete a covering of polysaccharide or polypeptides sometimes known as the **slime capsule** (Figure 3). The actual size and composition of this **glycocalyx** vary with the metabolic state of the organism. In many pathogens it is the first line of defense against host recognition. The slime capsule may also help the cell to resist drying out and aid in cell attachment. The interaction of the glycocalyx with other bacteria from the same or different species helps to form biofilms such as plaque on teeth and the bacterial filaments sometimes seen in aquatic environments. A more ordered structure is also present in some Bacteria and Archaea. A crystalline glycoprotein S-layer can form a mesh surrounding the cell and has been linked to attachment, biofilm formation, and the repulsion of host defenses.

Bacterial motility is often provided by the rotation of **flagella**, long filaments extending outside of the glycocalyx. These are highly complex structures, whose composition and role in chemotaxis are discussed further in Section C10. After suitable staining, these can often appear to be the dominant feature of cells such as those of *Salmonella* (Figure 3). Prokaryotes are also coated with other finer hair-like appendages that are generally classed as **fimbriae**. There are currently seven different types known, identified on the basis of their protein composition, length, and diameter. They are often seen as broken

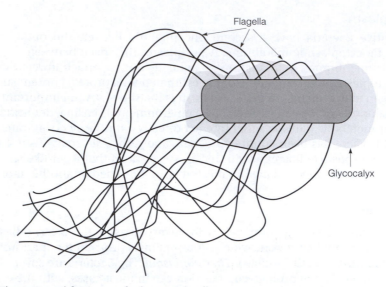

Figure 3. The external features of a bacterial cell.

filaments surrounding the cell in electron micrographs, suggesting that they are brittle and dispensable. Some are pathogenicity factors, but it is thought that most function in attachment. Finer still are the fibrils that coat the surface of bacteria such as *Streptococcus*. The limitations of the electron microscope make it difficult to say whether these are a separate type of hairy coat or an artifact from the drying of the glycocalyx. The largest filamentous structures outside the cell are the sex pili generated by F^+ bacteria before transduction (Section F8).

C8 The bacterial cell wall

Key Notes

Structure of the cell wall

Gram-positive Bacteria have a thick peptidoglycan layer outside the cell membrane to confer rigidity, whereas the Gram-negatives have a thinner layer within the periplasm. The Gram-negative outer membrane bounds the periplasm.

Peptidoglycan

Bacterial peptidoglycan is composed of a backbone of alternating sugars (*N*-acetylglucosamine and *N*-acetylmuramic acid) with a β1–4 linkage. This sugar polymer is cross-linked to similar molecules by an interbridge of amino acids and diaminopimelic acid. Lysozyme can break down peptidoglycan, which results in the internal osmotic pressure of the cell causing lysis. Teichoic and teichuronic acids are dispersed throughout Gram-positive peptidoglycan and may have a function in ion binding. Not all Archaea have an equivalent to peptidoglycan, but those that do possess pseudopeptidoglycan.

The Gram stain

This staining procedure was invented by Christian Gram in around 1894, and has since been shown to have phylogenetic relevance. Cells stained with crystal violet and iodine are washed with alcohol, which decolorizes Gram-negative cells. Counterstaining with carbol fuchsin gives a pale pink color to Gram-negatives, while Gram-positives retain the deeper purple of the crystal violet–iodine complex.

The Gram-negative outer membrane

The outer Gram-negative membrane is a lipid bilayer, but is modified to lipopolysaccharide (LPS). When purified from the rest of the cell this is sometimes called endotoxin due to its strong antigenic properties. The lipid part of LPS (lipid A) anchors the structure to the outer membrane and is covalently linked to the core polysaccharide (R antigen or polysaccharide). The core is topped by the O antigen, which is made of three to five repeating sugars, depending on the species. The outer membrane is less selective than the cell membrane, with exit and entry points provided by porins.

Related topics

(B2) Identification of Bacteria

Structure of the cell wall

The cell membrane (Section C7) of the prokaryotic cell alone is not sufficient to provide the rigidity that a free-living organism requires. In Gram-positive bacteria the shape and integrity of the cell is maintained by a thick single layer of **peptidoglycan** (Figure 1). However, the Gram-negative cell wall has evolved a greater level of complexity, with a

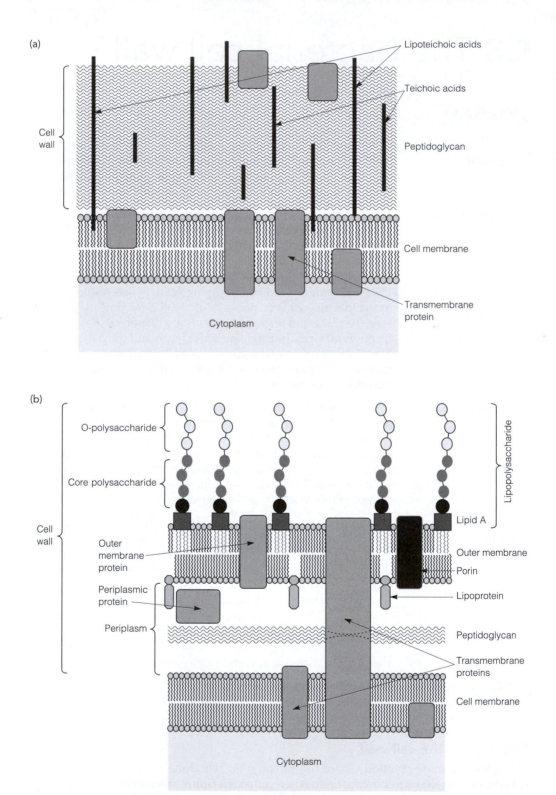

Figure 1. Bacterial cell walls. (a) Gram-positive Bacteria; (b) Gram-negative Bacteria.

second membrane outside the cell membrane to form a periplasm (Section C7), in which a chemically similar but thinner peptidoglycan layer appears. The outer membrane is composed mainly of **lipopolysaccharide** (LPS). Both Gram-negative and Gram-positive cell walls have many proteins embedded in them, penetrating through or attached to their surfaces. Some of these serve in the detection of environmental signals (Section F6). Growth substrates, metabolites, and secreted proteins are also allowed to pass through the thick outer layer by means of specific protein ports.

The main differences between Gram-positive and Gram-negative Bacterial cell walls are shown in Table 1.

Table 1. Comparison of the cell walls of Gram-positive and Gram-negative Bacteria

Gram-positive cell wall	Gram-negative cell wall	Comments
Thick peptidoglycan layer	Thin peptidoglycan layer	Absent in *Mycoplasma*
Polysaccharide	Lipopolysaccharide	
Teichoic acid	Absent	
Teichuronic acid	Absent	
Peptidoglycolipids	Absent	Found in *Corynebacterium*, *Mycobacterium*, *Nocardia*
Glycolipids	Absent	
Absent	Lipoprotein	
Absent	Phospholipid/phosphoprotein	
Absent	Porin	
Absent	Periplasm	

Peptidoglycan

Peptidoglycan or **murein** is made up of two sugar derivatives, *N*-**acetylglucosamine** (NAG) and *N*-**acetylmuramic** acid (NAM) with a β1–4 linkage. Chains of alternating NAM and NAG are cross-linked by amino acids such as L-alanine and D-glutamic acid as well as diaminopimelic acid (DAP). The way in which the cross links are formed differs in Gram-negative and Gram-positive Bacteria (Figure 2). In most Gram-negative Bacteria there is a direct **interbridge** between polypeptide side-chains emerging from adjacent NAG/NAM polymers, while in most Gram-positive Bacteria the interbridge that links side-chains is a glycine pentapeptide. The structure of peptidoglycan is highly conserved among the Bacteria as a whole, the only variations being slight changes in the interbridge. Peptidoglycan is resistant to many chemical challenges, but is easily broken down by **lysozyme**, which breaks the bonds between NAG and NAM. Without the constraining polymer, the osmotic potential is too much for the cell membrane to contain and the cell bursts. Lysozyme is present in many human bodily secretions as the first form of defense against Bacterial invasion. Those cells without peptidoglycan (particularly *Mycoplasma*) avoid this lysis.

The Gram-positive Bacteria have substances called **teichoic** and **teichuronic acids** interspersed with the peptidoglycan polymer. 'Teichoic' is a broad term covering polymers containing glycerophosphate or ribitol phosphate and the primary function of these

(a)

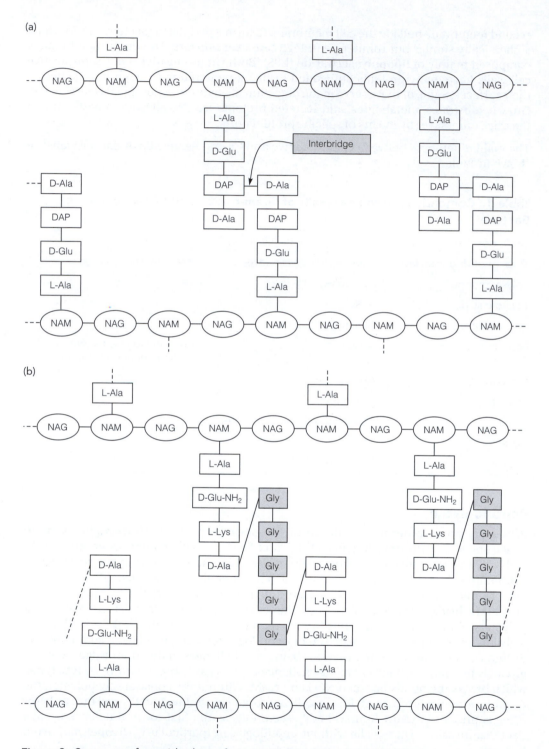

Figure 2. Structure of peptidoglycan from (a) *Escherichia coli* and (b) *Staphylococcus aureus*. NAG, *N*-acetylglucosamine; NAM, *N*-acetylmuramic acid; DAP, diaminopimelic acid. The remainder of the abbreviations refer to the ᴅ or ʟ forms of the amino acids listed in Section E1.

compounds appears to be to bind essential divalent cations such as Mg^{2+}, maintaining the local ionic environment of the cell. The gross effect of the presence of teichoic acids in the membrane is to give the cell a slightly negative charge.

It is extremely difficult to generalize on Archaeal cell wall structure, as the diversity is far higher than that of the Bacteria. Although our knowledge of the Archaeal cell wall is not as highly developed as that of Bacteria, we do know that some Archaea have a compound very similar to the murein, named **pseudopeptidoglycan**. This has a backbone containing NAG, but alternating NAM is replaced with N-acetyltalosaminuronic acid and is β1–3 linked instead of β1–4. However, many Archaea maintain rigidity with the use of a mixture of polysaccharide, protein, and glycoprotein.

The Gram stain

The primary method of distinction between the two main groups of Bacteria remains the Gram stain. Gram-positive and Gram-negative is used in identification, classification, and taxonomy immediately after morphological study. This important method is based on the ability of the cell wall to retain or lose certain chemicals. The staining procedure was devised in 1884 by Christian Gram. It has since proved to have first a biochemical and then a phylogenetic significance in the Bacteria.

The cells are first fixed to a glass microscope slide (normally by heating gently over a Bunsen flame) and **crystal violet** is used to stain the preparation. In order to complex the crystal violet with the cell wall, a solution of **iodine** is then added. If the slide is now washed with **alcohol**, the bound crystal violet stain will be washed out of the thin-walled Gram-negative cells, but cannot pass through the thicker Gram-positive wall. A counterstain of **carbol fuchsin** then stains all Gram-negative cells a pale pink, while the Gram-positive cells retain their deep violet color. These color differences can be seen clearly under a light microscope. Most Bacteria react true to their phylogeny with this stain, with only a few species such as *Paracoccus* behaving abnormally. Some attempt has been made to classify the Archaea with the Gram stain, but unfortunately in this kingdom the variability in staining begins at the subgenus level.

The Gram-negative outer membrane

In contrast to the Gram-positive Bacteria, the model Bacterium *E. coli* does not present a coat of peptidoglycan to the outside world (Figure 1). Instead the murein is suspended in the periplasmic space and there is a second outer membrane of LPS. The Gram-negative outer membrane has some features in common with the cytoplasmic membrane in that it is a lipid bilayer and is considered to be a fluid mosaic. However, rather than being composed of phospholipid alone, there are many lipids with polysaccharide attached. This alters the chemical and physical properties of the membrane, in that it is much more porous to much larger molecules. This porosity is enhanced by the presence of many **porin** and transport proteins.

Our detailed knowledge of the outer membrane structure of Gram-negative Bacteria has been driven by the antigenic properties of LPS, so the details purporting to represent Gram-negatives are in reality more relevant to pathogenic enteric Bacteria such as *E. coli*, *Salmonella*, and so on. The presence of LPS can provoke a strong immune response in mammals, so in genera including *Escherichia*, *Salmonella*, *Shigella*, *Pseudomonas*, *Neisseria*, and *Haemophilus*, the term LPS has become interchangeable with '**endotoxin**.' The chemistry of this LPS outer membrane is complex and variable between species. In *Salmonella* the lipid part of the molecule (**lipid A**) is phosphorylated glucosamine linked to long carbon chains which, in contrast to membrane lipids, may be branched. Lipid A of Gram-negative Bacteria is covalently linked to the **core polysaccharide** (Figure 3) but even if separated from the rest of the molecule it is frequently toxic to mammals, even if the bacterium of origin is nonpathogenic. Lipid A serves to anchor the LPS to the rest of the outer membrane.

The core polysaccharide (or R antigen or R polysaccharide) is the same in all members of a particular genus and is made up of heptose and hexose sugars plus ketodeoxyoctonate (KDO). Finally the **O antigen** (or somatic antigen or O polysaccharide) is made up of three to five repeating sugars. The composition of the repeat varies from species to species, and can be repeated from 1 to 40 times. This is the major antigenic determinant of pathogenic cells.

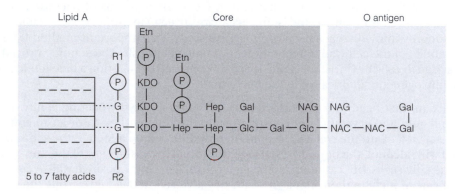

Figure 3. Lipopolysaccharide. P, phosphate; G, glucosamine; NAG, *N*-acetylglucosamine; NAC, N-acetylgalactosamine; Gal, galactose; Glc, glucose; Hep, heptose; Etn, ethanolamine; R1 and R2, phosphoethanolamine or aminoarabinose; KDO, ketodeoxyoctonate.

C9 Cell division

Key Notes

Binary fission

The ability of many prokaryotes to approximately double in biomass, before dividing, means that it is easy to model growth mathematically. The four stages of binary fission are characterized by: growth and replication; FtsK ring construction and chromosome separation; FtsK ring constriction and septation; and cell separation. Cell division must be accompanied by the efficient dispersal of at least two complete genomes between mother and daughter cells, which is attained by replication and site-specific recombination.

Sporulation

Gram-positive Bacteria, such as *Bacillus subtilis*, differentiate to form resting cells within the cytoplasm known as endospores. At the beginning of spore development, DNA moves to the central axis of the cell and becomes tightly coiled, which is quickly followed by DNA separation and protoplast formation. The mother cell engulfs the protoplast to make a forespore and coat synthesis begins. A primordial cortex forms while the mother cell cytoplasm becomes increasingly dehydrated. Cortex and spore coat synthesis is completed and the mature spore becomes increasingly resistant to heat, chemicals, and radiation. Finally the spore is liberated, as what remains of the mother cell undergoes autolysis. When the spore reaches a suitable environment, the process is reversed with activation of the spore followed by germination only a few minutes later.

Related topics

(C2) Prokaryotic diversity
(C6) The major prokaryotic groups

(D1) Measurement of microbial growth

Binary fission

Most Bacteria and Archaea cells accumulate biomass until a critical volume is reached when they divide into two identical daughter cells. This division occurs in a single plane of symmetry. To some extent division dictates the simplicity in cell shape (Section C2) and the prevalence of cell numbers of $2n$ when cells appear in chains or clusters. The link of **binary fission** to cell growth calculations is discussed in Section D1. The process of cell division is outlined in Figure 1 and can be divided into four steps: growth and replication; FtsK ring construction and chromosome separation; FtsK ring constriction and septation; and lastly cell separation. The process is poorly understood, involving many proteins and an extremely efficient control system. A central role is played by **FtsK**, a structural and functional analog of the eukaryotic tubulins. FtsK was named after the filamentous temperature-sensitive mutants that were first recognized in *E. coli*. At some

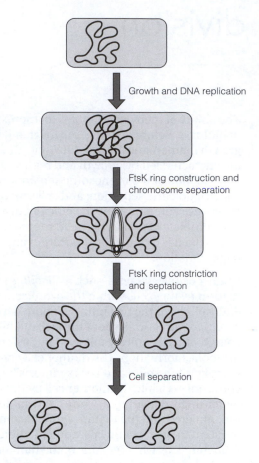

Figure 1. Bacterial cell division.

temperatures mutants were unable to septate and became long, multinucleate filaments, a phenomenon seen in some wild-type lactobacilli and in *Streptomyces*.

During the process of chromosomal replication, FtsK monomers begin to move to a point midway between the two ends of the cell. It is unclear how this happens, but inhibitors (the proteins MinC, D, and E) have been identified at the poles preventing the initiation of unwanted separation. The FtsK proteins form a ring and it is certain that at this point other proteins are recruited to the locality to deal with the **site-specific recombination** events (Section F9) that must take place to allow the separation of the two replicated chromosomes. The formation of a septum begins with constriction of the FtsK ring across the width of the cell and the septum forms completely, so that the daughter cell becomes separate (Figure 1). There are undoubtedly many other mechanisms governing the act of cell division, since the entire contents of the cell, including ribosomes, proteins, lipids, cell membrane, cell wall, and any extra-chromosomal material must also be divided between the two cells, as both daughter cells are immediately and fully functional. It is remarkable that this process is repeated every 40 minutes by *E. coli*. The universality of FtsK-like proteins across both the Bacteria and Archaea suggests that this is an evolutionarily conserved, fundamental mechanism for single-celled life.

Table 1. Endospore and vegetative forms of *Bacillus subtilis*

Property	Vegetative cell	Spore
Heat, radiation, and chemical resistance	Low	High
Metabolism	High	Low
mRNA content	High	Low
Lysozyme sensitivity	High	Zero
Calcium dipicolinate	Low	High
Water as % of volume	80	10–25
pH of soluble content	7	~ 5.7

Sporulation

The formation of endospores is a survival mechanism practiced by Gram-positive Bacterial genera (Section C8), the best studied of which is *Bacillus subtilis*. It can be regarded as a specialized case of cell division, with some of the mechanisms being similar and performed by protein orthologs (for example, the sporulation protein SpoIIIE is very similar to FtsK). The spore and vegetative cell have very different properties, outlined in Table 1.

The process of spore formation can be divided into seven stages (Figure 2):

1. *Beginning of spore development.* DNA moves to the central axis of the cell and becomes tightly coiled.

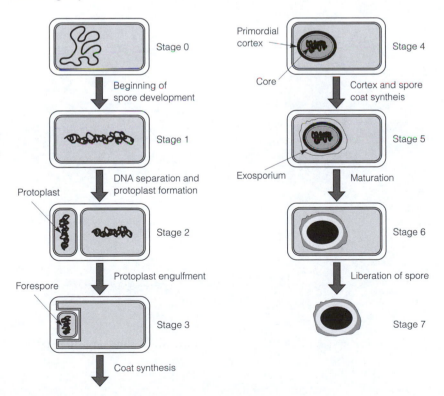

Figure 2. *Bacillus subtilis* sporulation.

2. *DNA separation and protoplast formation.* One complete chromosome becomes enclosed by the cell membrane and a **protoplast** is formed.

3. *Protoplast engulfment.* The mother cell engulfs the protoplast to make a **forespore**. The chromosome of what will become the spore is now surrounded by two membranes.

4. *Coat synthesis.* Wall material is deposited between the membranes and a **primordial cortex** forms around them. The material outside the cortex (the mother cell cytoplasm) becomes increasingly dehydrated.

5. *Cortex and spore coat synthesis.* Incorporation of calcium and dipicolinic acid.

6. *Maturation.* The spore becomes increasingly resistant to heat, chemicals, and radiation. The cortical layers become distinct.

7. *Liberation of spore.* The mother cell undergoes autolysis.

More than 50 genes and their products are specifically involved in sporulation, all coordinated both spatially and temporally over about 6–7 hours. The reverse of the process, first **activation** and then **germination**, takes only minutes.

C10 Bacterial flagella and movement

Key Notes

The flagellum	A single cell may have many flagella spread all over the surface of the cell (peritrichous flagella), the flagella may be polar, found at one (monotrichous if single) or both ends (amphitrichous if only two in total). Lipotrichous flagella grow in tufts from one position on the cell surface. All bacterial flagella are made up of the protein flagellin.
The structure of the Bacterial flagellum	The flagellar motor proteins are buried in the cytoplasmic membrane. The shaft of flagellin passes through the peptidoglycan via the P ring and the outer membrane via the L ring. Power for the motor is provided as part of the overall bioenergetics of the cell, with protons from the electron transport chain flowing through the motor. This flow is converted into rotation.
Movement with flagella	Bacterial movement is either in the form of a run (movement in one direction) or a tumble (rotation in a fixed position). The location of the flagella (peritrichous or polar) dictates which direction of rotation of the flagella cause runs or tumbles.
Other types of movement	Bacteria are small enough to be influenced by Brownian motion and convection currents from heat sources. Other more active forms of motion exhibited by different types of prokaryotes include gliding, secretion of slime, and a ratchet system of cell membrane proteins. Spirochetes have a unique form of twisting motion, generated by the rotation of axial filaments or endoflagella located between the cell membrane and the cell wall.
Related topics	(C8) The bacterial cell wall (F6) Signal transduction and environmental sensing

The flagellum

The most common means of locomotion in prokaryotes is via the rotation of flagella. These act as propellers, allowing the cell to swim through liquid media. A single cell may have many flagella spread all over the surface of the cell (**peritrichous flagella**), the flagella may be **polar**, found at one (**monotrichous** if single) or both ends (**amphitrichous** if only two in total). **Lipotrichous** flagella grow in tufts from one position on the cell surface. Each flagellum is not straight but is helix shaped, with the length, wavelength, and amplitude of the helix varying from species to species. Although the presence or absence, the properties, and the position of the flagella can be used as taxonomic characters, all bacterial flagella are composed of the same protein: **flagellin**. However, much like cell wall composition, the situation in Archaea is very different, in that there are many different proteins in this kingdom that perform analogous functions.

The structure of the Bacterial flagellum

The flagellin filament is only visible using the light microscope if it is complexed with another compound to increase its thickness. The filament is only 20 nm in diameter, but can be up to 20 μm in length. In Gram-negative Bacteria, the **motor proteins** (providing rotation) are buried in the cytoplasmic membrane, with a drive shaft passing through the peptidoglycan in the periplasm via the **P ring** and the outer membrane via the **L ring**. The drive shaft is attached to the flagellin filament by means of a flexible hook (Figure 1). Flagella are biosynthesized from the MS ring upwards, with motor proteins, P ring, L ring, and the hook sequentially added. The hook has a cap on top, and the flagellin filament elongates between these two substructures.

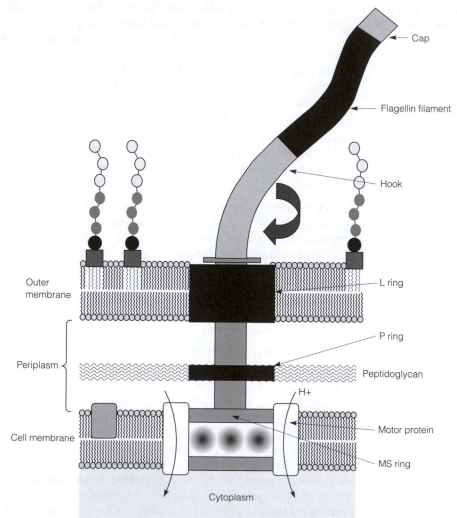

Figure 1. The Bacterial flagellum.

The flagellar motor is an adapted proton uniport (Section E2). The cell must have an electron potential gradient across the membrane and an ingress of protons through the motor causes the MS ring, and thus the flagellin, to rotate. This is an example of a simple **bioenergetic process** (Section E2).

Movement with flagella

Bacterial movement in solution can be divided into two main actions: the **run** is where the bacterium moves in a straight line towards an attractant. If the attractant moves or disappears, then the cell **tumbles** (moves around randomly on the spot) before beginning another run. Random tumbling is the only form of steering that bacteria have at their disposal. The movement towards an attractant is discussed more in Section F6. Depending on how the filaments are attached to the cell, flagella are used in three main ways to achieve locomotion:

- **Peritrichous**. Flagella rotate anticlockwise and bundle together. This allows the bacterium to move forward (run). Clockwise rotation causes the flagella to unbundle, and pushes the bacterium in every direction simultaneously (tumble).

- **Polar**. If the Bacterium has a reversible motor for its flagellum, anticlockwise rotation causes a run. Clockwise rotation can lead to either a run in the reverse direction, or a tumble. If the flagellar motor can only rotate anticlockwise, then the cell can only reorientate itself by stopping and allowing Brownian motion to randomly knock it into a suitable position to move in another direction.

- **Lipotrichous**. Runs and tumbles occur in the same way as for bacteria with peritrichous flagella.

Other types of movement

All prokaryotes appear to move under the light microscope, as they are sufficiently small to be shaken by random bombardment of local molecules (Brownian motion). In addition, convection currents generated by the light source will appear to make nonmotile bacteria move slightly. Although flagella are widespread throughout the Bacterial and Archaeal kingdoms, other forms of locomotion have evolved. Many Bacteria have a **gliding motility**, which allows flagella-free bacteria to move rapidly across surfaces. In the Cyanobacteria, this gliding is achieved by the secretion of slime, although the exact way in which this allows the Bacterium to move is poorly understood. *Flavobacterium johnsoniae* also glides, but appears to achieve this by using cell wall proteins to grab onto the surface and haul the Bacterium along a few nanometers at a time.

The spirochetes exploit their helical morphology to generate a unique form of movement, dependent on internalized flagella. These corkscrew-like cells have **axial filaments** or **endoflagella** located between the cell membrane and the cell wall. The rotation of these structures causes the whole cell to wriggle and rotate, and motion results.

C11 Prokaryotes and their environment

Key Notes

Prokaryotic niche diversity	Prokaryotes live in a wide variety of habitats at temperatures of around freezing to a maximum of 115°C. Their ability to use all naturally occurring carbon compounds as growth substrates allows them to grow in most places on earth. Despite their size, they are responsible for the cycling of many millions of tons of carbon, sulfur, and nitrogen through the atmosphere annually.
Cycling of elements through the biosphere	The biogeochemical cycling of elements, such as carbon, sulfur, and nitrogen, is mainly carried out by the Bacteria and Archaea, although the human population is making an increasing contribution. The turnover of elements allows the earth to function as a self-regulating entity, as proposed in Lovelock's Gaia hypothesis.
Detection of Bacteria in their natural habitats	For many years bacteriologists relied on isolation of laboratory cultures as a method of determining which Bacteria were present in a particular biotope. These methods are still used to some extent, but are now complemented by molecular methods, particularly fluorescence *in situ* hybridization (FISH) and what has recently been termed environmental genomics (essentially PCR-based methods).
Bacterial commensalism and microbial communities	The Bacteria and Archaea almost always grow with other species (including higher plants and animals). The majority are very difficult to grow in pure culture in the laboratory. Bacteria and Archaea grow in communities in the environment, with the individual strains and species exchanging metabolites (often in competition with their peers), so that most organic compounds are ultimately converted into biomass, carbon dioxide, and water.

Prokaryotic niche diversity

Bacteria and Archaea can grow in low temperature environments (for example, the Antarctic bacterium *Flavobacterium frigidarium* isolated from marine sediments can grow at 4°C, as can an archaeon isolated nearby, *Methanococcoides burtonii*) as well as high temperature ones. The highest known temperature allowing growth of a Bacterium is 95°C (*Aquifex pyrophilus*, isolated from an Icelandic marine thermal vent), while the archaeon *Pyrolobus fumarii* will grow at 113°C. Bacteria and Archaea are found in most places in the biosphere. They have become adapted to use a wide variety of compounds as sources of energy and carbon, and may be adapted to use those anthropogenically produced chemical compounds as well. The range of metabolic diversity means that microorganisms dominate nutrient cycling processes in almost every ecosystem. This role in the cycling of elements has led Bacteria and Archaea to form very close associations with their peers as well as with plants and animals. The close nature of their commensalism, parasitism, and symbiosis mean that pure cultures of many Bacteria and Archaea are extremely difficult to grow in the laboratory. These viable but nonculturable (VBNC) microorganisms can only be detected in biotopes using visualization techniques such as **fluorescence *in situ* hybridization** (FISH) or indirect PCR-based methods now known as **environmental genomics**.

Cycling of elements through the biosphere

James Lovelock's Gaia hypothesis portrays the earth as a self-regulating entity. The basis of this regulation is the involvement in the movement of millions of tons of chemical elements such as carbon, nitrogen, and sulfur through the atmosphere, marine, and terrestrial environments. The inter-conversion between solid and gaseous compounds of the elements is mostly accomplished by microorganisms, particularly the Archaea and Bacteria. The coupling of spontaneous chemical reactions with those catalyzed by microorganisms has given the study of these earth-scale transformations the name of **biogeochemistry**. An overall scheme showing the involvement of organisms in **biogeochemical cycling** is shown in Figure 1.

Sulfur cycle

The biogeochemical cycling of sulfur is extremely complex, and cannot be easily described diagrammatically. The complexity of the cycle is dictated by the relatively high number of oxidation states that sulfur can have in solution and in the presence of oxygen. The most significant carbon compound in the atmosphere is probably carbon dioxide, serving as a link between terrestrial and marine environments, but there is no direct sulfur equivalent. In prioritizing volatile atmospheric sulfur compounds, sulfur dioxide (SO_2), hydrogen sulfide (H_2S), carbonyl sulfide (COS), carbon disulfide (CS_2), dimethyl sulfide (CH_3SCH_3), dimethyl disulfide (CH_3SSCH_3), and methane thiol (CH_3SH) may be significant, depending on the geographical location examined. The nature of sulfur's chemistry may mean that nonvolatile molecules, such as methane sulfonate ($CH_3SO_3^-$) and sulfate (SO_4^-) may also be present in quantity in the atmosphere as dissolved aqueous ions.

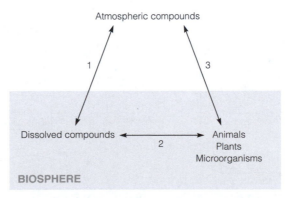

Figure 1. Biogeochemical cycling. Transformation (1) includes precipitation, occlusion at air/water interfaces; transformation (2) includes growth, fixation, secretion, and decay; transformation (3) includes decay, respiration, and combustion.

The biogeochemical cycling of sulfur serves to illustrate regulation on a global scale. The sulfur compound **dimethylsulfonium propionate** (DMSP) is produced intracellularly by marine photosynthetic algae. During and after algal blooms in the open oceans, this DMSP is released into solution on cell death, where it is metabolized by marine microorganisms to produce **dimethyl sulfide** (DMS). The generation of DMS is sufficiently high for some of the compound to enter the atmosphere as a gas above the ocean, where **photooxidation** by the sun's rays lead to the breakdown of DMS into two solid forms of sulfur: sulfate and methane sulfonate. These solids act as nuclei for the condensation of water, and clouds are formed. The clouds reduce the sunlight reaching the ocean surface, and the growth of the photosynthetic algae that produced the DMSP in the first place is reduced. This regulatory cycle stops algae covering the entire open ocean, although only in regions where man has not upset this cycle by the dumping of sewerage or other compounds that the algae might feed on.

Sulfur cycling also occurs through terrestrial ecosystems, where a combination of **sulfate-reducing Bacteria** (such as *Desulfobacter* and *Desulfovibrio* species), and **sulfate-oxidizing Bacteria** (such as *Thiobacillus* species) interconvert elemental sulfur, hydrogen sulfide, sulfate, thiosulfate, and polythionates. Sulfur is also assimilated by all organisms into the amino acids cysteine and methionine and so ultimately into proteins. Sulfur is also assimilated for the prosthetic protein groups known as iron-sulfur clusters.

Carbon cycle

Man's recent release of greenhouse gases such as carbon dioxide (CO_2) into the atmosphere is generally held to be responsible for global warming. In the absence of man, carbon dioxide serves as an atmospheric link between carbon released in the marine and terrestrial environments, with methane playing a secondary role. Outside the atmosphere, carbon forms a wide range of compounds, from simple metal cyanides to complex macromolecules such as starch, lignin, cellulose, and nucleic acids. Microorganisms take part in the ultimate conversion of all of these into carbon dioxide or methane, and these processes may be broadly divided into the aerobic and anaerobic (Figure 2a). Carbon compounds are broken down either aerobically by respiration or anaerobically by fermentation. The release of carbon dioxide is balanced by the fixation of CO_2 during photosynthesis by plants and microorganisms, or by a diversity of mechanisms described in Section E3.

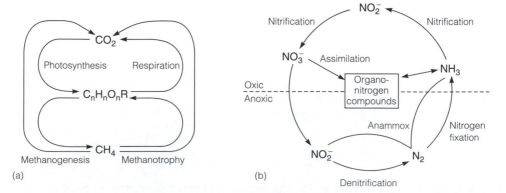

Figure 2. The biogeochemical cycling of the elements. (a) The carbon cycle. (b) The nitrogen cycle.

Nitrogen cycle

The nitrogen cycle is the easiest to define in terms of the number of participating major nitrogenous compounds. Proteins, prosthetic groups such as chlorophyll, nucleic acids, and other macromolecules can be regarded as fixed nitrogen, present in the marine or terrestrial environments. Ammonia, nitrate, and nitrite ions may participate either in the recycling of fixed nitrogen or in the generation of atmospheric molecular nitrogen (Figure 2b). The bulk of molecular nitrogen is reduced by Bacterial nitrogen fixation, while very small amounts are reduced by lightning strikes. The biologically mediated processes have been called **mineralization, nitrification, denitrification, nitrogen fixation**, and **nitrate reduction**.

Mineralization of fixed nitrogen is performed by many microorganisms during the degradation of organic matter and results in the generation of free ammonia. This ammonia may be re-assimilated or converted into nitrate by the process known as nitrification. Well-characterized nitrifiers include *Nitrosomonas europaea*. Nitrification is a two-step process via nitrite (Figure 2b).

The conversion of nitrate to molecular nitrogen (denitrification) is performed by a taxonomically undefined group of organisms including the Bacteria *Paracoccus denitrificans* and *Pseudomonas aeruginosa*, as well as the Archaea *Haloarcula marismortui* and *Pyrobaculum aerophilum*. Recent evidence suggests that some fungi can also denitrify. Nitrate is converted sequentially into nitrite, nitric oxide, and nitrous oxide before nitrogen is released from the organism. Each stage of this process is energetically beneficial to the organism and is generally coupled to the electron transport chain. For this reason it is also called **nitrate respiration** (Section E2).

Both Archaea such as *Methanosarcina barkeri* and Bacteria such as *Rhizobium* species are also capable of nitrogen fixation or **diazotrophy**. However, the rhizobia (bacteria classified as *Rhizobium* and *Bradyrhizobium*) have the unique ability to form nodules on the roots of certain plants, allowing nitrogen to be fixed almost directly from atmosphere to plant. The majority of diazotrophs are free living and employ various methods to protect the obligately anaerobic enzyme nitrogenase from the effects of molecular oxygen.

Detection of Bacteria in their natural habitats

The isolation and laboratory cultivation of a particular Bacterium from a biotope was at one time the sole evidence that that species was present. However, many studies have shown that there are many more prokaryotes in the environment than can be cultured using laboratory media, and nucleic acid analysis has complemented classical microbial ecology. Individual cells can be labeled with a dye attached to an oligonucleotide that hybridizes to ribosomal RNA. The organisms that bind the oligonucleotide will then show up under a fluorescence microscope. Careful choice of oligonucleotide can mean that these fluorescent probes can be species- or even strain-specific, allowing the microbial ecologist to examine and even count various sorts of microorganism in a particular niche. Various developments of this FISH technique have allowed the differentiation between metabolically active and inactive microorganisms.

The development of PCR means that the total ribosomal RNA (mRNA should only be produced by active cells) in a microbial community can be examined. By sequencing a library of ribosomal cDNA (complementary DNA produced from the action of reverse transcriptase on mRNA) from a particular biotope, the ecologist has some idea of the diversity of active microorganisms. Extending this idea of amplification of DNA to other genes has developed into the field of environmental genomics.

Both FISH and PCR of total genomic DNA suggest that there is a far greater diversity of Bacteria and Archaea than first thought, to the extent that the idea of discrete species barriers is beginning to break down. Although these new organisms are often held to be VBNC, new techniques and laboratory practices are finally yielding pure and mixed cultures suitable for physiological and biochemical studies.

Bacterial commensalism and microbial communities

Bacteria are classically studied as pure cultures in the laboratory. These cultures are regarded as clonal and genetically stable. However, the situation in the environment is somewhat different: Bacteria and Archaea rarely grow as pure cultures, more often forming associations with organisms around them. These associations can range from the symbiotic (Table 1) to the pathogenic. The diversity of life in a particular niche often makes **functional ecology** (the study of which organisms are responsible for which activities) challenging. Many ecosystems have evolved to include many Bacteria and Archaea exchanging and competing for metabolic intermediates to a state where no single chemical compound is completely metabolized by a single organism.

Table 1. Examples of associations between bacteria and higher organisms

Bacterial/Archaeal species	Partner	Description
Rhizobium leguminosarum	Peas and other legumes	Allows plant to fix nitrogen directly from the atmosphere
Various methanogenic Archaea	Ruminant animals	The presence of Archaea in the gut results in the Archaeal digestion of the cellulose content of plants
Symbiodinium species	Various coral species	Photosynthetic metabolism of the bacteria provides nutrients for the coral
Aeromonas veronii	Various leech species	Allows the digestion of blood
Vibrio fischeri	Squid	Confers a bioluminescent property on the squid

D1 Measurement of microbial growth

Key Notes

Types of microbial growth

Prokaryotes can use an enormous range of compounds as sources of carbon and energy. The few anthropogenic compounds they have yet to evolve to metabolize are known as xenobiotics. The mode of growth can be used as a system of classification.

History

Microorganisms have been cultured in the laboratory for more than 120 years and those who have made the greatest contributions to our understanding have included Kluyver, Monod, Novick, Szilard, and Pirt.

Growth of bacteria in liquid culture

When prokaryotes are grown in their planktonic form as free-living organisms in a liquid medium and not attached to any surface, they will divide in a regular manner until one growth factor runs out. The limiting growth factor may be its carbon source, energy source, or one of any number of micronutrients. Microbial growth can be measured conveniently by optical density (OD), a colligative property. If OD is plotted against time, the plot will be exponential during cell growth. Thus a plot of log OD against time results in a straight line with a slope equal to the specific growth rate. The growth rate can also be expressed as doubling time.

Growth phases of a planktonic culture

In a closed system to which no nutrients are added (batch culture), the growth curve of optical density and time can be divided into lag, log (or exponential), stationary (carbon-limited), and death phases. The phases between these main types are known as interphases and represent a switchover in the organism's metabolism from one mode to another.

Yield

The maximum amount of biomass per mole of a particular growth substrate is known as the yield. This can also be expressed as Y_{ATP}, the yield according to energy generated.

Interpretation of growth curves

When a microorganism is presented with more than one growth substrate, it will preferentially metabolize the one with the best energy yield first, before using the remaining substrates (diauxic growth). The growth curve will always follow the pattern of a high growth rate followed by a lower one.

Primary and secondary metabolism

A compound that is formed during the log phase of microbial growth is known as a primary metabolite. Those compounds that are formed at the very late stages of growth or during stationary phase are known as secondary metabolites. Secondary metabolites have a nonessential role in growth, are dependent on growth conditions for production, and may be produced as one of a family of similar compounds.

Related topics	(C2) Prokaryotic diversity (C6) The major prokaryotic groups (C9) Cell division	(D2) Batch culture in the laboratory (D3) Large-scale and continuous culture

Types of microbial growth

Microbes can convert an enormous range of compounds into biomass – from foodstuffs that humans would consider eating, right the way through to compounds, such as cyanides, that are considered highly toxic. Some man-made chemical compounds are new to the natural environment, so microbes have yet to evolve pathways to deal with them. These compounds are called **xenobiotics**, but eventually microbes will evolve systems to metabolize these compounds too.

We can give a very broad classification to microbes based on how they grow (Table 1) but also at which temperature and pH they do it (Figure 1). These terms can be combined, so *Thermus aquaticus* can be described as a thermophilic heterotroph, or *Paracoccus versutus* could be described as a mesophilic chemolithoheterotroph (Section C2). Many other terms are used to describe the overall modes of growth of organisms (Table 2) and these are frequently used to describe whole populations as well as individual species.

Table 1. Classification of microorganisms according to carbon and energy sources

Carbon or energy source	Description
Energy from a chemical	Chemotroph
Energy from light	Phototroph
Electrons from an organic compound	Organotroph
Electrons from an inorganic compound	Lithotroph
Carbon from carbon dioxide	Autotroph
Carbon from an organic compound	Heterotroph

Table 2. Other terms used to describe the growth of microorganisms

Growth property	Description
Obtain nitrogen from N_2 gas rather than 'fixed' nitrogen sources	Diazotroph
Ability to exist at low nutrient concentrations	Oligotroph
Uses reduced C1 compounds as sole source of carbon and energy	Methylotroph
Uses reduced methane as sole source of carbon and energy	Methanotroph
Able to use a mixture of heterotrophic and autotrophic growth mechanisms	Mixotroph
Only able to grow in the presence of salt	Halophile

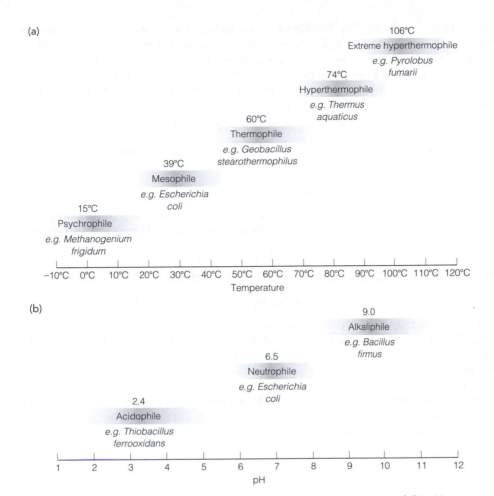

Figure 1. Classification of microorganisms according to (a) temperature and (b) pH.

History

Although microbes have been cultivated in pure and mixed cultures for more than 120 years, a mathematical interpretation of microbial growth is a relatively recent innovation. A summary of the timeline is shown below.

- **Kluyver (~ 1930)** Shake-flask technique.

- **Monod (1942)** Bacterial growth can be formulated in terms of growth yield, specific growth rate, and the concentration of the growth-limiting substrate.

- **Monod (1950); Novick and Szilard (1950)** Mathematics of continuous flow cultures.

- **Pirt (1975)** Publishes seminal book *Principles of Microbe and Cell Cultivation*, summarizing the previous 40 years' work and developing these themes.

Growth of bacteria in liquid culture

When Bacteria are grown in their **planktonic** form as free-living organisms in a liquid medium and not attached to any surface, they will divide in a regular manner until one growth factor runs out. The **limiting growth factor** may be the carbon source, energy source or one of any number of micronutrients. As Bacteria (and Archaea) grow by a process of binary fission (Section C9), at the end of a complete round of cell division there will be twice as many cells compared with the beginning of growth. If we consider a single cell, after division, this becomes two cells, those two cells become four, and so on, in other words:

$$1 \rightarrow 2 \rightarrow 4 \rightarrow 8 \rightarrow 16 \rightarrow 32 \dots$$

This could be expressed mathematically as 2 to the power of n, where n is the number of times the cell has divided.

$$2^1 \rightarrow 2^2 \rightarrow 2^3 \rightarrow 2^4 \rightarrow 2^5 \rightarrow 2^6 \dots$$

Using a method to determine the number of cells at any time (N_0) during the growth of the culture, and then the number of cells after n rounds of cell division, this becomes:

$$N = N_0 2^n.$$

The equation can be rearranged to be more useful in the lab to:

$$n = 3.3(\log N - \log N_0)$$

This is not such a useful equation as such because it is difficult to count micron-sized Bacteria into a pot. Fortunately, it is easy to measure the numbers of cells in solution using a spectrophotometer. The spectrophotometer is normally set to read light at wavelengths between 500 and 600 nm – the actual figure used depending on the species being examined. As the spectrophotometer measures the amount of light scattering in a solution (the **optical density**, the more the scattering the higher the **OD**), the readings obtained reflect the number of bacteria in solution. If the OD/number of cells is plotted against time, a graph is produced similar to the one in Figure 2.

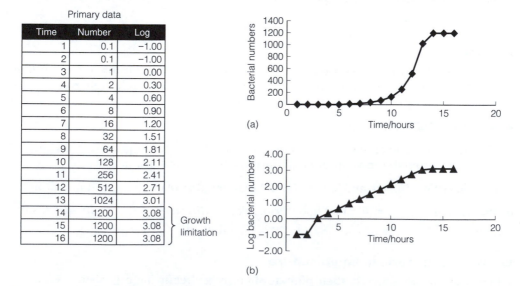

Figure 2. The relationship between cell number and time is exponential.

A curve using the primary cell number data is **exponential** for most of the curve, reflecting $N = N_0 2^n$. Exponential curves are hard to compare, so to make better sense of how the organism is growing a more useful plot is the log OD versus time. This now means that most of the curve is a straight line, the slope of which is the **specific growth rate**, μ. So for an organism grown under defined conditions (temperature, growth substrate, etc.), the specific growth rate can be defined as:

$$\mu = \frac{(\log N - \log N_0)}{t} \quad \text{or}$$

$$\mu = \frac{(\log OD - \log OD)}{t}$$

It is not always convenient to plot a complete graph of the growth of an organism, so growth rate is sometimes expressed as **doubling time**. The doubling time (t_d) of a culture is the time it takes for the OD to double, i.e. the cell numbers to multiply by two.

$$t_d = t_{2N} - t_0$$

The doubling time of *Escherichia coli* is around 20 minutes when it is growing under optimal conditions, but the doubling time of some organisms can be relatively large – the planctomycete used in the Anammox process (*Brocadia anammoxidans*, Section C6) has a doubling time in excess of 3 weeks.

Since OD is related to the number of particles in solution but not directly to concentration, it is thus a **colligative property**. If two species have different cell volumes, this means that for the same OD reading the grams biomass per ml may be different. It is possible to make a direct estimation of biomass by either taking samples and measuring the dry weight of the culture or measuring total protein or total carbon. Although these methods are accurate, none of them are rapid, so a compromise is to establish the relationship between OD and dry weight once, then extrapolate to other OD readings.

Growth phases of a planktonic culture

With the data that can be derived from OD measurements, it is now possible to draw a complete **growth curve** of a bacterium relating biomass to the changes in the state of the culture. If no new medium or medium components are added to the container, this **batch culture** will give us information over the maximum amount of biomass that can be made using the medium components, the specific growth rate, and other information.

The growth curve of a bacterium growing in a shake-flask batch culture will have the form illustrated in Figure 3.

The phases of growth can be divided into the **lag**, **log** (or **exponential**), **stationary**, and **death** phases. The phases in between these main types are known as **interphases** and represent a switchover in the organism's metabolism between one mode and another. If time = 0 is the point at which the culture was inoculated, the shape of the growth curve can be explained:

● **Lag phase** During this period the organism adapts to the medium into which it has been introduced. If it was grown up in a medium containing a different carbon source, of a less complex type, or grown at a different temperature, the lag phase can be long. The cells in the inoculum need to switch on new sets of genes and express new proteins to cope with the change in environment.

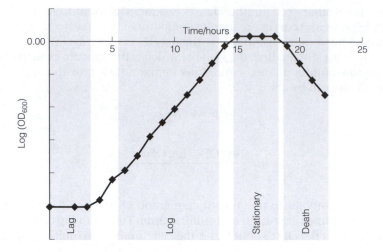

Figure 3. Growth of a bacterium in batch culture.

- **Log phase** Once all the necessary proteins are available, the cells will start dividing as quickly as they can. It is only during this phase of **exponential growth** that the equations relating to specific growth are applicable.

- **Stationary phase** Log phase cells will continue to divide, but eventually one component of the medium will run out. Media are normally designed to be **carbon limited** (i.e. the first thing to stop growth is the absence of a carbon source such as glucose or succinate), but cultures may become limited by nitrogen, phosphorus, trace elements or essential amino acids. The cells are still metabolically active but do not have the resources to divide. This stationary phase seen in laboratory cultures is indicative of the state of organisms in oligotrophic environments.

- **Death phase** After some time metabolism stops and the cells begin to die and lyse. The rate of lysis is constant over time. OD readings in this phase become unreliable, as during lysis the cell debris now contributes to light scattering. Occasionally this is manifested in a slight increase in OD as the number of particles in solution increases.

Yield

The amount of growth that can be obtained from a particular compound is of great interest to microbiologists, especially when the production of the maximum amount of biomass is required for the minimum cost. Substrates, such as pure glucose, are suitable for use on a small scale, but when tens or hundreds of liters of medium are to be made up the most efficient use of resources must be employed. Yield can be calculated simply by growing the same organism under the same conditions but with two different carbon and energy sources. The maximum OD at the end of each experiment signifies the best yield. However, this comparison is only valid if there is some way of normalizing the amount of carbon source added. An industrial microbiologist might look at yield in terms of grams of biomass per unit of currency spent on the medium. Biochemically, it would be more suitable to look at grams biomass produced per mole of carbon. If enough is known about the biochemical route by which the carbon source is utilized, it would be possible to estimate how much ATP should be generated or consumed during metabolism – for example, in comparison of the metabolism of glucose and lactose (glucose will yield more ATP). This gives the concept of Y_{ATP}, the yield according to energy generated.

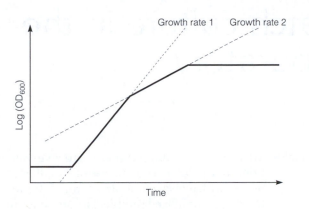

Figure 4. Diauxic growth of bacteria.

Interpretation of growth curves

Apart from the times at which the various phases of growth occur and comparing the rates of growth, we can also derive other information from growth curves. If the organism is given more than one substrate to grow on, we can see which substrate it uses preferentially. Most substrates are used sequentially by Bacteria and not at the same time. For example if *Escherichia coli* is grown in a minimal medium containing glucose and lactose, it will use up all the glucose first (growth rate 1 in Figure 4) before switching on the *lac* operon (Section F4) and using up the lactose (growth rate 2 in Figure 4). This can be explained in terms of the energetics of the compounds: the bacterium derives more ATP from the metabolism of glucose than lactose, so has a faster growth rate and uses this compound preferentially.

From this it can be seen that **diauxic growth** always follows the pattern of a high growth rate followed by a lower one.

Primary and secondary metabolism

A compound that is formed during the log phase of microbial growth is known as a **primary metabolite**. Those compounds that are formed at the very late stages of growth or during stationary phase are known as secondary metabolites. As these two definitions overlap, there is some confusion over how some microbial products might be classified, but a clear primary metabolite is alcohol from *Saccharomyces cerevisiae*. Here the production of alcohol is intrinsically linked to energy production, so accumulation of alcohol is parallel with the appearance of biomass.

Many significant industrial bioproducts (for example, almost all nonrecombinant antibiotics) are secondary rather than primary metabolites. To enhance the differences between these and the primary metabolites, secondary metabolites are also characterized as:

● Having a nonessential role in growth

● Being dependent on the growth conditions for production

● Being produced as one of a family of similar compounds (e.g. complex heterocycles differing in methylation)

The biochemical pathways of secondary metabolite synthesis tend to be longer than primary. Erythromycin is made by the action of 25 separate enzymes and tetracycline by more than 70.

D2 Batch culture in the laboratory

Key Notes

Principles of shake-flask culture

Microorganisms are commonly grown in batch culture in Erlenmeyer flasks. The process of inoculating these flasks begins with revival of the strain from long-term storage as a frozen or lyophilized sample. The strain is grown in a small quantity of medium before transfer to the final culture vessel. The inoculum of the final flask must only be 10% of the total volume to minimize carry over, overcome any quorum sensing limitations, and reduce stress. Once inoculated, the pattern of growth is subject to a number of growth-limiting factors, principal among which is oxygen availability.

Entrainment of oxygen

Shake flasks can only have a limited culture volume based on the mass transfer of oxygen to the medium they contain. The oxygen mass transfer rate of shake flasks is limited because the system relies on the agitation of the surface of the medium to allow air to form small bubbles in the medium, and the oxygen to diffuse from those bubbles to the bulk medium. This entrainment of air can only happen at the surface of the medium. Better mass transfer rates can be obtained in stirred tank reactors, although the addition of structures such as baffles can maximize entrainment.

Limitations of shake-flask culture

The main limitation of batch culture is the difficult balance that must be struck between the space available to incubate Erlenmeyer flasks and the oxygen mass transfer rate. However, the medium in a batch culture cannot be regarded as being of constant composition during the period of incubation, as microorganisms secrete many small molecules during growth. Metabolic products can change the pH of even strongly buffered media and toxic products can even induce the early onset of the stationary phase.

Fed-batch

To minimize down time, it is sometimes possible to provide a batch culture with fresh growth medium after an initial growth cycle has completed. However, the possibility still remains that toxic metabolites will not be sufficiently diluted by the fresh medium.

Related topics

(C3) Culture of bacteria in the laboratory
(D1) Measurement of microbial growth

(D3) Large-scale and continuous culture

Principles of shake-flask culture

The most common method of growing planktonic bacteria (i.e. those suspended in solution rather than attached to surfaces) in the laboratory is the shake or **Erlenmeyer** flask (see Figure 1). This type of culture is referred to as a **batch culture** as all the components are used only for a single cycle of growth, in contrast to **continuous culture** (Section D3) and fed-batch cultures (see below).

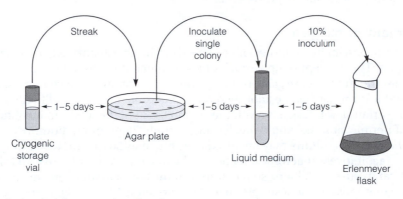

Figure 1. A simple laboratory batch culture.

The inoculation process for batch cultures is not just a question of adding a few cells to the medium. Frequently the bacterial strain must be **revived** from storage, either from a **lyophilized** (freeze-dried) or frozen state. Bacterial and Archaeal cells will withstand freezing at −70°C for many years, provided that the solution they are stored in contains a compound that prevents the formation of ice crystals. Crystals of ice are thought to puncture the cell membrane, rendering the frozen culture useless. However, a solution of 50% v/v glycerol or 20–50% DMSO (dimethyl sulfoxide) can prevent this from happening. To bring the cells back from their frozen state, a rich undefined medium (Section C3) is used. This often takes the form of an agar streak, as this can simultaneously be used to check for the purity of the stored cells. An individual colony is then picked from the revival medium and inoculated into between 1 and 10 ml of the liquid medium to be used in the experiment. This first growth in a small-scale batch culture helps the strain to adapt to the medium conditions as well as to generate sufficient biomass for the next inoculation.

The 10 ml inoculum is then used to inoculate a larger, Erlenmeyer flask. Many microorganisms seem not to grow well unless the inoculum size is between 1 and 10% of the final experimental volume. The reasons for this are obscure and vary from culture to culture but may be related to one or more of the following factors:

- **Carry over** of an essential nutrient from the inoculum medium to the experimental flask.

- Some form of **quorum sensing**. Many Bacteria have mechanisms to detect the numbers of the same species in their immediate vicinity.

- Reduction in stress. On the macro level there would appear to be little difference between adding 1 ml inoculum to 100 ml of medium and adding 10 ml of inoculum to the same volume. On the microscopic scale, the inoculum and fresh medium are not perfectly and immediately mixed, so a larger inoculum may briefly form a gradient between established and new conditions, giving the individual cells slightly longer to adapt.

Once inoculated, the Erlenmeyer flask can be placed on a rotary shaker, held at constant temperature, and growth of the organisms can begin. How the organisms grow is subject to a number of **growth-limiting factors**. In batch culture with Erlenmeyer flasks, the greatest limiting factor is always the concentration of oxygen, which will frequently dictate whether the organism grows at all.

Entrainment of oxygen

Shake flasks have a limited culture volume based on the transfer of oxygen to the medium they contain. The concept of how much of a substance it is possible to move from one phase to another (in this case gaseous to aqueous) in a system is known as **mass transfer**. By trial and error it has been established that an aerobic organism will only grow optimally if the Erlenmeyer flask has a volume 10 times that of the medium it contains. Thus a 250 ml Erlenmeyer flask should ideally contain only 25 ml of medium. When growing *Escherichia coli*, this volume may be raised to as much as 50 ml, but this is only possible as *E. coli* is facultatively anaerobic. Practically, the maximum size of standard Erlenmeyer flask that can be used is 5 liters, so if culture volumes of more than 500 ml are required, alternative methods such as a simple **stirred tank reactor** (Section D3) must be used.

The oxygen mass transfer rate of shake flasks is limited because the system relies on the agitation of the surface of the medium to allow air to form small bubbles in the medium, and the oxygen to diffuse from those bubbles to the bulk medium. This **entrainment** of air can only happen at the surface of the medium, so if the level of the medium is too far up the conical flask, the surface area available is too small for mass transfer of oxygen to the medium beneath it. Entrainment of air can be increased by making the mixing of the liquid more vigorous, frequently by increasing the speed of agitation of the flask. Those organisms with a very high demand for oxygen can demand adapted flasks in which baffles in the sides of the flasks disrupt the smooth mixing of the medium further still.

Limitations of shake-flask culture

As outlined above, the main limitation of batch culture is the difficult balance that must be struck between the space available to incubate Erlenmeyer flasks and the oxygen mass transfer rate. In addition, batch culture should not be regarded as solutions of defined composition. Before the Bacteria are introduced into the medium, it may be possible to define all the components of the medium even in exact terms of the elemental concentration of carbon, nitrogen, phosphorus, sulfur, and so on. As soon as the organism starts to grow, the composition of the medium changes, and will continue to change until the organism stops growing. We assume that biomass increases and the carbon and energy sources decrease in a regular manner. However, microorganisms secrete a variety of small molecules into the medium as they grow, from protons to heterocyclic carbon compounds. Thus during batch culture, the pH as well as the concentrations of oxygen and many other compounds not only change but can fluctuate in an unpredictable manner. To reduce the effect of pH fluctuations, the growth medium is **buffered** (normally with a phosphate buffer), but even so a change of one or more units of pH during batch culture is not uncommon. The onset of stationary phase (Section D1) is taken to be due to the lack of a suitable carbon source, but is frequently the result of the accumulation of toxic metabolites in batch culture. Despite these limitations, batch culture is a simple, quick, and for the most part reproducible method of growing small quantities of microorganisms in liquid culture.

Fed-batch

For small-scale laboratory experiments, batch culture is ideal. However, relative to the time that the organism is growing, the time taken to prepare the equipment for another experiment (the **down time**) is long. The flask must be sterilized, cleaned, refilled with medium, autoclaved, and another inoculum prepared before another experiment can begin. This time can be minimized if the majority of the biomass and spent medium is poured away or removed by pumping, and fresh medium is added directly to the flask. The residual biomass serves as the inoculum for the next growth cycle. Although this is attractive in terms of reducing down time, it increases the number of interventions in the experiment, and thus the possibility of contamination increases as well. In situations where the culture itself is undefined, such as during enrichment of microorganisms with a particular property, then contamination can be seen to be a smaller problem. However, the possibility still remains that toxic metabolites will not be sufficiently diluted by the fresh medium to allow exactly the same growth parameters after feeding compared with the primary culture.

D3 Large-scale and continuous culture

Key Notes

The simple stirred tank fermenter

Some of the limitations of shake-flask culture can be overcome by the construction of a vessel with pH control, agitation via an impeller and baffles, and aeration via a sparger. This increases the oxygen transfer rate (OTR) and allows the entire volume of a vessel to be filled with medium. The OTR can be monitored with an oxygen probe, and the culture warmed and cooled *in situ*.

Other fermenter types

Depending on the application, a number of other fermenter designs can be employed in the growth of microorganisms. These include the airlift, fluidized bed, and fixed bed reactors.

Continuous culture

If a stirred tank reactor is fed with fresh medium at a constant rate, and excess is allowed to flow to waste, a culture will be created with constant growth rate. This is known as steady-state culture. If limited by carbon source, this is called a chemostat, but biomass (turbidostat) and redox (potentiostat) variations are possible. The defined and constant nature of the culture allows the calculation of many parameters, including replacement time, maximum growth rate, or biomass and growth-limiting substrate concentrations.

Related topics

(D1) Measurement of microbial growth

(D2) Batch culture in the laboratory

Shake-flask batch culture provides a simple and convenient method of growing small amounts of microorganisms. However, if grams of biomass are required for protein purification or liters of medium are needed for product recovery then the limitations of the shake flask quickly become apparent. The most significant problem is getting enough oxygen to the culture: to grow a 5 liter culture in a single flask would require a 50 liter glass Erlenmeyer flask if the ratio of one-tenth volume of culture to the total volume of container is maintained. Furthermore, the shake flask is a highly dynamic system. As the organism grows, it excretes primary and secondary metabolites into the medium. Some of these metabolites might actually prevent efficient use of the substrate, but more significantly there may be a change in pH by many units. If a batch culture of 1 liter or more is to be grown efficiently, more controlled conditions are required.

The simple stirred tank fermenter

A culture will grow in a reproducible form if it is supplied with the same medium under the same conditions of temperature, pH, and oxygen concentration. A shake flask will provide constant aeration if the medium does not exceed 500 ml, but above this volume insufficient oxygen can be entrained at the surface of the culture. The stirred tank fermenter overcomes this primary obstacle to microbial growth in larger volumes by providing agitation via an impeller, rotating at the bottom of a circular vessel. Additional mixing of the culture may be provided by one or more baffles on the sides of the vessel (Figure 1). Oxygen, normally as sterile air, enters the fermenter underneath the impeller via a sparger. The sparger breaks up the flow of air into bubbles, which are broken into yet smaller bubbles when hitting the impeller. The combination of an impeller running at > 250 rpm, baffles, and sparger enable a highly efficient oxygen transfer to the culture. If the speed of the impeller and the rate of flow of oxygen to the fermenter are linked via a processor to an oxygen probe in the culture, the oxygen saturation of the culture can be monitored and regulated continuously.

At high biomass loads, a vessel is subject to some cooling to the atmosphere, but may even be warmed by the metabolic energy of the culture itself. Shake-flask temperature is regulated by the air temperature of an incubator, but a fermenter is often too large both spatially and in terms of heat capacity to be regulated in this way. Instead, a jacket surrounds the fermenter vessel linked to a thermostatically controlled water supply (Figure 1). This can be supplemented with warming/cooling coils in the culture itself. Finally the pH of the fermenter is kept constant with a probe linked to alkali and acid pumps.

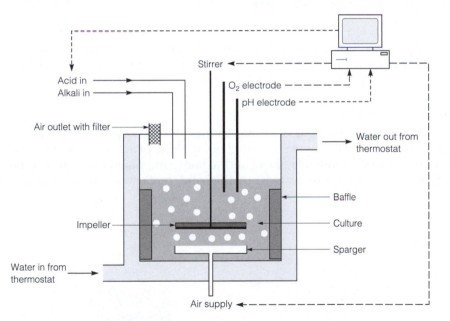

Figure 1. A simple fermenter with temperature, pH, and oxygen control.

Other fermenter types

The stirred tank fermenter provides an efficient means for the growth of most bacteria. However, as culture volumes approach hundreds of liters, the power demands of turning the impeller fast enough to ensure sufficient oxygen transfer can become too high. The **airlift reactor** (Figure 2) has no internal moving parts and stirs the culture via the passage of the air itself. This design of reactor is also useful for cells that are prone to lysis by mechanical shear.

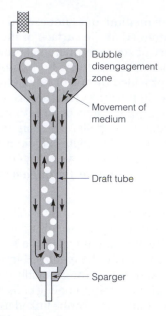

Figure 2. An airlift fermenter with draft tube.

When oxygenation is not so important (e.g. when using anaerobic or microaerophilic cells) or when immobilized enzymes are used, much simpler reactors can be employed (Figure 3), such as fluidized or fixed bed reactors.

Continuous culture

A stirred tank reactor regulated for temperature, pH, and oxygen so that all conditions remain constant can be adapted to a continuous mode of action. If the continuous culture grows so that it is limited by one of the medium components, it is called a **chemostat** (Figure 4), but limitations on biomass (**turbidostat**) or electron potential (**potentiostat**)

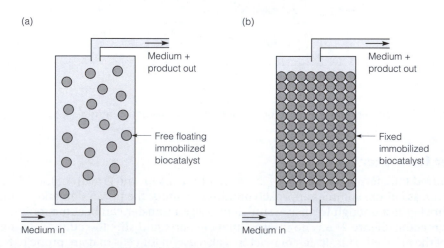

Figure 3. (a) Fluidized and (b) fixed bed bioreactors.

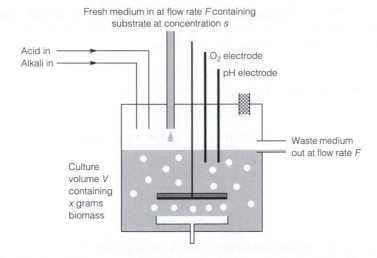

Figure 4. The chemostat. Temperature, pH, and oxygen control are omitted for clarity. x = biomass concentrations; s = growth-limiting nutrient concentration; F = flow rate of culture; V = volume.

can be imposed. In an industrial context, continuous culture is useful as it provides a constant production of biomaterials for downstream applications. In the research laboratory, the kinetics of the chemostat provide additional insight into the physiology and biochemistry of pure and mixed cultures.

We assume that the action of the impeller means that the medium coming into the culture vessel is mixed instantaneously. This is known as the **replacement time** (t_r, the time for one complete volume change of the vessel).

If V is the volume of the reactor and F is the flow rate:

$$t_r = V/F$$

If the flow rate is too high, the organism cannot divide quickly enough to maintain growth in the vessel, and eventually is diluted away (washout). The flow rate can be adjusted so that the rate of washout of biomass equals the **maximum growth rate** (μ_m). At this point any change in substrate concentration will have a direct effect on biomass. The set of conditions under which biomass remains constant over several volume changes of the chemostat is known as a **steady state**.

The **specific growth rate** of a chemostat culture is equal to the dilution rate, D

$$D = F/V$$

This is always the case as the net increase in biomass equals growth minus output. Over an infinitely small period, this becomes:

$$V.dx = V.\mu x.dt - Fx.dt$$

Dividing throughout by $V.dt$

$$dx/dt = (\mu - D)x$$

So at the steady state when $dx/dt = 0$, then $\mu = D$.

We can also calculate the **biomass and growth-limiting substrate concentrations** by rewriting the balance for the growth-limiting substrate (net increase in biomass = growth – output) for the case over an infinitely small period:

$$V.dx = F.s_r.dt - Fs - V\mu x.dt/Y$$

where Y is the growth yield. Divide throughout by $V.dt$

$$ds/dt = D(s_r - s) - \mu x/Y$$

In the steady state, $dx/dt = ds/dt = 0$, then the steady-state values of x and s (written as $\tilde{x}$ and $\tilde{s}$) are given by:

$$(\tilde{\mu} - \tilde{D})\tilde{x} = 0$$

and

$$D(\tilde{s}_r - \tilde{s}) - \mu\tilde{x}/Y = 0$$

To obtain $\tilde{x}$ and $\tilde{s}$, we can use the equation for specific growth rate

$$\mu = \mu_m s/(s + K_s)$$

where K_s is the equivalent of the Michaelis–Menten constant and is inversely proportional to the affinity of the organism for the substrate. As $\mu = D$, then

$$\tilde{s} = K_s D(\mu_m - D)$$

or

$$\tilde{x} = Y(\tilde{s}_r - \tilde{s}) = Y\{\tilde{s}_r - K_s D/(\mu_m - D)\}$$

The **critical dilution rate**, the dilution rate above which the culture begins to **washout**, is obtained from the maximum growth rate when $s = s_r$. Inserting this value in $\mu = \mu_m s/(s + K_s)$ results in:

$$\mu = D_c = \mu_m \tilde{s}_r/(\tilde{s}_r + K_s)$$

If s_r is very much greater than K_s, then it follows from the above equation that D_c is approximately equal to μ_m.

E1 Heterotrophic pathways

Key Notes

The 'bag of enzymes concept'
Many prokaryotes have complex subcellular structure, differentiation and compartmentalization of proteins, as described in other sections of this book. However, in more macrobiological systems, they can be adapted to overproduce proteins so that a single type can account for up to 15% of total protein.

High-energy compounds
Heterotrophy refers to the breaking down of organic molecules to obtain energy. This energy is generally stored in the form of high-energy compounds, such as ATP and NAD^+. The formation of such compounds relies on balanced redox reactions that generate organic molecules containing oxygen and phosphate groups.

Glycolysis and alternative pathways
There are a number of hexose monophosphate pathways (including the Entner-Doudoroff pathway, the phosphoketolase pathway, and the pentose phosphate pathway) that can be used as alternatives to glycolysis for the oxidation of glucose. These pathways yield less ATP per molecule of glucose than glycolysis, but they generate important metabolic intermediates including NADPH and pentose sugars for nucleic acid synthesis.

Citric acid cycle
The citric acid cycle occurs in the cytoplasm of aerobic Bacteria and in the mitochondria of aerobic eukaryotes. Important intermediates for fatty acid synthesis, nucleotide synthesis, and amino acid synthesis are also generated by the citric acid cycle.

Fermentation
Fermentation is the incomplete oxidation of an organic substrate and it occurs under anaerobic conditions. Energy yields from fermentation are lower than comparative yields from respiration. The products of incomplete oxidation can include pyruvate, lactate, formate, and ethanol.

Related topics
(C2) Prokaryotic diversity

The 'bag of enzymes concept'

Bacteria have been regarded by other biologists as extremely simple organisms, with no cellular or subcellular structure. As can be seen from other sections of this book, this is far from the case. Protein targeting, post-transcriptional processing, and post-translational processing are all part of the biochemistry of the bacteria. However, our detailed knowledge of this biochemistry also allows us to manipulate transcription and translation for the overproduction of proteins. An *Escherichia coli* cell can be manipulated to make more than 15% of all its protein as a single recombinant form, and as such *E. coli* is the primary expression system for most protein studies from crystallography through to Michaelis–Menten kinetics.

High-energy compounds

The ability to produce high-energy compounds for metabolism and storage is a prerequisite for cell survival. Energy is acquired by cells through a series of balanced **oxidation-reduction (redox)** reactions from organic or inorganic substrates. The simplest redox reaction can be seen in the reaction below.

$$H_2 + \tfrac{1}{2}\,O_2 \rightarrow H_2O$$

H_2 = reductant (electron donor) that becomes oxidized
O_2 = oxidant (electron acceptor) that becomes reduced

The energy that is released in redox reactions is stored in a variety of organic molecules that contain oxygen atoms and phosphate groups. **ATP**, adenosine triphosphate, is a **high-energy compound** found in almost all living organisms. It is synthesized in catabolic reactions, where substrates are oxidized, and utilized in anabolic, biosynthetic reactions. Intermediates called **carriers** participate in the flow of energy from the electron donor to the terminal electron acceptor. The co-enzyme **nicotinamide adenine dinucleotide** (NAD⁺) is a freely diffusible carrier that transfers two electrons plus a proton and a second proton from water, to the next carrier in the chain.

$$NAD^+ + 2H^+ + 2e^- \rightleftharpoons NADH + H^+$$

The reactions for the phosphorylated derivative (NADP⁺) are similar. NAD⁺ is usually used in energy-generating reactions and NADP⁺ in biosynthetic reactions.

All protozoa, all fungi, and most prokaryotes synthesize ATP by oxidizing organic molecules. This can be either via **respiration** or by **fermentation**. Respiration requires a terminal electron acceptor. This is usually oxygen, but nitrate and sulfate are among the compounds used in anoxic conditions. Fermentation requires an organic terminal oxygen acceptor.

Microorganisms can be grouped according to the source of energy they use, and by the source of carbon which may either be an organic molecule or from CO_2 (carbon dioxide fixation) (Table 1).

Glycolysis and alternative pathways

The reactions termed **glycolysis (Embden-Meyerhof-Parnas)** take place in the cytoplasm of many prokaryotes and eukaryotes. This is well described in many biochemical textbooks, including *Instant Notes in Biochemistry*.

Some important groups of Bacteria, for example some Gram-negative rods, do not use glycolysis to oxidize glucose. They use a different mechanism, the **Entner-Doudoroff** pathway (Figure 1), which yields one mole of ATP, NADPH, and NADH from every mole of glucose. This is a **hexose monophosphate pathway** (HMP) and in this pathway only one molecule of ATP is produced per molecule of glucose metabolized.

Another HMP is the **phosphoketolase** pathway, which is another method for glucose breakdown found in *Lactobacillus* and *Leuconostoc* spp. when grown on five-carbon sugars (pentoses). The pathway produces lactic acid, CO_2, and either ethanol or acetate (Figure 2).

Table 1. Classification of microorganisms by energy and carbon source utilized

	Type	Electron donor	Energy source	Carbon source	Examples
Organotrophs	Chemo-organotroph	Organic compounds	Redox reactions of organic compounds	Organic compounds	All fungi, protists, most terrestrial bacteria
	Photo-organotroph	Organic compounds	Light	Carbon dioxide and organic compounds	Nonsulfur Bacteria
Lithotrophs	Chemo-lithotrophs	Inorganic compounds	Redox reactions of inorganic compounds	CO_2	*Thiobacillus Nitrosominas, Nitrobacter, Hydrogenimonas Beggiatoa*
	Photo-lithotrophs	Inorganic compounds	Light	CO_2	Photosynthetic green and purple Bacteria, photosynthetic protists

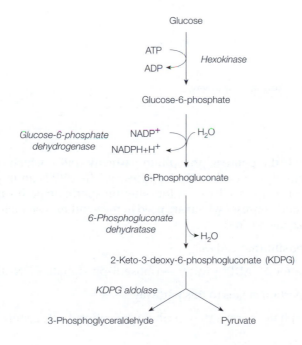

Figure 1. Entner-Doudoroff pathway.

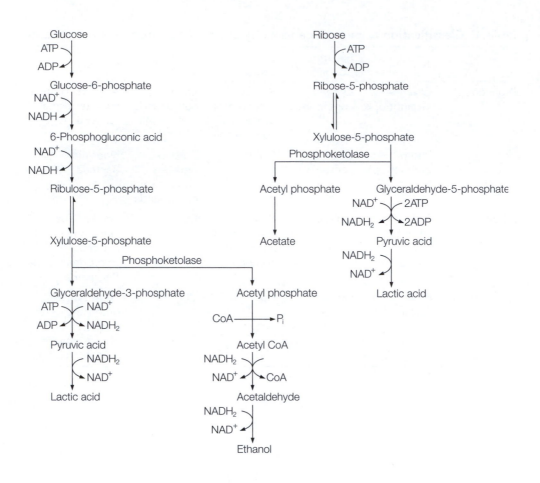

Figure 2. The phosphoketolase pathway.

An important HMP is the **pentose phosphate pathway** (PPP), which often operates in conjunction with glycolysis or other HMP pathways. The PPP is an important provider of intermediates that serve as substrates for other biosynthetic pathways. This pathway yields NADPH + H+ and pentoses which are used in the synthesis of nucleotides including FAD, ATP, and coenzyme A (CoA).

The reactions can be summarized as:

glucose-6-phosphate + 2 NADP+ + water → ribose-5-phosphate + 2 NADPH + 2H+ + CO_2

There are three important stages to this pathway.

1. Glucose-6-phosphate is converted to ribulose-5-phosphate, generating two NADPH + 2H+.

2. Ribulose-5-phosphate isomerizes to ribose-5-phosphate.

3. Excess ribose-5-phosphate is converted to fructose-6-phosphate and glyceralde-hyde, via a series of reactions, to enter glycolysis.

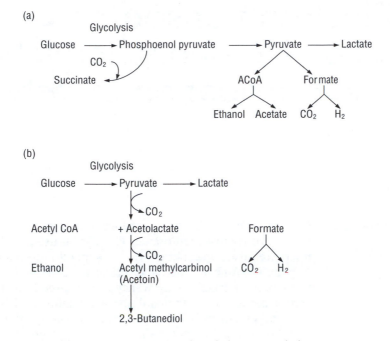

Figure 3. Products of fermentation. (a) Mixed acid. (b) Butanediol.

Citric acid cycle

The **citric acid cycle** is found in the cytosol of aerobic prokaryotes and in the mitochondria of eukaryotes. Anaerobic organisms have incomplete cycles while facultative aerobic organisms only have a functional citric acid cycle in the presence of oxygen. This is well described in *Instant Notes in Biochemistry*. Important intermediates for fatty acid synthesis, nucleotide synthesis, and amino acid synthesis are generated by the citric acid cycle.

Fermentation

Fermentation is an **incomplete oxidation** of an organic substrate. During fermentations an electron donor becomes reduced and energy is trapped by **substrate-level phosphorylation**. Fermentation products include pyruvate if the **glycolytic** pathway is used, or lactate, formate, 2,3-butanediol, and ethanol from the **butanediol pathway** (used by *Klebsiella, Erwinia, Enterobacter,* and *Serratia* spp.) or succinate, ethanol, acetate, and formate from a **mixed acid fermentation** (found in *Escherichia, Salmonella,* and *Shigella* spp.). See Figure 3 for details.

E2 Electron transport, oxidative phosphorylation, and β-oxidation of fatty acids

Key Notes	
Electron transport	Electron transport is used to create a proton motive force (PMF) across membranes. This PMF is used by all microorganisms to generate ATP via a membrane-bound ATPase. The Archaea and Bacteria also use PMF to drive the movement of flagella, allow transport of charged substrates across membranes, and maintain their osmotic potential. In eukaryotic microbes, PMF is established across the inner membrane of the mitochondrion. In the electron transport chain, a series of balanced oxidation and reduction reactions drives the movement of electrons through the carrier series from NADH to oxygen. During this process energy is released and ATP is synthesized.
Anaplerotic pathways	Lost intermediates from glycolysis and the citric acid cycle are replenished by anaplerotic reactions, where carbon dioxide is fixed into three-carbon compounds by carboxylation reactions.
Glyoxylate cycle	Some substrates that microbes can utilize as carbon sources, for example the two-carbon compound acetate, can lead to the depletion of citric acid cycle intermediates. Reactions that result in the loss of CO_2 during the cycle can be avoided by using the glyoxylate cycle.
Fatty acid oxidation	Fatty acids can be used as substrates by microorganisms through the fatty acid or β-oxidation pathway. This is located in the mitochondria of eukaryotes and the cytoplasm of prokaryotes.
Anaerobic respiration	Many microbes live in low or no oxygen environments. Alternative electron acceptors, such as nitrate and sulfate, can be utilized instead of oxygen by these organisms to complete the electron transport chain.
Related topics	(C11) Prokaryotes and their environment (H2) Eukaryotic cell structure

Electron transport

Peter Mitchell theorized that the generation of ATP only occurred because mitochondria and Bacteria could pump protons across a membrane. These primary pumps lead to generation of a charge across the membrane, known as the **proton motive force** (PMF). As the protons try to move back across the membrane, the energy of their movement can be harnessed in a number of ways. The most important use of the PMF in many aerobic organisms is the **generation of ATP**, normally using the enzyme f_1f_0 ATPase. This ATP-generating enzyme is found in the cytoplasmic membrane of Bacteria and the inner membrane of the mitochondria and in the absence of the PMF actually cleaves ATP into ADP plus phosphate. However, when PMF is applied, the ATPase works essentially in reverse, generating ATP from ADP and phosphate. It is thus known as a **secondary pump**. For all this to happen efficiently, the organism or organelle must conform in several ways to the Mitchell hypotheses, which are as follows:

- Protons are pumped across mitochondrial and bacterial membranes in such a way as to generate an **electric potential** across the membrane. A membrane-bound enzyme (ATPase) couples synthesis of ATP to the flow of protons down the electric potential gradient.

- Solutes can accumulate against a concentration gradient by the coupling of proton flow to the movement of the solute by a **transmembrane protein**. These cotransporters may act as **symports**, **antiports** or **uniports**.

- The flow of protons through the flagellar transmembrane proteins rotates components of the flagellum and allows a prokaryote to move.

There are five types of component molecule.

1. Enzymes that catalyze transfer of hydrogen atoms from reduced NAD$^+$ to flavoproteins (NADH dehydrogenases).

2. Flavoproteins. Flavin mononucleotide (FMN) and flavin adenine dinucleotide (FAD). The flavins are reduced by accepting a hydrogen atom from NADH and oxidized by losing an electron.

3. Electron carriers, cytochromes. Cytochromes are porphyrin-containing proteins each of which can be reduced or oxidized by the loss of a single electron:

$$\text{cytochrome-Fe}^{2+} \rightarrow \text{cytochrome-Fe}^{3+} + e^-$$

4. Iron-sulfur proteins. These are carriers of electrons with a range of reduction potentials.

5. Quinones. These are lipid-soluble carriers that can diffuse through membranes carrying electrons from iron-sulfur proteins to cytochromes.

Protons for the final reduction in the transfer chain are supplied by the disassociation of water, providing the build up of hydroxyl ions on the inside of the membrane.

Protons flow back into the mitochondrial matrix or bacterial cell through the enzyme ATP synthase, driving ATP synthesis. This enzyme is in two parts, one localized on the Bacterial cytoplasmic or mitochondrial matrix side and the other which spans the membrane to the outside of the bacterial cell or the intermembrane space of the mitochondrion.

The rate of **oxidative phosphorylation** is set by the availability of ADP, electrons only flow down the chain when ATP is needed. When there are high levels of ATP and energy-rich compounds like NADH[+] and FADH$_2$, accumulation of citric acid inhibits the citric acid cycle and glycolysis.

However, an alternative theory, the **conformational change hypothesis**, proposes that ATP synthesis occurs because of conformational changes created in the ATP synthase enzyme by electron transport. This theory is currently being intensively researched.

Anaplerotic pathways

The intermediates of glycolysis and the citric acid cycle are used as precursors of biosynthetic pathways. To maintain the energy-yielding processes of glycolysis and the citric acid cycle these lost intermediates must be replenished by **anaplerotic reactions**. Three-carbon compounds are carboxylated to form oxaloacetate. Both pyruvate and phosphoenol pyruvate (PEP) can be used in these reactions, for example, oxaloacetate can be formed from the enzymatic addition of carbon dioxide to pyruvate or PEP.

Glyoxylate cycle

A number of organic acids can be used by microorganisms as electron donors and carbon sources. Those that are common to the citric acid cycle – citrate malate, fumarate, and succinate, for example – can be metabolized using the enzymes of the citric acid cycle. However, utilization of acetate via the citric acid cycle will cause the depletion of oxaloacetate. If this occurs the citric acid cycle could not operate. To compensate for the loss of oxaloacetate the **glyoxylate shunt** occurs (Figure 1).

In this pathway, which shares many of the reactions of the citric acid cycle, reactions that give rise to the release of CO_2 are bypassed and instead isocitrate is split into succinate plus glyoxylate by the enzyme isocitrate lyase. Succinate can be used in biosynthetic reactions, while glyoxylate is combined with acetyl CoA via malate synthase to form malate, which enters the citric acid cycle.

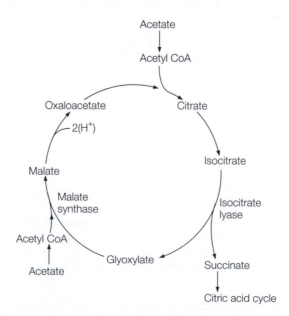

Figure 1. Glyoxylate shunt.

Fatty acid oxidation

Fatty acids can be used as substrates for microbial metabolism. The metabolic process is called β-**oxidation** and it occurs in the mitochondria of eukaryotes or the cytoplasm of prokaryotes. Two-carbon units are removed from the fatty acid to yield their acyl CoA. The pathway begins with the activation of the fatty acid by CoA. There then follow two separate dehydrogenation reactions where electrons are transferred to FAD and NAD^+, finally yielding CoA and an activated fatty acid to restart the cycle. The complex series of reactions for fatty acids with odd or even numbers of carbons and saturated or unsaturated bonds between them are described in detail in many biochemistry textbooks, including *Instant Notes in Biochemistry*.

Anaerobic respiration

Anaerobic respiration occurs in prokaryotes that are unable to use oxygen as a terminal electron acceptor. These organisms are termed **obligate anaerobes**. Other prokaryotes can use anaerobic respiration facultatively, if oxygen happens to be unavailable. Less energy is generated during anaerobic respiration than in aerobic respiration. Nitrate, sulfate, and carbon dioxide can be used as **alternative electron acceptors**.

Nitrate respiration uses the most common alternative electron acceptor, nitrate (Figure 2). The first step of the reaction is catalyzed by the enzyme nitrate reductase, an enzyme that is only synthesized under anaerobic conditions. The product is nitrite, which is excreted by most Staphylococci and Enterobacteria. Other Bacteria will reduce nitrite further to ammonia or nitrogen gas. This reaction, and the enzyme that catalyzes it, are termed **dissimilatory** because nitrogen is reduced during the biological breakdown of organic compounds. This type of respiration leads to **denitrification** (Section C11).

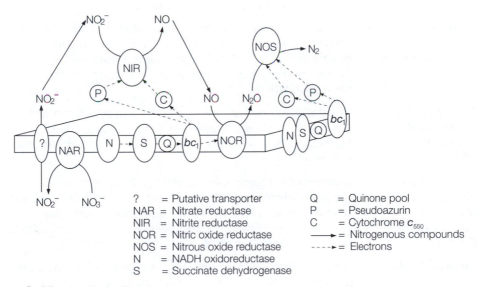

Figure 2. Nitrate respiration.

E3 Autotrophic reactions

Key Notes

Chemolithotrophy

Autotrophic microorganisms can survive in the absence of organic carbon sources by fixing atmospheric or dissolved CO_2 to form carbohydrates. Chemolithotrophs have the ability to fix CO_2 using the Calvin cycle, and the energy required to drive the reactions comes from the oxidation of inorganic substrates such as ammonia.

Photosynthesis

Photosynthesis can be divided into two sets of reactions, those that are light-dependent (light reactions) and those that are light-independent (dark reactions). The light reactions convert light into chemical energy through the synthesis of ATP, which is then used to drive the Calvin cycle (dark reactions). Photosynthesis may be described as oxygenic if oxygen is generated (as in the Cyanobacteria and the photosynthetic eukaryotes) or as anoxygenic if it is not (as in the green and purple Bacteria). The light reaction can be driven by photosystems I and II in eukaryotes, but may only be driven by photosystem I in some prokaryotes.

Light reactions in bacterial photosynthesis

Photosynthetic green and purple Bacteria contain chlorophyll A and B, and carry out anoxygenic photosynthesis that utilizes only photosystem I.

Light reactions in eukaryotic photosynthesis

In eukaryotes, photosynthesis occurs in the chloroplasts and involves photosystems I and II. The light-dependent reactions generate $NADPH + H^+$ and the resulting proton gradient is used to generate ATP by noncyclic phosphorylation.

Dark reactions in eukaryotic and prokaryotic photosynthesis

The dark (light-independent) reactions of photosynthesis are called the Calvin cycle and use the energy generated from light-dependent reactions to synthesize carbohydrates from CO_2 and H_2O.

Other light-independent mechanisms for fixing carbon dioxide

Autotrophic heterotrophs cannot use intermediates generated during metabolism of some classes of compounds as carbon sources. Instead these compounds are oxidized to CO_2 and water completely, and then intracellular CO_2 is used directly in the Calvin cycle.

Related topics

(C6) The major prokaryotic groups	(J2) Archaeplastida, Excavata, Chromalveolata, and
(I2) Fungal nutrition	Amoebozoa: nutrition and metabolism

Chemolithotrophy

Chemolithotrophy is found in a limited number of microorganisms. Chemolithotrophs obtain their energy by the oxidation of inorganic substrates and their carbon from CO_2. However, these reactions yield less energy than oxidation of glucose to CO_2, so large quantities of substrates have to be metabolized to generate enough energy for sufficient ATP and NADH generation. An example of this is in the process whereby ammonia is oxidized to nitrate (nitrification, Section C11) and ATP can be generated from this reaction. However, electrons cannot be donated directly for NADH production from ammonia or nitrate because they have a more positive redox potential than NAD^+. Instead a process termed 'reversed electron flow' allows electrons from the oxidation of nitrite to be used to generate small but sufficient amounts of NADH (Figure 1) for growth. A similar process is used by sulfur-oxidizing bacteria, where the oxidation of sulfite (or thiosulfate) to sulfate yields electron reducing power for the production of NADH and ATP.

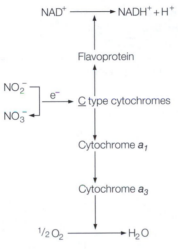

Figure 1. Reversed electron flow.

Photosynthesis

A large number of microorganisms have the ability to use sunlight to generate ATP by photophosphorylation. This process may not generate oxygen, a reaction termed anoxygenic, as found in the green and purple Bacteria, or it may generate oxygen, termed oxygenic, by the photolysis of water, as found in the blue green Bacteria and algae (Table 1). The reactions are complex but can be divided into two sets of reactions, the light reactions where light energy is converted into chemical energy (ATP) and the dark reactions where ATP is used to synthesize glucose.

Light reactions in bacterial photosynthesis

Photosynthetic Bacteria contain **bacteriochlorophylls** a and b, with absorption maxima of 775 and 790 nm, respectively. These pigments are contained within sac-like extensions of the plasma membrane called **chlorosomes** in green sulfur and nonsulfur bacteria and **intracytoplasmic vesicles** in purple Bacteria. Bacterial photosynthesis is an **anoxygenic** (nonoxygen-producing) photosynthesis that relies on photosystem I only, and is termed a cyclic phosphorylation (Figure 2).

Table 1. Classification of photosynthetic microorganisms according to hydrogen (reductant) and carbon source

Nutritional classification	Examples	Carbon source	Hydrogen source	Oxygen evolution
Primarily photolithotrophs	Green sulfur Bacteria (Chlorobiaceae)	CO_2, acetate, butyrate	H_2, H_2S, $S_2O_3^{2+}$	Negative
	Purple sulfur Bacteria, (Chromatiaceae)	CO_2, acetate, butyrate	H_2, H_2S, $S_2O_3^{2+}$	Negative
Photo-organotrophs	*Purple nonsulfur Bacteria (Rhodospirillaceae)	Organic (CO_2)	H_2, organic	Negative
	*Green gliding Bacteria (Chloroflexaceae)	Organic (CO_2)	H_2, organic	Negative
	Halobacteria (Archaea)	Organic	Organic	–
Photolithotrophs	Blue green Bacteria (Cyanobacteria)	CO_2	H_2O	Positive
Photolithotrophs	Photosynthetic protista	CO_2	H_2O	Positive

*Can grow as chemoorganotrophs aerobically in the dark.

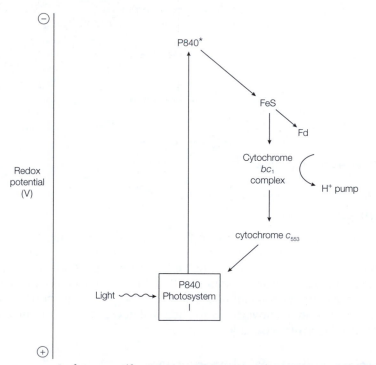

Figure 2. Photosystem I of green sulfur Bacteria. FeS, iron sulfur protein; Fd, ferredoxin.

There is no net change in the numbers of electrons in the system. ATP synthesis occurs during the generation of a protein motive force during photosynthesis, which allows ATP synthase to synthesize ATP. The electrons expelled from the reaction center return to the bacteriochlorophyll via the electron transport chain. The photosynthetic apparatus consists of four membrane-bound pigment–protein complexes, plus an ATP synthase. For NADPH synthesis, Bacteria must use electron donors like hydrogen, H_2S, sulfur, and organic compounds with a more negative reduction potential than water. In this case direct transfer can occur via ferredoxin.

The purple sulfur Bacteria cannot synthesize NADPH directly by photosynthetic electron transport. This is because their acceptor molecules are more positive than the $NADP^+/$NADPH couple (–0.32 volts). In this case electrons enter the cytochromes from the electron donors, and ATP from the light reactions is used to reduce $NADP^+$ to NADPH, by energy-dependent reverse electron flow (Figure 3).

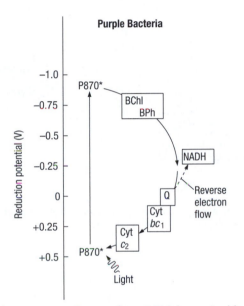

Figure 3. Energy-dependent reverse electron flow. BChl, bacteriochlorophyll; BPh, Bacterial pheophytin.

Light reactions in eukaryotic photosynthesis

Photosynthesis occurs within **antenna complexes** and **reaction centers** in the thylakoid membrane of chloroplasts in eukaryotic microorganisms. Antenna complexes are formed from several hundred chlorophyll molecules plus accessory pigments. Light excitation of the chlorophyll molecule results in an electron in a chlorophyll molecule being excited to a higher orbit, and this energy is transferred between chlorophylls until it is channeled into the chlorophyll molecules of the reaction center.

The reaction center contains two photosystems, called **photosystem I (PS I)** and **photosystem II (PS II)**, with different light-energy absorption maxima. PS I absorbs at 700 nm, and PS II at 680 nm. The reaction centers are linked by other electron carriers, and if the components are arranged by their redox potentials they assume a Z shape, so the scheme is called the **Z scheme** (Figure 4).

The reactions of the Z scheme generate NADPH from NADP. ATP is generated by noncyclic phosphorylation reactions because of the creation of a **proton gradient** between the thylakoid space and the stroma by the reactions of PS I and PS II. H^+ is pumped from the stroma into the thylakoid space. ATP is generated as protons return to the stroma via an ATP synthase present in the thylakoid membrane (Figure 5). PS I may operate without

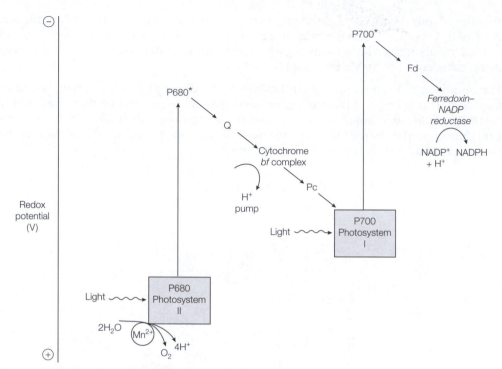

Figure 4. The Z scheme for noncyclic photophosphorylation. From Hames D & Hooper N (2011) *Instant Notes Biochemistry*, 4th ed. Garland Science.

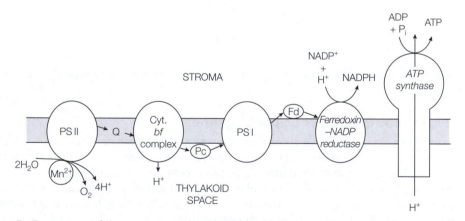

Figure 5. Formation of the proton gradient and ATP synthesis. From Hames D & Hooper N (2011) *Instant Notes Biochemistry*, 4th ed. Garland Science.

PS II in some circumstances, and in this reaction no O_2 is produced; only ATP is produced via the proton gradient.

Dark reactions in eukaryotic and prokaryotic photosynthesis

The light-independent reactions of photosynthesis use the NADPH and ATP generated from the light reactions to synthesize carbohydrates from CO_2 and water. This is called the **Calvin cycle**. A key enzyme in this cycle is **RuBisCo** (Figure 6) (ribulose bisphosphate

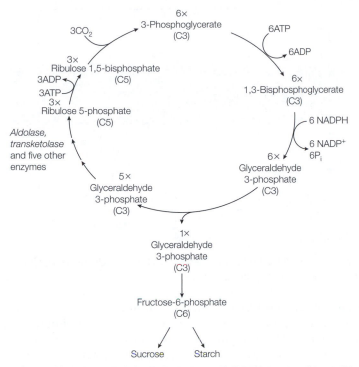

$3CO_2$

6×
3-Phosphoglycerate
(C3)

6ATP

6ADP

3×
Ribulose 1,5-bisphosphate
(C5)

3ADP

3ATP

3×
Ribulose 5-phosphate
(C5)

6×
1,3-Bisphosphoglycerate
(C3)

6 NADPH

6 NADP+
6Pᵢ

*Aldolase,
transketolase*
and five other
enzymes

6×
Glyceraldehyde
3-phosphate
(C3)

5×
Glyceraldehyde
3-phosphate
(C3)

1×
Glyceraldehyde
3-phosphate
(C3)

Fructose-6-phosphate
(C6)

Sucrose   Starch

Figure 6. The Calvin cycle. From Hames D & Hooper N (2011) *Instant Notes Biochemistry*, 4th ed. Garland Science.

carboxylase), a large, multi-component enzyme. This enzyme incorporates CO_2 into ribulose 1,5-bisphosphate to form first a six-carbon compound, which then splits to form two three-carbon molecules (3-phosphoglycerate). Subsequent reactions regenerate ribulose 1,5-bisphosphate from one of the 3-phosphoglycerate molecules, to continue the cycle. The other molecule of 3-phosphoglycerate is transported to the cytosol and used in respiration and to produce storage sugars.

Other light-independent mechanisms for fixing carbon dioxide

Some α-proteobacteria (notably species of *Rhodobacter* grown in the dark and *Paracoccus*) do not have the metabolic machinery to utilize some carbon compounds directly for anabolic reactions. When growing as methylotrophs (Section E4) the organisms cannot use C1 intermediates as a source of carbon. C1 compounds are instead oxidized completely to carbon dioxide and water, and then intracellular carbon dioxide is used directly in the Calvin cycle as a means of fixing carbon. Organisms with this ability could be called **autotrophic heterotrophs**. Some chemolithotrophs fall into this category too as nitrification and sulfur oxidation can be regarded as waste products of other bacteria. If both hydrogen and oxygen are present, the chemolithotroph *Cupriavidus necator* (formerly *Ralstonia eutropha*) uses a membrane-bound hydrogenase to chemiosmotically generate ATP (Figure 7). A separate cytosolic hydrogenase is used to generate NADH so that both NADH and ATP can be used in a normal Calvin cycle (see *Instant Notes in Biochemistry* for further details). This aerobic oxidation of hydrogen should not be confused with the anaerobic processes used by the sulfate-reducing Bacteria (Section E4).

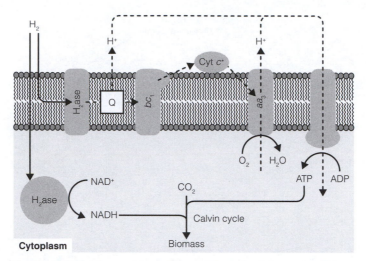

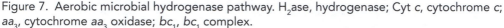

Figure 7. Aerobic microbial hydrogenase pathway. H$_2$ase, hydrogenase; Cyt c, cytochrome c; aa_3, cytochrome aa_3 oxidase; bc_1, bc_1 complex.

The problem of synthesizing carbon-containing compounds in environments rich in CO$_2$ but poor in organic compounds has also been overcome in a different way by *Chloroflexus* and some Archaea. These organisms use carboxylases to convert acetyl CoA or propionyl CoA to malyl CoA and glyoxylate. The glyoxylate can then be used for anabolism, while acetyl CoA is regenerated to complete the '**hydroxypropionate cycle**' (Figure 8).

The concept of using what we regard as central catabolic pathways to anabolic reactions is taken to a logical extreme in the **reverse citric acid cycle**. The citric acid (TCA) cycle normally evolves CO$_2$ and produces reducing equivalents. However, by using very reduced compounds such as sulfide, thiosulfate or hydrogen as electron donors, carbon can be fixed by species of *Chlorobium* using CO$_2$ and water and the TCA cycle in reverse. This process requires an additional 2 ATP to yield acetyl CoA so that this molecule can act as an anabolic intermediate (Figure 9).

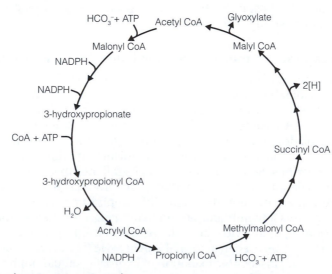

Figure 8. The hydroxypropionate cycle.

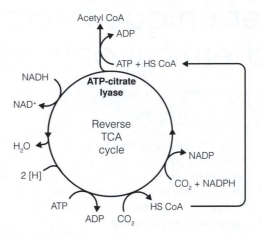

Figure 9. The reverse citric acid (TCA) cycle.

E4 Other unique microbial biochemical pathways

Key Notes

Nitrogen fixation

Some Bacteria can fix atmospheric nitrogen when fixed nitrogen sources (nitrate, nitrite or ammonia) are not available. The key anaerobic enzyme for this process is nitrogenase. Microorganisms use a variety of strategies to keep nitrogenase oxygen-free while normal aerobic metabolism takes place.

Methylotrophy

Methylotrophs are organisms that can use reduced C1 compounds such as methane, methanol, and methylamine as a source of carbon and energy. C1 compounds are first oxidized at the expense of NADH or ATP to methanol, and the energy is recovered by subsequent oxidations all the way to carbon dioxide and water. The key enzyme of methylotrophy, methanol dehydrogenase, uses an unusual prosthetic group (pyrroloquinoline quinone) to oxidize methanol to formaldehyde. Methanol toxicity is kept to a minimum by using tetrahydromethanopterin as a carrier molecule for the aldehyde group.

Iron oxidation

Bacteria can use reduced iron as an electron donor at low pH when growing as autotrophs. The energy from the oxidation of iron (II) to iron (III) is enough to drive ATP synthesis, and the appearance of iron (III) salts in acid mine run off is of a characteristic deep red color.

Anaerobic hydrogen oxidation

Acetogenic and methanogenic prokaryotes oxidize hydrogen to water anaerobically, using the energy to translocate protons and so generate ATP chemiosmotically.

Methanogenesis

Archaea have many biochemical pathways in common with the Bacteria. A unique property of the Archaea is the possession of a pathway that converts hydrocarbons into methane with the release of energy.

Sulfate reduction

Desulfovibrio and *Desulfotignum* can reduce sulfate and assimilate the sulfur as cellular biomolecules, or harvest energy from the reaction via a dissimilatory pathway. Sulfur itself can act as an electron acceptor and is reduced to hydrogen sulfide directly by many Bacteria.

Related topics

(C11) Prokaryotes and their environment
(I2) Fungal nutrition

(J2) Archaeplastida, Excavata, Chromalveolata, and Amoebozoa: nutrition and metabolism

Nitrogen fixation

Assimilation of nitrogen in microbes is very variable. Only one group can utilize atmospheric nitrogen, in a process called **nitrogen fixation**. This reaction is only seen in *Proteobacteria* such as *Azotobacter* or *Rhizobium* as well as in Gram-positive Bacteria such as some species of *Clostridium*.

Nitrogen fixation is mediated by an oxygen-sensitive enzyme called **nitrogenase**

$$N_2 + 6H^+ + 12\,ATP + 12\,H_2O \rightarrow 2NH^{3+} + 12\,ADP + 12\,P_i$$

Nitrogenase plays a pivotal role in the global nitrogen cycle (Section C11). This reaction is an extremely energy expensive one and is carried out anaerobically. However, most nitrogen fixers are aerobes, so a variety of simple differentiation techniques are used to create anaerobic compartments. The heterocysts of filamentous Cyanobacteria are found in normal chains of aerobic cells, but are the main sites of nitrogen fixation. *Azotobacter* uses high rates of partly uncoupled aerobic metabolism for high turnover of oxygen near the cell wall. This allows the cytoplasm to become slightly anaerobic and so nitrogen fixation can occur.

Methylotrophy

Methylotrophs are organisms that can use reduced C1 compounds such as methane, methanol, and methylamine as a source of carbon and energy. This is an oxidative form of growth that usually has methanol, formaldehyde, and formate as intermediates (Figure 1). Methylotrophs are ubiquitous in our environment, but specialists such as the methanotrophs (obligate methylotrophs growing on methane such as *Methylococcus capsulatus*) are found in the aerobic parts of marshes or hot springs near where methanogenesis takes place.

Compounds such as methane are first oxidized at the expense of NADH or ATP to methanol, and the energy is recovered by subsequent oxidations all the way to carbon dioxide and water. Although organisms such as *Paracoccus* can use the CO_2 generated this way for autotrophic assimilation of carbon via the Calvin cycle, non-autotrophs such as *Methylobacterium extorquens* can assimilate carbon at the level of formaldehyde. These heterotrophic methylotrophs must balance the dissimilatory energy generation steps with assimilation. Heterotrophs use complex pathways (the serine, xylulose monophosphate or ribulose monophosphate pathways) to transfer the methyl group in formaldehyde onto molecules compatible with central anabolic pathways.

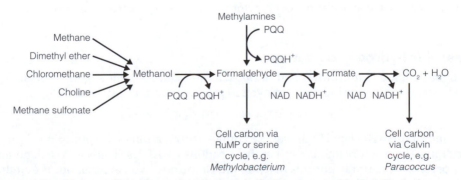

Figure 1. Microbial transformations during growth on C1 compounds. RuMP, ribulose monophosphate.

The key enzyme of methylotrophy, methanol dehydrogenase, oxidizes methanol to form-aldehyde using an unusual prosthetic group. Pyrroloquinoline quinone (PQQ) interacts with the electron transport system during its reduction and concomitant production of formaldehyde. Formaldehyde is toxic to most organisms and it is unusual to see it as a cytoplasmic intermediate. Methylotrophs keep formaldehyde levels very low, using a tet-rahydromethanopterin transferase to move the aldehyde group to a less toxic compound for further metabolism.

Iron oxidation

Bacteria such as *Acidithiobacillus ferrooxidans* and *Leptospirillum ferrooxidans* and the Archaea *Ferroplasma* are able to grow using reduced iron as an electron source, gaining carbon from the atmosphere. These iron oxidizers grow at low pH (from 1 to 3) and are frequently found in high numbers in acid mine run off. Here, oxygen is low and iron high. The energy from the oxidation of iron (II) to iron (III) (Figure 2) is enough to drive ATP synthesis when coupled to some of the components of the electron transport system. The iron oxidation system is responsible for the deep red color (caused by iron III salts) found in many mine-polluted rivers and streams. Phototrophic bacteria (such as *Chlorobium*) can also perform a similar reaction at higher pH, coupling iron oxidation to sulfide oxida-tion or even denitrification.

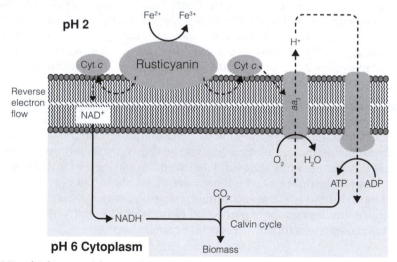

Figure 2. Microbial iron oxidation. Cyt *c*, cytochrome *c*; *aa$_3$*, cytochrome *aa$_3$* oxidase.

Anaerobic hydrogen oxidation

Acetogens such as *Clostridium* and *Acetobacterium* can ferment sugars to acetate, but can also use carbon dioxide and hydrogen to produce acetate and water:

$$2HCO_3^- + 4H_2 + H^+ \rightarrow CH_3COO^- + 4H_2O$$

This membrane-associated hydrogenase couples the reaction to the pumping of protons or sodium. The re-entry into the cell of these cations via ATPase leads to ATP generation. Most organisms carrying out this reaction are Gram-positive Bacteria, but the Archaeal methanogens (see below) can also oxidize hydrogen anaerobically to yield methane and water.

Methanogenesis

Energy metabolism in the Archaea has many similarities to either the Bacteria or the Eukarya, subject to slight modifications (e.g. the Entner-Doudoroff pathway for glucose catabolism). Pathways cannot be generalized because of the metabolic diversity of the Archaea, but the metabolism of carbon sources resulting in the release of methane (methanogenesis) is unique. Substrates such as carbon monoxide, formate, and carbon dioxide are metabolized anaerobically in the presence of hydrogen:

$$CO_2 + 4H_2 \rightarrow CH_4 + 2H_2O$$

C1 compounds such as methanol, methylamine, and dimethylsulfide are used when hydrogen is an electron donor in the following way:

$$CH_3OH + H_2 \rightarrow CH_4 + H_2O$$

However, some methanogens can use methanol in the absence of hydrogen:

$$4CH_3OH + CH_4 \rightarrow CO_2 + 2H_2O$$

Compounds such as acetate are cleaved in an acetotrophic process:

$$CH_3COO^- + H_2O \rightarrow CH_4 + HCO_3^-$$

All these reactions are chemiosmotically linked to ATP synthesis. The acetotrophic reaction yields the least amount of energy per mole substrate, whereas the CO_2-type substrates yield the most.

Sulfate reduction

The bioconversion of sulfate to hydrogen sulfide and organic sulfur compounds is a crucial part of the biogeochemical sulfur cycle. Bacteria such as *Desulfovibrio* and *Desulfotignum* perform this function along with a single genus in the Archaea (*Archaeoglobus*). These organisms use sulfate (SO_4^{2-}) as a source of energy (dissimilatory metabolism) and as a source of sulfur (assimilatory metabolism). Although little energy can be derived from the reduction, assimilatory and dissimilatory reactions can be linked to the use of hydrogen, pyruvate or lactate as electron donor. Dissimilation of sulfate uses ATP and proceeds via an intermediate known as APS (adenosine phosphosulfate, Figure 3). A further reduction produces sulfite (SO_3^{2-}) and the release of adenosine monophosphate. A final reduction yields hydrogen sulfide, which is then excreted to be used by other microorganisms or released into the atmosphere. Bacteria can assimilate the sulfur by a further ATP-consuming reaction to produce PAPS (phosphoadenosine-5'-phosphosulfate). Reduction to sulfite is this time accomplished by using NADPH. Assimilation into compounds such as the amino acid cysteine is again via hydrogen sulfide (Figure 3).

Sulfur itself can act as an electron acceptor and is reduced to hydrogen sulfide directly by many Bacteria including species of *Desulfuromonas*, *Wolinella*, and *Campylobacter* when growing anaerobically. A wide variety of electron donors are used and the bioenergetics of the pathway are still not fully understood.

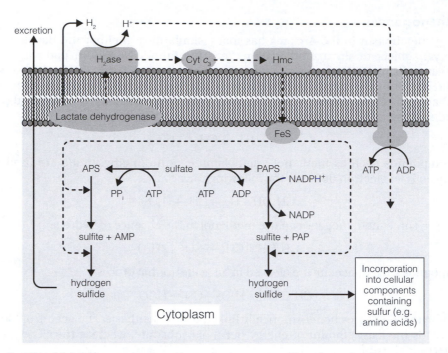

Figure 3. Microbial sulfate reduction. H_2ase, hydrogenase; Hmc, hexadecaheme cytochrome; Cyt c_3, cytochrome c_3; FeS, iron-sulfur protein.

SECTION F – PROKARYOTIC DNA AND RNA METABOLISM

F1 DNA – the primary informational macromolecule

Key Notes

Microbiology and the discovery of DNA

Microbiology has played a pivotal role in the discovery of the mechanisms involved in the transfer of information within and between cells.

DNA transforms bacteria

The earlier discovery of DNA by Miescher was finally put in context by the discovery that this was the same molecule as Griffith's "transforming principle". Avery, MacLeod, and McCarty showed that nonvirulent *Streptococcus pneumoniae* could be transformed into a pathogen by incubation with DNA from virulent *S. pneumoniae*.

Genetic information is nucleic acid, not protein

Hershey and Chase differentially labeled protein and DNA to show that information was passed from T2 bacteriophage by DNA and not by protein.

The relationship of purines to pyrimidines

Chargaff also showed that:

- The ratio of A to C to G to T varied between different species but not between individuals of the same species.

- The ratio of A to C to G to T did not vary between tissues of the same individual, and does not change with environmental conditions or age.

- In all DNA samples, the concentration of A was always the same as that of T, and the concentration of G was always the same as that of C.

The DNA double helix

The experiments performed by Chargaff, Avery *et al.*, and others, allowed Watson and Crick to not only propose a structure for DNA, but also to postulate that the replication of DNA must be semi-conservative, with separation of the strands before synthesis of complementary DNA. They also proposed that codons might explain the interaction of DNA, RNA, and protein, with an adaptor hypothesis. Later, all these suppositions were verified experimentally.

Related topics

(F5) Messenger RNA and translation   (F10) Bacteriophages

Microbiology and the discovery of DNA

Microbiology has played a pivotal role in the discovery of the mechanisms involved in the transfer of information within and between cells. The biggest milestone in molecular genetics was the elucidation of the structure of DNA by Watson, Crick, Franklin, and Wilkins in the early 1950s, but the cellular role for this macromolecule had already begun to be established.

DNA transforms bacteria

Although biochemistry was well advanced by 1930, the identity of the molecule responsible for the transfer of genetic information between mother and daughter cell was not agreed upon. Protein was held to be the most likely candidate as it held sufficient diversity in sequence to code for the many cellular variations. DNA had been purified as early as 1868 (by **Friedrich Miescher**, from pus in surgical bandages) and many of its characteristics determined, but had not been explicitly linked to information flow. A breakthrough came in 1944 when **Avery**, **MacLeod**, and **McCarty** showed that DNA could confer virulence on non-pathogenic *Streptococcus pneumoniae*. This finding was built on earlier work by **Griffith**, who had shown that the same effect could be achieved with heat-killed bacteria (Figure 1).

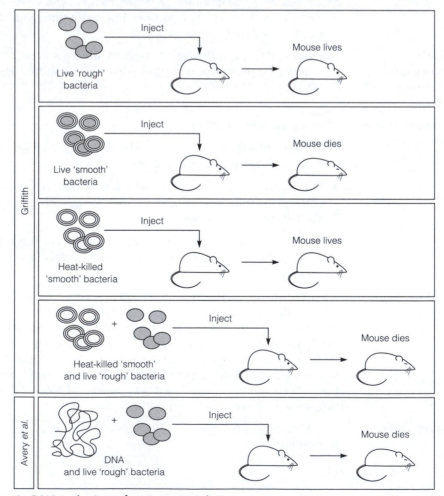

Figure 1. DNA is the 'transforming principle.'

Griffith described virulent *S. pneumoniae* as smooth due to the appearance of the colonies on agar plates. The virulence factor is a capsule that causes the smooth colonies and allows the Bacterium to cause pneumonia in mice. The finding that live noncapsulate Bacteria could cause the death of mice when mixed and incubated with heat-killed capsulate bacteria was not fully explained at the time. Griffith did not fractionate *S. pneumoniae* but did suggest that a subcomponent of the capsulate Bacteria was responsible, a component he called the '**transforming principle**.' The fractionation was performed by Avery, MacLeod, and McCarty, who went on to not only find that the principle was DNA, but also propose that DNA was a carrier of genetic information. In many ways these experiments were a series of fortunate events, in that Griffith could have chosen many other Bacteria to study, but chose one that was naturally competent (Section F8) under the conditions he used. Furthermore, *S. pneumoniae* can be transformed by linear fragments of double-stranded DNA (dsDNA), and had Avery, MacLeod, and McCarty carried out their experiments with *E. coli*, they would not have obtained such easily explained results. Meanwhile, the debate over whether DNA was the only repository for genetic information continued.

Genetic information is nucleic acid, not protein

Another microbiological experiment was used to determine if DNA alone could be responsible for the persistence of genetic information in a daughter cell. The experiment supposed that labeled DNA would be detectable in all progeny of a cell, and conversely that labeled protein would also persist, if either were the primary informational macromolecule. This sounds simple in theory, but practically bacteriophages (Section F10) were the only way to manufacture a genetic entity with either DNA labeling only or protein labeling only. **Hershey** and **Chase** ^{32}P-radiolabeled T2 bacteriophage DNA, and then allowed the self-assembly of virions (Section F10) with unlabeled protein. Concurrently they ^{35}S-radiolabeled T2 bacteriophage coat protein, and made virions with unlabeled DNA (Figure 2).

By interrupting the infection of *E. coli* by T2, by using a blender, they were able to separate the heads from the cells by centrifugation. In this way they demonstrated that the DNA entered the cell and caused a change but not the T2 protein, conclusively demonstrating that DNA carries genetic information.

The relationship of purines to pyrimidines

The composition of DNA had been rigorously examined by **Chargaff** and colleagues in the late 1940s. They established that DNA had a set of properties, commonly known as **Chargaff's rules**. These were as follows:

- The ratio of A to C to G to T varied between different species but not between individuals of the same species.

- The ratio of A to C and G to T did not vary between tissues of the same individual, and does not change with environmental conditions or age.

- In all DNA samples, the concentration of A was always the same as that of T, and the concentration of G was always the same as that of C, commonly summarized as **A + G = T + C.**

The structure of DNA could not have been worked out without these rules, and the implications of the structure would not have been fully appreciated without the work on the transforming principle and interrupted T2 infection.

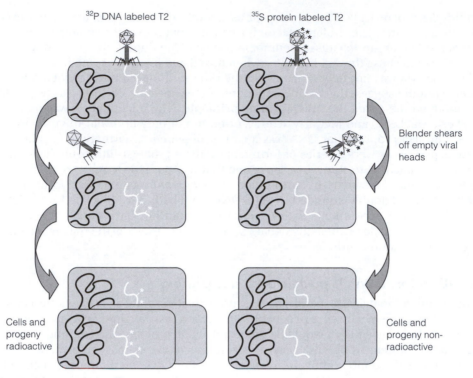

Figure 2. The Hershey–Chase experiment.

The DNA double helix

The process by which the structure of DNA was found to be a double helix has been well documented elsewhere. It is important to see the structure in the context of the previous chemical and biological evidence – for example, the double helix was only regular in the way that X-ray diffraction experiments suggested if G was paired with C and T with A. More remarkably, Watson and Crick were able to suggest much more about the biochemistry of DNA than the mere shape of the molecule alone. They proposed that the replication of DNA must be **semi-conservative**, with separation of the strands before synthesis of **complementary** DNA. Subsequently, the very nature of DNA as an informational macromolecule led them to suppose the existence of **codons** and thus to an **adaptor hypothesis**. In conjunction with the discovery of mRNA (Section F5) they proposed that, in order for amino acids to reflect DNA sequence, a molecule must intervene between the two. This led Crick to draw a molecule with many similarities to tRNA. The combination of these hypotheses surrounding DNA structure has led to the elucidation of most of our understanding of molecular biology, from gene to mRNA to protein (via tRNA).

F2 Genomes

Key Notes

Comparative genomics	The information available from completely and partially sequenced genomes allows comparison of gene function across phylum boundaries. The relatively small size of microbial genomes means that many more species have complete sequences compared with higher organisms.
Generalized structure of the Bacterial genome	The singular circular model of a Bacterial genome is applicable to *E. coli*, but other bacteria have multiple chromosomes, some of which are linear. Genome size seems to be related to the organism's ecological niche, with obligate intracellular pathogens having the smallest genomes and free-living organisms the largest. A prokaryote's genome may include one or more plasmids, which can provide specialized genetic information for activities not normally associated with cell growth (e.g. antibiotic production). Eukaryotes and prokaryotes differ in the arrangement of genes. Relatively speaking, Bacterial genomes are composed mainly of genes, whereas regulatory systems for genes comprise most plant and animal genomes.
Generalized structure of Archaeal genome	A typical Archaeal genome is very similar to that of a Bacterial one. Generally the chromosome is single and circular, of a similar size to the Bacteria and may be complemented by the presence of plasmids. However, Archaea have many genes and genetic systems in common with eukaryotes as well as Bacteria.
Eukaryotic genomes	Eukaryotes have many more chromosomes than prokaryotes (between 4 and 105 per cell) and do not generally have plasmids. Extrachromosomal DNA does appear in the form of the mitochondrial and chloroplast genomes, which are held outside the cell nucleus.
Related topics	(C2) Prokaryotic diversity (H2) Eukaryotic cell structure (C6) The major prokaryotic groups (K3) Virus genomes (C7) Composition of a typical prokaryotic cell

---

Comparative genomics

With genomic data, via DNA sequencing, becoming more easily available, it is becoming increasingly relevant to consider the whole of a microorganism rather than its individual genes. In this way subtle differences can be examined (for example, what makes *Bacillus anthracis* the causative agent of anthrax compared to the genetically very similar *Bacillus cereus*). The arrangement of genes relative to one another, their presence or absence, the sequence of the genes, and intergenic regions are all used not only to compare species of Bacteria, but also to examine more distant relationships, for example, between higher animals and microorganisms. The size of higher organisms' genomes (the human genome is 3200 million bp (Mbp) has made widespread whole genome sequencing a task performed by a few consortia of public and private laboratories worldwide. However, the relatively small size of Bacterial and Archaeal genomes (3–8 Mbp) has led to the release of new complete genome sequences on an almost weekly basis. The way in which DNA sequences are obtained and assembled into genome-sized pieces is beyond the scope of this text, but a primary consideration in approaching genomics is how the microorganism's genome is arranged. The structure of the viral genome is considered in Section K3.

Generalized structure of the Bacterial genome

The Bacterial genome is often portrayed as a stable, single, circular molecule. However, the genomes of most Bacteria are fluid (constantly changing in response to external stimuli) and composed of several molecules including extra chromosomes, megaplasmids, and plasmids.

The model organism for molecular biology, *Escherichia coli*, is considered to be the paradigm for all Bacterial and Archaeal genomes. However, its single haploid circular chromosome, consisting of around 4.6 million bp, is rather unusual compared with other genera, but is by far the best studied. Other Bacterial genomes comprise several chromosomes, some of which are circular and some of which are linear (Table 1).

The size of a Bacterial genome is related to the ecological niche in which the organisms live. Obligate pathogens, such as the causative agent of epidemic typhus (*Rickettsia*

Table 1. Chromosomal structures of Bacterial genomes*

(Total size bp)	Number of circular chromosomes	Number of linear chromosomes	Extrachromosomal DNA
Escherichia coli (4 639 221 bp)	1	0	Plasmids may be present
Mycoplasma pneumoniae (816 394 bp)	1	0	Some species in the genus have a single plasmid
Rickettsia prowazekii (1 111 523 bp)	1	0	No known plasmids
Paracoccus denitrificans (~3 740 000 bp)	3 (2 + 1.1 + 0.64 Mbp)	0*	Plasmids and megaplasmids may be present
Cupriavidus necator (5 810 922 bp)	2	0	Plasmids may be present
Deinococcus radiodurans (3 284 156 bp)	2	0	Plasmids may be present
Streptomyces coelicolor (6 667 507 bp)	0	1	Linear and circular plasmids may be present

*An additional linear chromosome has been found in the closely related species *Paracoccus pantotrophus*.

prowazekii), seem to have minimized their genomes to such an extent that they rely on host proteins and metabolites in order to replicate. This is taken to the extreme in the smallest known genome, that of *Carsonella ruddii*, which is composed of only 159 663 base pairs of DNA. In comparison, free-living organisms, such as the metabolically versatile *Pseudomonas aeruginosa* and *Streptomyces coelicolor*, have to cope with changes in temperature over tens of degrees, varying carbon and energy sources in the space of minutes, and other environmental challenges. As a consequence they have a larger complement of genes regulated by a more complex sensing apparatus, and thus a larger genome.

Another strategy used by microorganisms to cope with transient environmental change is the acquisition of plasmids. Plasmids are small circular extrachromosomal pieces of DNA, which replicate independently of the genome. In contrast to the singular genome, there may be between 10 and 100 000 complete copies of a plasmid in a Bacterial cell. Plasmids may carry genes that allow the microorganism to become pathogenic (one of the main differences between species of *Salmonella* is the presence of plasmid(s) carrying pathogenicity factors), resist antibiotics (resistance to kanamycin, streptomycin, and many other antibiotics may be carried on plasmids) or metabolize a particular set of compounds (for example, the proteins making up the *xyl* pathway used by *Pseudomonas putida* for the degradation of toluene). Occasionally these plasmids are integrated into the genome and only exist as extrachromosomal DNA in the presence of certain physiological stimuli. While the plasmids that are used in molecular biology are in the range of 2.5–10 thousand bp (Kbp), naturally occurring plasmids can be many hundreds of thousands of base pairs in size, bringing into question the philosophical difference between these megaplasmids and the chromosomes themselves.

The characteristics that distinguish Bacterial genomes from the eukaryotes lie mainly in how the genetic information is arranged. Relatively speaking, the Bacterial genome is information-rich, containing many regions coding for proteins and RNA but comparatively few regions involved with the regulation of expression. Genes of similar function tend to be clustered together, and often genes in a single metabolic pathway or all involved in the synthesis of a complex multi-subunit protein are found in operons (Section F4). Genes in an operon are sometimes so tightly packed together that they overlap. The fluidity of the Bacterial genome is reflected in gene order found in different Bacterial genera: there is no similarity in the arrangement of genes among the major phyla, and often gene order is very different in species of the same genus.

Different Bacterial genomes have varying composition in terms of nucleotides. The G+C content of the Bacteria ranges from 25 to 75%, and this is often reflected in the more frequent use of certain codons for certain amino acids (termed codon usage). While Bacterial genomes do contain repeating elements, they are often long repeats of >10 bp and may be associated with pathogenicity islands, insertion sequences or the remnants of excised lysogenic bacteriophage.

Generalized structure of Archaeal genome

A typical Archaeal genome is very similar to that of a Bacterial one (Table 2). Generally the chromosome is single and circular, of a similar size to the Bacteria, and may be complemented by the presence of plasmids (Section F11). The main differences are in the fine structure of the arrangement of genes and the proteins that associate with the genomic DNA.

While the Archaea have operons and tend to exhibit clustering of genes according to function, the arrangement of the genes has elements in common with both the eukaryotes

Table 2. Chromosomal structures of Archaeal genomes

(Total size bp)	Number of circular chromosomes	Number of linear chromosomes	Extrachromosomal DNA
Methanococcus jannaschii (1 664 970 bp)	1	0	Species characteristic plasmids of 58 407 bp and 16 550 bp
Aeropyrum pernix (1 669 695 bp)	1	0	No known plasmids
Pyrococcus horikoshii (1 738 505 bp)	1	0	Plasmids may be found

and the Bacteria. An Archaeal operon may contain genes that have close relatives in both the other kingdoms, and rarely the genes themselves may be made up of domains that may have origins in different kingdoms. However, about a third of the genes in any archaeon are unique to this kingdom.

The replication origin of the Archaeal genome has many features in common with the eukaryotes and this similarity in the gross chromosomal features is apparent through the use of histone-like proteins to stabilize the chromosomal tertiary structure.

Eukaryotic genomes

The smallest eukaryotic genome is that of the parasite *Encephalitozoon cuniculi* (2.5 million bp), and many eukaryotic microorganisms have smaller genomes than the larger more differentiated organisms. Eukaryote genomes are characterized by having a large number of chromosomes (between 4 and 105 in the haploid state, Table 3) but

Table 3. Features of eukaryotic genomes

Organism	Description	Genome size (bp)	Haploid number of chromosomes
Encephalitozoon cuniculi	Human pathogen	2 507 519	11
Saccharomyces cerevisiae	Budding yeast	12 495 682	4
Cyanidioschyzon merolae	A unicellular red alga	16 520 305	10
Plasmodium falciparum	Causes malaria	22 853 764	7
Neurospora crassa	Fungus	38 639 769	7
Caenorhabditis elegans	Microscopic worm	100 258 171	6
Arabidopsis thaliana	Flowering plant	115 409 949	5
Drosophila melanogaster	Fruit fly	122 653 977	4
Homo sapiens	Human	$\sim 3.3 \times 10^9$	23
Fugu rubripes	Pufferfish	$\sim 3.65 \times 10^8$	21
Oryza sativa	Rice	$\sim 4.3 \times 10^8$	18
Amphibians	Various	$10^9 - 10^{11}$	
Equisetum arvense	Horsetail	$> 10^{11}$	105

do not generally have stable extrachromosomal DNA as plasmids. As well as the nuclear chromosomes, some of the cell organelles (mitochondrion, chloroplast) have their own chromosomes, which code for proteins specific for the function of the organelle. For detailed information on the human genome, consult *Genomes 3* (Brown TA (2007) Garland Science).

Fungal genomes are characterized by their lack of introns (only 43% of *Saccharomyces pombe* genes contain introns of a total of 4730). These introns are small, being only 50–200 bp in size compared with the introns of >10 kb in mammals. Although the genes are not as tightly packed as in Bacteria or Archaea, fungal genomes are information-rich and contain little repetitive DNA (50–60% of the *S. cerevisiae* nuclear genome is transcribed, compared with 33% of *Schizophyllum commune* and only 1% in *Homo sapiens*).

F3 DNA replication

Key Notes

DNA replication

DNA replication in all organisms proceeds via similar mechanisms. Although the speed of replication can vary, it proceeds by common rules:

- Replication is semi-conservative.
- All DNA and RNA polymerases synthesize DNA in a 5′ to 3′ direction.
- All cellular DNA and RNA polymerases initiate synthesis with an RNA primer.
- All DNA and RNA polymerases require magnesium ions to function.

Synthesis of nucleic acids by polymerases

All the bacterial nucleic acid polymerases share a common hand-like structure, with the template polynucleotide binding to the palm (the active site) and nucleotides entering between the thumb and forefinger. Bacteria have several DNA polymerases, but only one core RNA polymerase, whereas the eukaryotes have many specialized variants of both.

Initiation of DNA replication

DNA inside the cell is highly supercoiled, so must be unwound and rewound by topoisomerase I or DNA gyrase as replication proceeds. Bacterial DNA is always replicated from the origin of replication. Two primosomes are formed here and DNA replication occurs bidirectionally.

DNA replication fork

The DNA replication fork is formed at the origin of replication by the action of DNA helicase and re-annealing of the single strands is prevented by single-stranded binding protein. The opened helix is termed the replication bubble, in which the primosome forms. DNA primase adds a few complementary RNA nucleotides to the template strand, which acts as a primer for the main DNA replication complex, DNA polymerase III. The leading strand is synthesized continuously, but the lagging strand is made in short lengths that are eventually joined together. Primase adds short RNA fragments (Okazaki fragments) to the lagging strand, allowing DNA polymerase III to extend from 5′ to 3′ until it meets the next RNA primer, where extension stops. The short stretches of RNA are then removed by DNA polymerase I and any breaks in the phosphodiester bonds are mended by DNA ligase.

Proofreading	DNA polymerases I and III both have a separate active site that allows these enzymes to check for perfect complementarity as DNA is synthesized.
Termination of synthesis and resolution of replicated circular genomes	The movement of the two replisomes is stopped at 180° from *ori* by the binding of terminator utilization substrate (Tus). This leaves the two complete circular chromosomes intertwined, a situation that is resolved by the action of XerC and XerD at the *dif* site.
Linear genomes	The full details of the replication of Bacterial linear genomes have yet to be elucidated but, in *Borrelia burgdorferi*, a central *ori* is replicated bidirectionally and terminated by unknown mechanisms. In the bacteriophage φ29 replication is initiated at either end of the chromosome and is terminated in the center by the collision of the two replisomes.
Mutation and adaptation	Stable inheritance of genetic information is maintained by systems that check and remediate DNA. Microorganisms need these systems to cope with factors such as ultraviolet light and mutagenic chemicals, to prevent induced mutagenesis. Errors in the DNA replication machinery can also cause spontaneous mutagenesis, although in some genomes (at mutational hot spots) this type of mutagenesis seems to be allowed.
	Any strain derived from this wild type with any change in its genomic make up compared with this wild type is known as a mutant. If the difference in genotype results in an observable change to the properties of the organism, this is a change in phenotype. The mutation of the wild type with a nutritional requirement results in an auxotroph.
Related topics	(F1) DNA – the primary informational macromolecule (F7) DNA repair (F9) Recombination

DNA replication

DNA replication in all organisms takes place via similar mechanisms, but much of the detail has been gained from our understanding of the molecular biology of the bacterium *Escherichia coli*. In all bacteria, DNA replication is triggered by cell mass, but the speed of genomic duplication varies from species to species. In *E. coli*, new bases are added at approximately 1000 nucleotides per second (nt s^{-1}), but in *Mycoplasma capricolum* this rate falls to only 100 nt s^{-1}. This means that, although the mycoplasma has a considerably smaller genome, both bacteria can completely replicate their genomes in around 45 minutes.

There are two rules that govern how DNA replication can occur. The first is that replication is **semi-conservative**, in that double-stranded DNA is separated and each strand acts as a template for synthesis of a new strand composed of bases exactly complementary to the template. The second is that all nucleic acid polymerases (DNA and RNA polymerases) synthesize DNA in a 5′ to 3′ direction using a primer to initiate synthesis. This primer is

RNA, whether or not DNA or RNA is to be synthesized. Nucleotides (also known as bases) are added to the exposed 3′-hydroxyl group of the primer so that they are exactly complementary to the template strand (A matched with T or U; C matched with G).

Synthesis of nucleic acids by polymerases

All enzymatic extension of polynucleic acids requires a template and the synthesis of nucleic acids proceeds in a 5′ to 3′ direction. The RNA and DNA polymerases, in common with many other enzymes associated with nucleic acid metabolism, also require Mg^{2+} or a similar divalent ion to function. This ion is not strictly a prosthetic group for these enzymes, but helps in the binding of nucleotides to the active site. All the bacterial nucleic acid polymerases share a common structure, in that the protein around the active site folds to form a structure that could be said to be like a hand. The template polynucleotide binds to the palm (the active site) and nucleotides enter between the thumb and forefinger.

Bacterial cells have a variety of DNA polymerases (Table 1) each with specialized functions, but most only have one RNA polymerase. In contrast most eukaryotes have a plethora of both specialized DNA and RNA polymerases. Unfortunately our knowledge of the cell biology of the Archaea is not sufficiently developed to be able to draw many inferences on this topic, but we do know that their primary replicative polymerase resembles one class of eukaryotic DNA polymerase more than any of the bacterial types.

Table 1. Properties of bacterial and Archaeal DNA polymerases

Polymerase	Function	Origin	Direction of synthesis
DNA polymerase I	Aids in both replication and repair	Bacteria	5′ → 3′
DNA polymerase II	Primary repair complex	Bacteria	5′ → 3′
DNA polymerase III	Primary replication complex	Bacteria	5′ → 3′
Family B polymerase	Primary replication complex	Archaea	5′ → 3′

Initiation of DNA replication

Our knowledge of Bacterial chromosomal replication is mostly limited to circular chromosomes composed of double-stranded DNA. Rather than being arranged as a simple circle, the DNA is **supercoiled** – twisted and folded in on itself. This is to compact the DNA into the very small volume of the cell. The model microorganism, *E. coli*, is only 2 or 3 μm long, yet its 4.6 million bp chromosome is about 1 mm in circumference. The number of supercoils in the genome is determined by the competitive action of two winding and unwinding enzymes, topoisomerase I and DNA gyrase. The equilibrium between the two is established according to the physiological state of the organism, so is influenced by external environmental factors. In addition, topoisomerase and gyrase have to act locally to allow DNA replication to begin. The supercoiled DNA must first be relaxed to allow replication proteins access to the template strands.

Bacterial DNA is always replicated from a single location on the chromosome, known as the **origin of replication**. This origin (abbreviated as *ori*) is a sequence of around 300 base pairs and here a group of enzymes collectively called the **primosome** acts (Figure 1). Two primosomes are formed at the origin of replication of circular genomes, so DNA replication occurs **bidirectionally**.

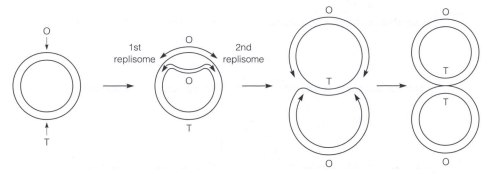

Figure 1. Replication and resolution of circular bacterial genomes. O, *ori*; T, *dif*.

DNA replication fork

Once DNA replication has been initiated, a large number of enzymes must function to allow DNA polymerase access to the template strand. The DNA is separated into single strands by **DNA helicase**, and spontaneous re-annealing is prevented by **single-stranded binding protein** (SSB). The primosome is then formed in the developing replication bubble, allowing the binding of **DNA primase**. This primase adds a few complementary RNA nucleotides to the template strand, which acts as a primer for the main DNA replication complex, DNA polymerase III (Figure 2). The extension of DNA complementary to the template strand means that the replication bubble can be extended quickly. However, it was less clear for many years how the other strand (often called the lagging strand) was replicated at the same time and yet still synthesized from 5′ to 3′.

An elegant explanation of the extension of the replication bubble was that it was discontinuous, again using DNA polymerase III extending from RNA primers. It was found that primase regularly adds short RNA fragments (**Okazaki fragments**) to the lagging strand, allowing DNA polymerase III to extend from 5′ to 3′ until it meets the next RNA primer, where extension stops. The lagging strand is, for a short time, a DNA:RNA heteroduplex, before DNA polymerase I functions in its repair mode and removes RNA with an **exonuclease** activity, replacing it with complementary DNA nucleotides. Any gaps in the newly synthesized lagging strand are then resolved by DNA ligase. In this way, both the template (or leading) and lagging strands of the replication bubble are quickly and accurately replicated (Figure 2).

Proofreading

At each stage of DNA replication, the sequence of the newly synthesized DNA is checked to see that it exactly matches the template strand, a function known as proofreading. First the incorporation of each nucleotide is dictated so that it exactly matches the complementary template base just by the shape of the DNA polymerase III active site. A separate active site in the enzyme (in the epsilon subunit) then checks the base pairing again after incorporation as the polymerase moves down the template strand. Similarly, DNA polymerase I has a proofreading domain in many organisms (though not in *E. coli*) to ensure that the replacement of RNA with DNA is accomplished without error (Figure 3). DNA polymerase I also participates in the continuous checking of the replicated double-stranded genome as part of the mutation repair systems described in Section F7.

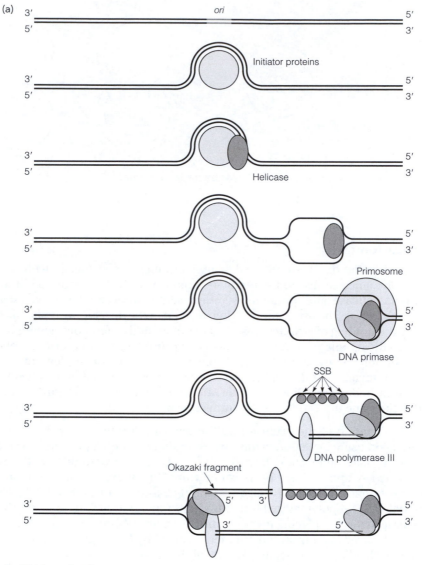

Figure 2. DNA replication.

Termination of synthesis and resolution of replicated circular genomes

The termination of synthesis of a copy of a circular Bacterial genome is signaled by a region coded within the DNA. This site is at a 6 o'clock position on the genome, if the *ori* is at 12 o'clock. A terminator utilization substrate (Tus) binds at this site, stopping the movement of the two replisomes. This might appear to be the end of the replication process, but the cell is now faced with the challenge of separating the two complete chromosomes from one another. Without an effective means of resolving the two chromosomes, they will remain intertwined with one another. Two intertwined genomes close to one another are prone to generalized recombination (Section F9) and must be separated as quickly as possible. A short DNA sequence known as *dif* is found in the middle of the termination region, and here the proteins XerC and XerD bind both double helices and, via a transient DNA/protein complex known as the Holliday junction, cut and resplice the

(b)

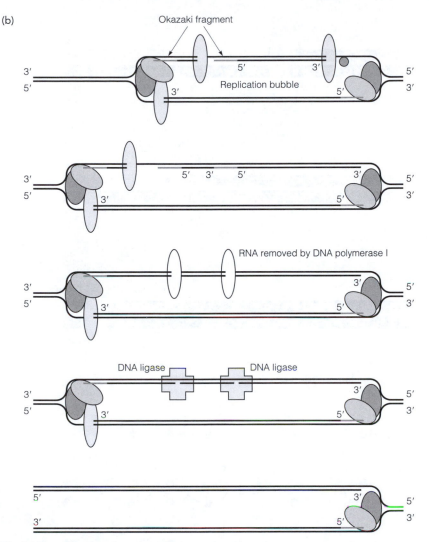

Figure 2. DNA replication (continued).

DNA to form two separate circular chromosomes. Thus a process of site-specific recombination (Section F9) stops more widespread generalized recombination.

Linear genomes

A Bacterial chromosome is typically presented as a single circle of double-stranded DNA. In reality, the situation is far more complex. Most importantly Bacteria may have one or more chromosomes (Section F2), each of which may be linear or circular. However, the full details of the mechanisms of bacterial linear DNA replication have yet to be fully elucidated. The linear chromosome of *Borrelia burgdorferi* has an origin of replication in the center, and presumably replicates bidirectionally (Figure 4), although the mechanisms of termination are unclear. More unusual mechanisms of linear chromosome replication are exhibited by the bacteriophage, such as the *Bacillus* phage φ29, where replication is initiated at either end of the chromosome and is terminated in the center by the collision of the two replisomes. Similar systems may exist in other Bacteria and Archaea.

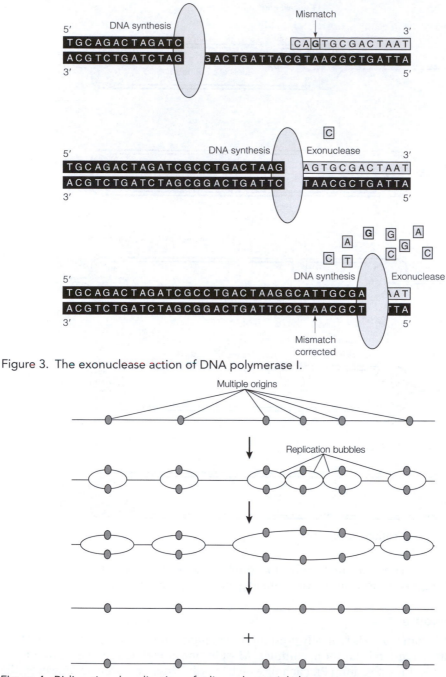

Figure 3. The exonuclease action of DNA polymerase I.

Figure 4. Bidirectional replication of a linear bacterial chromosome.

Mutation and adaptation

As with all organisms, microorganisms have mechanisms to preserve their DNA genome so that their genetic complement is stably passed from generation to generation. However, in contrast with higher organisms, Bacteria and Archaea have no method of replication-based recombination so have to rely on processes of direct genomic change to enable diversity in their populations. Microorganisms face many environmental factors that could possibly cause mutation in their genomic DNA. Permanent changes in the DNA caused by external factors are known as **induced mutagenesis**. Those changes caused by errors in the DNA replication machinery, or other mistakes made by the cellular DNA metabolizing enzymes are known as **spontaneous mutagenesis**. The rate of error in DNA replication is between 10^{-7} and 10^{-11} per base pair per round of replication, equivalent to around 10^{-4} to 10^{-8} errors per gene, per generation in a bacterium the size of *E. coli*. Spontaneous change can also include those genomic rearrangements that an organism may make to its own genome, as exhibited by *Streptomyces*. These mutation **hot spots** on the genome are normally associated with an abundance of short inverted repeats. It is thought that the repetitive nature of the sequence causes polymerase to stutter and make more errors at these particular points.

A strain isolated from its environment and held in pure culture in the laboratory is defined as the **wild type** of a particular strain. Any strain derived from the wild type with any change in its genomic make up compared with this wild type is known as a **mutant**. This change can be referred to as a change in **genotype**. If the difference in genotype results in an observable change to the properties of the organism, such as a sudden inability to use a particular carbon source, this is said to be a change in **phenotype**. Changes in dependence on medium components are frequently used in the elucidation of biochemical pathways or in the deduction of regulatory systems, so the mutation of the wild type with a nutritional requirement is a common experiment. These strains are known as **auxotrophs**. For example, if *E. coli* is subjected to a chemical mutagen and as a result of mutation now requires the vitamin B_{12} to be present in its medium for the strain to grow, the strain is **auxotrophic** for B_{12}, or can be called a B_{12} auxotroph.

F4 Transcription

Key Notes

Transcription

The first stage of converting the primary genetic information from the stable DNA code into protein is to transcribe the DNA into messenger RNA (mRNA). The RNA polymerases that carry this out bind to a DNA-coded signal (promoter) upstream of the gene to be transcribed. Genes in Bacteria can be transcribed alone (monocistronically) or with others as part of a polycistronic operon.

Promoters

RNA polymerase is composed of four polypeptides ($\alpha\beta\beta'\sigma$). The core polymerase ($\alpha\beta\beta'$) has a high processivity but low DNA affinity. The presence of sigma factor (σ) makes the holo-polymerase and confers DNA sequence specificity to the promoter. Once the RNA polymerase has bound, the sigma factor dissociates. The core polymerase then synthesizes RNA complementary to the lower (template) DNA strand. In *E. coli*, the promoter is made up of two conserved regions, the '–10' or Pribnow box, and '–35'. The numbers –10 and –35 refer to the number of bases the sequence is from the base where transcription starts. The –10(TATAAT)/–35(TTGACA) consensus promoter is not the only sequence of promoter in the genome, other sequences allow the binding of alternative sigma factors, which can control specialized groups of genes. These regulons can thus be switched on and off according to external changes in the cell's environment.

Termination of transcription

The signal for termination of transcription is provided by a structure on the mRNA itself. Termination can be signaled by a stem-loop structure, or by the action of the protein Rho.

Regulation of transcription

All genes are regulated in some way at some stage during the cell cycle. In Bacteria this mainly happens in two ways: by derepression (where a protein bound to a promoter stopping transcription is removed and the gene is switched on) and attenuation (where the presence or absence of a substrate necessary for the function of the gene product governs the transcription of the gene itself). Less common is activation, where the presence of protein is used to switch a gene on.

The *lac* operon

The *lac* operon is made up of a promoter, an operator (a site on the DNA where regulatory proteins bind) and the genes *lacZ*, *lacY*, and *lacA*. Another promoter controls the transcription of *lacI*, coding for the regulator (Lac repressor protein). When LacI is bound to the *lacZYA* operon operator, transcription is blocked and the cell is unable to produce β-galactosidase. Allolactose (a by-product of the action

	of β-galactosidase on lactose) is the primary inducer of the operon. In this derepressed state, the *lac* promoter is relatively weak and only achieves full strength if the protein CRP binds as well.	
The *trp* operon	The tryptophan (*trp*) operon is made up of five genes (*trpEDCBA*), regulated by the binding of the *trp* repressor complexed to tryptophan. A series of stem-loop structures can form, which regulate transcription of the operon by attenuation.	
Related topics	(E2) Electron transport, oxidative phosphorylation, and β-oxidation of fatty acids	(F5) Messenger RNA and translation (F6) Signal transduction and environmental sensing

Transcription

The first stage of converting the primary genetic information from the stable DNA code into protein is to **transcribe** the DNA into messenger RNA (**mRNA**). This process is carried out by RNA polymerase, which has many common features with the DNA polymerases. The structure of the RNA polymerase is a little similar to the DNA polymerases in the active site region, and also requires Mg^{2+} to function and synthesizes nucleic acid in a 5′ to 3′ direction. However, the RNA polymerases do not require a primer to initiate synthesis. Instead, their signal for the initiation of transcription is a specific sequence on the DNA, known as the **promoter** region.

A gene's promoter is said to be 'upstream', in that the promoter region is situated to the 5′ end of the coding region. The promoter region allows the RNA polymerase to bind and begin transcription so that the resulting mRNA contains not only the coding region itself but also all the signals to start and stop the synthesis of the polypeptide. How RNA polymerase works is intrinsic to the concept that one gene makes one polypeptide. In eukaryotes, genes are arranged so that the promoter region is in such a position that when transcription occurs a single mRNA molecule is produced that can be used to code for a single polypeptide (**monocistronic**). Genes that code for polypeptides that have a common purpose (such as the manufacture of a multi-polypeptide protein) are placed in many different parts of the genome, frequently on different chromosomes. In prokaryotes (both the Bacteria and the Archaea) genes are more likely to be arranged so that the coding regions for enzymes involved in a single pathway are clustered together. Furthermore, several genes may be arranged so close to one another that they are transcribed from a single promoter (Figure 1). This **polycistronic** arrangement is called an **operon** (see the *lac* operon below).

Promoters

RNA polymerase is composed of four different polypeptides, called beta, beta prime, alpha, and the sigma factor ($\alpha\beta\beta'\sigma$). The **core polymerase** ($\alpha\beta\beta'$) has a high processivity (RNA polymerizing activity) but low DNA affinity. The addition of the sigma factor to make **holo-polymerase** confers DNA sequence specificity, forcing the RNA polymerase to bind at the promoter region. Once the RNA polymerase has bound, the sigma factor dissociates. The core polymerase then synthesizes RNA complementary to the lower

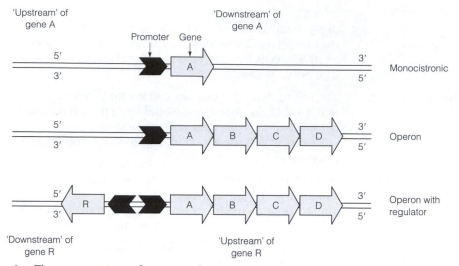

Figure 1.   The arrangement of genes and promoters in Bacteria.

(template) DNA strand. The promoter region tends to have a slightly higher A and T content than the surrounding DNA, and the most abundant form of promoter in *Escherichia coli* has two fairly well conserved sequences, TATAAT (the **Pribnow** box or –10 sequence) and TTGACA (the –35 sequence). The higher A/T content means that there are fewer hydrogen bonds to hold the double-stranded DNA together and thus the double helix is easier to force apart, a process termed melting. Melting of the promoter region allows access of the RNA polymerase, which specifically targets the –10 and –35 regions by the use of the sigma factor. The numbers –10 and –35 refer to relative position in relation to the number of bases the sequence is from the base where transcription starts (known as the transcription start site, Figure 2).

The –10(TATAAT)/–35(TTGACA) **consensus promoter** is not the only sequence of promoter in the genome. There are many promoters with a sequence very similar to –10/–35 and the greater the difference in the base sequence is to this consensus, the weaker the promoter. A strong promoter (such as that of the *lac* operon) forms a very tight bond with the sigma factor, and transcription is very likely to be initiated from such a promoter. A weak promoter binds the sigma factor by only a few bases, and is concomitantly less likely to initiate RNA synthesis. Other promoters have completely different sequences that bind **alternative sigma factors**. A good example of this is the alternative sigma factor produced in response to low oxygen in *E. coli* and some other bacteria. The sigma factor still allows RNA polymerase to recognize a –10 site, but instead will only allow the holo-polymerase to bind where there is an additional specific sequence (FNR site) at around 42 bases upstream of the transcription start site. The use of alternative sigma factors allows a whole group of genes and operons, known as a **regulon**, to be switched on and off according to external changes in the cell's environment (Section F6). Induction of the FNR regulon allows the cell to induce all the genes that are useful to cope with low oxygen, principally alternative electron acceptors (Section E2).

Termination of transcription

Confusingly, the signal for termination of transcription is provided by a structure on the mRNA itself. After the last gene in the operon has been completely transcribed, the RNA

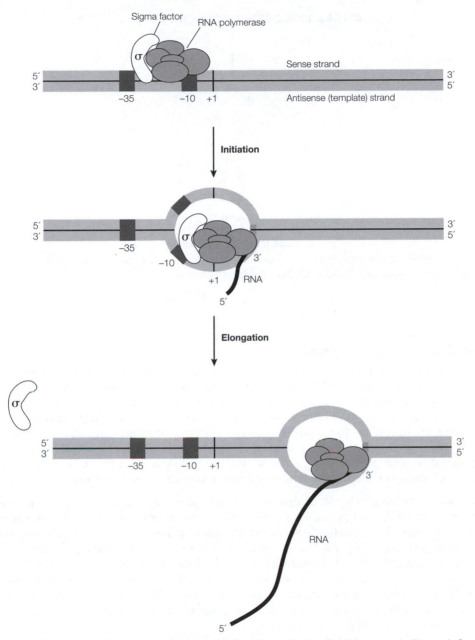

Figure 2. Formation of the transcription complex. From Turner P, McLennan A, Bates A & White M (2005) *Instant Notes Molecular Biology*, 3rd ed. Garland Science.

polymerase continues transcription past the last gene's termination codon. This part of the mRNA folds up into a **stem-loop** structure (Figure 3) that causes the RNA polymerase to pause and cease transcription. The RNA polymerase then dissociates from the DNA and the mRNA it has generated, leaving the complete transcript ready for the ribosomes to translate it into protein.

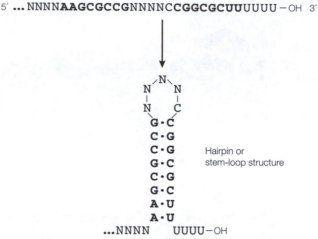

Figure 3. RNA hairpin structure. From Turner P, McLennan A, Bates A & White M (2005) *Instant Notes Molecular Biology*, 3rd ed. Garland Science.

Regulation of transcription

The strength of a promoter and alternative sigma factors represent two ways in which the cell can adjust the rate of mRNA transcription from a particular promoter, and thus the amount of individual proteins in the cell. There are some genes that seem to be switched on most of the time, particularly those genes involved in 'housekeeping' functions of the cell, such as central metabolic pathways, and those that account for gene products, such as the ribosomal components. However, we have come to recognize that all genes are **regulated** in some way at some stage during the cell cycle, and the bacterial cell has a variety of means of altering the flow of information from the genome to the proteome.

The two most common forms of regulation cited in the study of bacteria are **derepression** (where a protein bound to a promoter stopping transcription is removed and the gene is switched on) and **attenuation** (where the presence or absence of a substrate necessary for the function of the gene product governs the transcription of the gene itself). The former is exemplified by the *E. coli lac* operon, the latter by the *trp* operon from the same organism. It should be noted that **activation** (where the presence of a protein is used to switch a gene on) is used much less frequently in Bacteria than derepression. The contrary is true in the Archaea and in eukaryotes, where transcription factors are the main regulators in activating gene expression.

The *lac* operon

The most commonly used model for transcription and the regulation of transcription is the *lac* operon of *E. coli*. The operon comprises a promoter, an **operator** (a site on the DNA where regulatory proteins bind), and three genes. These three genes allow the Bacterium to use lactose instead of glucose as a carbon source (Figure 4), *lacZ* coding for the enzyme β-galactosidase (the gene product LacZ – note the difference in italicization and capitalization between the gene and the protein it makes), *lacY* coding for a permease (gene product LacY) that allows lactose through the membrane, and *lacA*, a gene of poorly understood function that codes for a transacetylase.

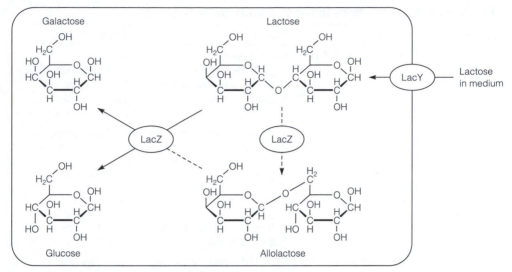

Figure 4. Lactose metabolism in *Escherichia coli*.

Upstream of the *lacZYA* promoter is another gene called *lacI*, which has its own promoter (Figure 5). The product of this gene, the protein LacI, is the primary regulator of the *lac* operon and is sometimes called the **Lac repressor**. When LacI is bound to the *lacZYA* operon operator, transcription is blocked, the cell is unable to produce β-galactosidase, and thus is unable to use lactose as a carbon source.

The functions of all the genes were found as a result of analysis of mutations. Changes in the DNA sequence in the genes themselves can knock out individual genes (Table 1).

The finding of a mutation that meant an inactive LacI was produced, and this allowed constitutive expression of *lacZ* (i.e. in the presence and absence of lactose), was significant and was one finding that allowed Jacob and Monod to begin to elucidate the mechanism of LacI repression. They proposed that LacI was always produced from its own promoter, but was structurally altered by the presence of lactose itself, which stopped it binding to the *lacZYA* promoter (Figure 6). It was found that the molecule that caused this change and induced the *lac* operon was allolactose (Figure 4). Allolactose is the primary **inducer** but there are other molecules such as XGAL that can also induce the operon.

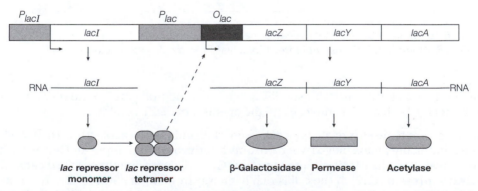

Figure 5. Structure of the lactose operon. From Turner P, McLennan A, Bates A & White M (2005) *Instant Notes Molecular Biology*, 3rd ed. Garland Science.

Table 1. Effect of mutations in *lac* operon structural genes

Mutation	Genotype	Phenotype	Notes
Inactivation of LacI gene	*lacI⁻ lacZ⁺ lacY⁺ lacA⁺*	In the presence of lactose: β-galactosidase⁺ In the absence of lactose: β-galactosidase⁺	LacZ LacY LacA produced in both the presence and absence of lactose (constitutively)
Inactivation of LacZ gene	*lacI⁺ lacZ⁻ lacY⁺ lacA⁺*	In the presence of lactose: β-galactosidase⁻ In the absence of lactose: β-galactosidase⁻	No β-galactosidase (LacZ), so lactose cannot be metabolized
Inactivation of LacY gene	*lacI⁺ lacZ⁺ lacY⁻ lacA⁺*	In the presence of lactose: β-galactosidase⁺ In the absence of lactose: β-galactosidase⁻	No permease (LacY), so lactose cannot enter the cell to be metabolized
Inactivation of LacA gene	*lacI⁺ lacZ⁺ lacY⁺ lacA⁻*	In the presence of lactose: β-galactosidase⁺ In the absence of lactose: β-galactosidase⁻	No transacetylase, but no effect on lactose metabolism

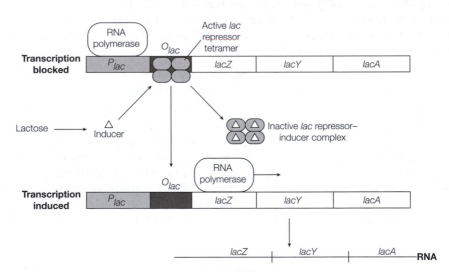

Figure 6. Binding of the inducer activates the *lac* repressor. From Turner P, McLennan A, Bates A & White M (2005) *Instant Notes Molecular Biology*, 3rd ed. Garland Science.

The model was backed up with studies in the nonstructural parts of the operon, in the promoters of both *lacZYA* and *lacI*, and the operator of *lacZYA* (Table 2).

The *lac* promoter itself is relatively weak, even when induced by allolactose. High levels of transcription are only achieved when not only is the promoter derepressed but also when it is activated by the **cAMP receptor protein** (**CRP**). This is sometimes called **catabolite activator protein** (**CAP**). If the cell is rich in energy from other sources, it may not need to utilize lactose at all. The precursor of ATP, AMP, is in equilibrium with a cyclic form, cAMP (Figure 7). When glucose is low, there is an abundance of AMP, which in turn means

Table 2. Effect of mutations in *lac* operon promoters and operators

Mutation	Genotype	Phenotype	Notes
Inactivation of *lacZ* promoter e.g. mutation in –10 region	*lacI⁺ lacZ⁻ lacY⁻ lacA⁻*	In the presence of lactose: β-galactosidase⁻ In the absence of lactose: β-galactosidase⁻	With no promoter and no permease or β-galactosidase, lactose cannot be metabolized
Change in *lacZYA* operator means LacI can no longer bind	*lacI⁺ lacZ⁺ lacY⁺ lacA⁺*	In the presence of lactose: β-galactosidase⁺ In the absence of lactose: β-galactosidase⁺	β-galactosidase and permease produced constitutively
Change in *lacI* promoter to make it stronger	*lacI⁺ lacZ⁺ lacY⁺ lacA⁺*	In the presence of lactose: β-galactosidase⁻ In the absence of lactose: β-galactosidase⁻	As too much LacI is produced, all the allolactose is titrated out Thus there is always some LacI bound to the *lacZYA* operator, and β-galactosidase is never produced

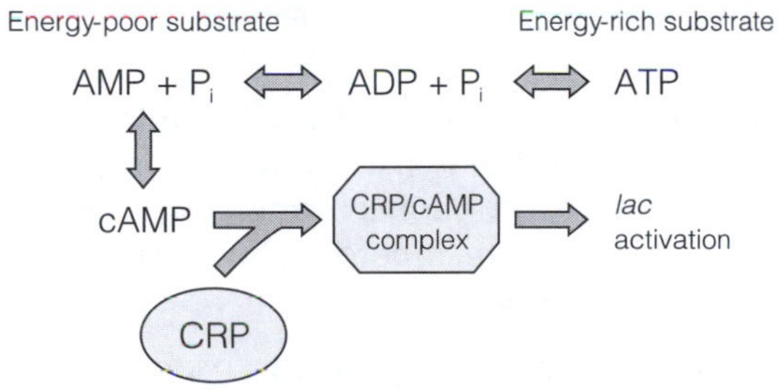

Energy-poor substrate   Energy-rich substrate

$$AMP + P_i \Longleftrightarrow ADP + P_i \Longleftrightarrow ATP$$

cAMP → CRP/cAMP complex → *lac* activation

CRP

Figure 7. cAMP receptor protein (CRP) responds to cAMP.

there is a lot of cAMP. The cAMP binds to CRP, which can then activate the *lac* operon by as much as 40-fold.

The *trp* operon

The tryptophan operon comprises five genes (*trpEDCBA*), which, when transcribed and translated, enable the cell to synthesize tryptophan from glutamine and chorismate (Figure 8). Tryptophan itself inhibits the first step of the pathway at the protein level, but the cell also imposes regulation during both transcription and translation.

It is important to remember that the cell does not need the tryptophan synthesis enzymes when there is an abundance of tryptophan itself, a contrast to the *lac* operon where lactose abundance requires that the cell induces a metabolism system. The transcription of the *trp* operon is thus stopped by the binding of the *trp* repressor complexed to tryptophan. As the levels of tryptophan fall, the operon is derepressed. The difference between repression and derepression is about 70-fold, much less than that of the *lac* operon in its 'on' and 'off' states, and this gave a clue that there was more than one mechanism regulating *trpEDCBA* expression.

The number of bases between the *trp* promoter/operator and triplet 1 of *trpE* is unusually long. The 162 nt (nucleotides) form a Rho-independent terminator site. When

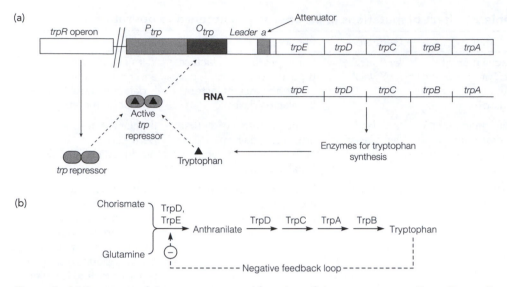

Figure 8. (a) Structure of the *trp* operon and function of the *trp* repressor. From Turner P, McLennan A, Bates A & White M (2005) *Instant Notes Molecular Biology*, 3rd ed. Garland Science. (b) The pathway of tryptophan synthesis in *E. coli*.

transcribed, the site consists of a short GC-rich palindrome followed by eight uracil bases. If this palindrome can hybridize to itself, it forms a stem-loop structure (or hairpin), which is a highly efficient transcriptional terminator, allowing the RNA polymerase to synthesize only the first 140 bp of the operon. The mechanism by which these short transcripts are made is termed **attenuation**.

In bacteria, transcription by RNA polymerase and translation by ribosomes often occur in close proximity, with the ribosome binding to its mRNA binding site as soon as the RNA polymerase has transcribed it. To ensure that the RNA polymerase and ribosome are as close to one another as possible, the initial transcription of the first hundred bases results in the formation of a stem-loop between regions 1 and 2 of the mRNA (Figure 9). This causes the RNA polymerase to pause, and while this takes place, a ribosome loads itself onto the mRNA at the 5' end.

When tryptophan is absent or at very low levels, the ribosome begins to translate the 5' end of the *trp* mRNA, until it encounters two codons coding for tryptophan itself (UUGUUG, region 1 in Figure 9). As tryptophan is at such low levels, there are few tRNA-tryptophan molecules around for the ribosome to incorporate into the growing peptide and, while the ribosome is waiting, a hairpin between regions 2 and 3 forms which reinforces this pause and allows the RNA polymerase to transcribe the entire operon. Contrast this with the situation when tryptophan is abundant: again the RNA polymerase begins transcription, pauses, and the ribosome loads, but now there is no barrier to the ribosome incorporating tryptophan, so it can move forward, unfolding the hairpin between regions 2 and 3. This causes region 3 to be available to form another secondary structure, this time with region 4 (Figure 9). However, the combination of regions 3 and 4 coupled to the presence of poly U a little further downstream is a Rho-independent transcription termination signal. The RNA polymerase falls off after only having manufactured the first 140 nt of the *trp* mRNA. In this way, the transcription of the *trp* operon is attenuated in the presence of tryptophan, but permitted in its absence, allowing the cell to conserve energy by avoiding the wasteful transcription of unneeded mRNA.

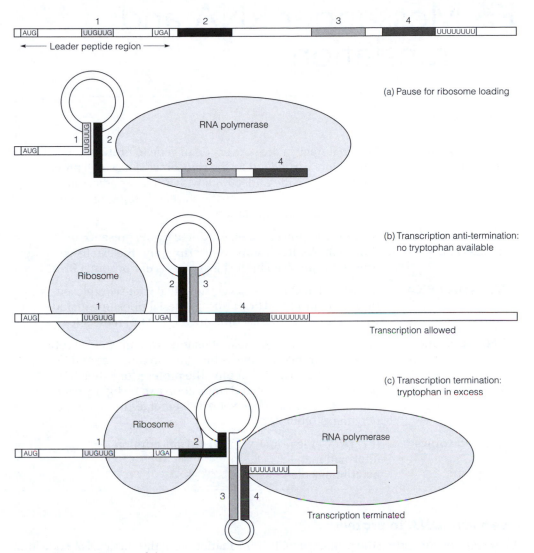

Figure 9. Attenuation of the *trp* operon.

F5 Messenger RNA and translation

Key Notes

Overview: DNA to protein	The final stage in the conversion of gene to protein is translation, carried out by the ribosome. The overall process can only be clearly understood with a clear appreciation of which molecules possess the start and stop signals for transcription and translation.
Messenger RNA processing	Bacteria can process RNA, exhibited by the presence of short poly-A tails on some transcripts, and the extensive processing involved in the biosynthesis of ribosomal RNA.
Ribosomal RNA	Ribosomal RNA is extensively modified while complexed as a ribonucleoprotein. The mature Bacterial ribosome includes 16S (1541 bp), 23S (2904 bp), and 5S (119 bp) molecules.
RNA to protein	The Bacterial and Archaeal ribosomes resemble eukaryotic ribosomes in many respects but differ in total size and in the sizes of the rRNA that bind the proteins together. In Bacteria transcription and translation can be tightly coupled, with mRNA translated as soon as it emerges from the RNA polymerase complex.
Related topics	(F1) DNA – the primary informational macromolecule (F4) Transcription

---

Overview: DNA to protein

A key stage of converting the primary genetic information from the stable DNA code into protein is to **translate** the messenger RNA (mRNA) into a polypeptide. This process is carried out by the ribosome. The ribosome can only bind to messenger RNA, at sites to the 5′ of the AUG start codon. Confusion often arises over the location of the promoter, the ribosome binding site (RBS), and the transcript start site, as these are often all marked on schematics showing the features of a gene in its genomic location (Table 1).

Table 1. Location of nucleic acid signals for transcription and translation

Conversion	Start signal	Stop signal
DNA to RNA	Found on DNA	Found on mRNA
TRANSCRIPTION	PROMOTER	TERMINATOR
mRNA to protein	Found on mRNA	Found on mRNA
TRANSLATION	RIBOSOME BINDING SITE	TERMINATION CODON

Messenger RNA processing

A common misconception about prokaryotes is that they are unable to perform any processing on their RNA, a false distinction often drawn between them and the eukaryotes. Although transcript processing is less common in prokaryotes, it still does occur. There is evidence that some bacterial transcripts have short poly-A tails added, and there is evidence of intron/exon-like processing in a very few genes. The Archaea possess many of the transcript processing features associated with eukaryotes, including extensive intron/exon structures, but neither the Bacteria nor the Archaea rely heavily on mRNA splicing to generate protein diversity. Instead they rely on population diversity to cope with environmental change.

One function in which extensive RNA processing is always present in Bacteria is in the manufacture of the ribosomal RNA. In *Escherichia coli* there are seven different operons for ribosomal RNA (***rrn* operons**), scattered throughout the genome. The number, amount of duplication, location, and order of the genes within these operons varies in different species. At the E. *coli* K12 *rrnA* genomic locus the 16S rRNA gene (*rrsA*) is separated from the 23S rRNA gene (*rrlA*) by genes coding transfer RNA (*ileT* and *alaT*). Downstream of *alaT* at the end of the operon is the gene for 5S rRNA (*rrfA*) (Figure 1). The operon is transcribed as a single polycistronic RNA of about 5000 nucleotides but is quickly processed by RNase III, plus other site-specific RNA-hydrolyzing enzymes called M5, M16, M23, and M25. The processing takes place in the order outlined in Figure 1, and begins just as the unprocessed RNA emerges from the RNA polymerase, i.e. the 5′ end of the operon is processed before the 3′.

Ribosomal RNA

The immature *rrn* transcript can form stable secondary structures made up of stems of RNA:RNA complementary duplexes, and loops of single-stranded RNA. The structure allows the protein to bind to the rRNA, initially to form a **ribonucleoprotein** (**RNP**) complex. Some of these proteins remain to become part of the mature ribosome, but some

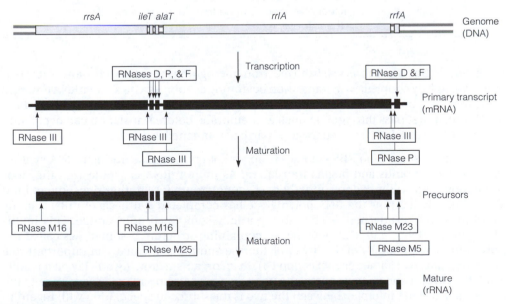

Figure 1. Processing of the E. *coli* K12 *rrnA* primary transcript.

are lost as 24 base-specific methylations of the rRNA take place. It is also at this stage that the first RNase III stage of maturation occurs. Further processing of the precursors by M5, M16, and M23 RNases generates three ribosomal RNA species, named 16S, 23S, and 5S. The 'S' stands for Svedberg units and refers to the sedimentation coefficient – an older measure of molecular weight and size. With the advent of sequencing, the sizes of the *rrn* genes are equally well known, 16S being 1541 bp, 23S 2904 bp, and the smallest 5S is only 119 bp in *E. coli*.

RNA to protein

The Bacterial and Archaeal ribosomes resemble eukaryotic ribosomes in many respects. The prokaryotes lack any compartmentalization of the processes involved, in contrast to the many membranous structures the eukaryotes have evolved. Key differences between the prokaryotes and the eukaryotes are listed in Table 2. Prokaryotic ribosomes move around in the cytoplasm freely, much as those in the eukaryotic mitochondrion do. The actual mechanism by which mRNA is translated into the polypeptide is very similar in all organisms. A detailed description of the process can be found in other books within the *Instant Notes* series.

Table 2. Differences between eukaryotes and prokaryotes displayed during translation

Property	Eukaryotes	Prokaryotes
Ribosomal RNA size	28S, 18S, 5.8S, 5S	23S, 16S, 5S
Ribosomal subunit size		
Total protein complex size	80S	70S
Large subunit	60S	50S
Small subunit	40S	30S
Location of ribosomes	Bound to endoplasmic reticulum or nuclear envelope. Can be free floating in the cytoplasm	Free floating in the cytoplasm.

In Bacteria it has been observed from electron micrographs that transcription and translation are tightly coupled, with translation beginning as soon as a nascent mRNA emerges from the RNA polymerase complex. This is in marked contrast to eukaryotes, where mRNA must first pass through the nuclear membrane before translation can occur, and in many cases mRNA is targeted towards particular organelles.

In Bacteria, as soon as an RBS emerges from an RNA polymerase during mRNA synthesis, a ribosome binds and begins translation. As more mRNA is synthesized the ribosome moves down the codons until there is enough room for another ribosome to bind at the RBS. By the time the RNA polymerase has completed synthesis of mRNA, many ribosomes will be attached, each with a partially completed polypeptide. This multi-ribosome-, mRNA-, and RNA polymerase-containing structure is sometimes called the **polysome**. The proximity of the RNA polymerase and ribosome plays an important role in the regulation of some genes (Section F4). In an organism growing quickly on a readily available carbon and energy source, the polysome may also be spatially very close to the replisome, and the interplay between the two is the source of speculation with regard to regulatory mechanisms.

F6 Signal transduction and environmental sensing

Key Notes

Cellular response to environmental stimuli

Free-living, single cells must be able to adapt quickly to relatively large changes in their immediate environmental conditions. Signal transduction mechanisms allow these external changes to be reflected in differential gene expression. Some molecules, such as lactose, diffuse directly into the cell and act on an inducer or repressor, but other compounds are prevented from doing so. In common with physical changes (pH, temperature, etc.) these molecules can be detected via sensors in the cell membrane, and then a signal sent to the genome via phosphorylation of cytoplasmic proteins. The process of translocation via this phospho-relay pathway can be branched to include activation or deactivation of many genes or operons.

Two-component sensors

The simplest sensor consists of a sensor kinase in the cell membrane and a response regulator in the cytoplasm. Binding of an external signal results in the hydrolysis of ATP at the cytoplasmic side of the sensor kinase, and activation of the response regulator by transfer of a phosphate group.

Oxygen sensing

Facultative anaerobes are able to grow in atmospheric oxygen concentrations as well as less aerobic conditions. Accordingly they must have mechanisms to sense and respond to oxygen concentrations to employ different biochemical pathways appropriate to aerobiosis. An example of an oxygen-sensing system is FNR (after fumarate and nitrate reduction). The group of genes and operons controlled by FNR (the regulon) allow *E. coli* to use fumarate and nitrate as alternative electron acceptors to oxygen. There is some evidence that there is crosstalk between sensors with similar but not identical function.

Chemotaxis

A control mechanism to efficiently move towards food or away from toxins is essential to motile microorganisms. Chemotaxis towards an attractant (inducer) is initiated by the stimulation of a cell membrane-located transducer. The signal is translocated to the flagellar motor by means of the Che proteins, and the balance between attraction and repulsion is maintained by methylation of the transducer.

Related topics

(E2) Electron transport, oxidative phosphorylation, and β-oxidation of fatty acids

(F4) Transcription

Cellular response to environmental stimuli

Single-celled organisms, such as Bacteria, can face enormous changes in their external environment over very short periods of time. If we consider *Escherichia coli* and its life cycle, we can see how significant these changes are. In its normal habitat of the colon, the *E. coli* cell is kept at a more or less constant temperature of 37°C, is surrounded by other microbes and nutrients (in the form of partly or completely digested food), and is in a low-oxygen environment. After excretion, and before the cell re-enters the digestive system of the same or a different host, *E. coli* is suddenly thrust into a colder, well oxygenated, more aqueous, and nutrient-poor situation. To be able to cope with all these changes in lifestyle, the cell must quickly turn off some genes and turn on others. At the gene level the switches are repressor and activator proteins (Section F4 on the *lac* and *trp* operons), while at the protein level enzymes can be switched on or off by the presence or absence of metabolites (see Section F4 on the role of allolactose). These cytoplasmic responses have come into effect due to changes outside the cell, and the way in which an indicator that the outside world has changed is carried from the cell wall and to the genes or proteins that might be involved in a response is called **signal transduction**.

Some molecules that have a significant effect on the cell are small and can diffuse or be actively transported. Once in the cell they can have a direct effect on their target, for example, sugars such as lactose, and alcohols such as methanol. However, some useful growth components must be isolated from the cytoplasmic contents (cyanide, formaldehyde, and even oxygen can be used for growth but are also cytotoxic) or are unable to pass easily through the cell wall and membrane (large molecules such as polydextrans). Other changes on the outside of the cell must be transformed into a signal that enzymes can recognize. Change in pH, ultraviolet light or heat are pertinent examples. Even the enzymes of extremophiles (Section C2) can only tolerate relatively small ranges of heat and pH, yet the cell as a whole can survive a greater range of temperature or acidity. Rather than let environmental changes into the cell directly, at their external concentrations, the exterior of the cell has developed systems to sense change and then relay that change to a target. Frequently this relay involves the transfer of phosphate groups from a sensor protein onto one or more relay proteins and finally to the effector protein. For this reason it is sometimes called a **phospho-relay pathway**. When we examine cell response pathways it is convenient to think of them as linear pathways in which a sensor relays a series of phosphate molecules via a defined set of proteins to an effector. In the cytoplasm, the **translocation** of a signal from sensor to effector is certainly much more branched, with overlapping sets of relay proteins accepting phosphates from many different sensors. For this reason our understanding of the entire process of cellular sensing has become intertwined with computational techniques such as neural networking. To gain access to the entire branched cellular response, we must first understand simple examples such as two-component regulatory systems, and then develop it to look at a more global response such as the effect of oxygen on the cell or chemotaxis.

Two-component sensors

Taken in isolation, the simplest bacterial sensing system allows the cell to adapt to an external factor by means of two protein components: a **sensor kinase** and a **response regulator**. The sensor kinase is buried so that there is an environmental signal binding site outside the cell membrane, and an ATP binding cassette on the cytoplasmic side of the cell membrane (Figure 1). Binding of an external molecule causes conformational change across the whole protein, and the protein on the inner membrane is now able to transfer a phosphate group from ATP to a relay protein. In some cases this relay protein is a transcriptional regulator.

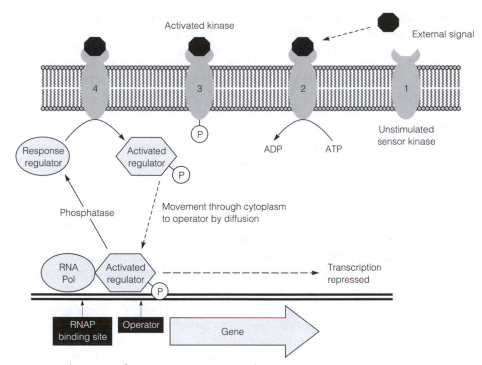

Figure 1. A schematic of a two-component regulatory system.

Oxygen sensing

One of the more frequent changes all microorganisms face is a change in the external oxygen tension. Some organisms can adapt between oxygen concentrations equivalent to atmospheric levels and almost completely anoxic conditions. These **facultative anaerobes**, such as *E. coli*, achieve this by changing the terminal electron acceptor (Section E2) used in oxidative phosphorylation. Under aerobic conditions they use water as a terminal electron acceptor and cytochrome aa_3 oxidase. In the absence of oxygen and the presence of fumarate, cytochrome aa_3 oxidase is switched off and the fumarate/nitrate system is switched on. The primary sensor for this oxygen-dependent switch is a regulatory protein called **FNR** (after fumarate and nitrate reduction). This is a global regulator in that it alters the transcriptional properties of hundreds of *E. coli* genes, acting as an inhibitor of some while elevating the transcription of others. The group of promoters and genes that are regulated by FNR are known collectively as the **FNR regulon**, and are all characterized by a distinct sequence in the relevant operator, an FNR binding site. Where FNR acts as an activator, the FNR binding site replaces the –35 sequence of the standard s^{70} promoter, so that transcription cannot take place unless active FNR is present. The FNR binding site is found in a similar place to the LacI binding site (Section F4) when FNR is acting as a repressor.

The way in which FNR senses the presence of molecular oxygen is based more on chemistry than molecular biology. In the absence of oxygen, a cluster of iron and sulfur molecules forms at the N-terminal end of the protein, which allows FNR to dimerize and interact with DNA at its specific binding sites. When oxygen is present, the $[2\text{Fe-2S}]^{2+}$

cluster is destroyed and the protein can no longer function in DNA binding. This oxygen-sensing capability is carried out in concert with a number of other regulons, such as the Arc cluster. CRP (Section F4) participates in another regulon as well, and it is interesting to note that the binding sites for FNR and CRP differ by only one base pair, raising the possibility that a considerable amount of **crosstalk** occurs between regulatory systems, allowing the cell to simultaneously adapt to many conditions at once.

Chemotaxis

Microorganisms, by definition, exist on a microscopic scale, and this means that, in relative terms, separation from a food source by only a few millimeters presents an enormous challenge. Most bacteria have some form of motility, using pili (Section C10) or gliding, or merely by extending colony size. To conserve energy, movement must not be random, depending on a chance encounter with a food source, but directed. The sensing process provides this direction. If movement is towards an organic compound, this is named **chemotaxis**, but other forms of taxis exist. These include geotaxis (movement towards or away from gravity) and phototaxis (movement towards or away from light). The best understood chemotactic system is formed by the Che proteins, present in many Gram-negative Bacteria.

In response to the presence of a suitable carbon source (the **inducer**), a **transducer** (MCP) forms a complex with a sensor kinase, CheA. The transducer functions as a medium to pass a signal from the periplasm and across the cytoplasmic membrane. Unlike the more simple two-component sensors, the transducer and sensor kinase interaction is mediated by a coupling protein, CheW. However, the result is still the phosphorylation of CheA. The signal is carried through the cytoplasm by transfer of the phosphate group to CheY. Phosphorylated CheY-P has a direct interaction with the flagellar motor switch, changing the direction of rotation. Once CheY-P has acted on the motor, it is dephosphorylated by CheZ (Figure 2) and rotation continues in the original anticlockwise direction.

As the microbe approaches the food source, and the concentration of the inducer changes, the cell must adapt its chemotactic response so as not to overshoot. This adaptation is

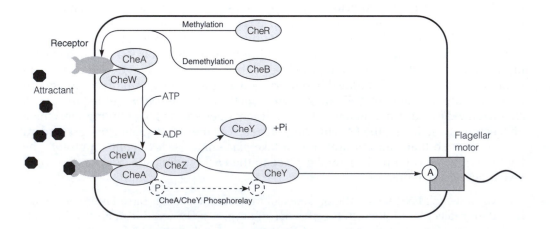

Figure 2. Chemotaxis is controlled by the Che proteins.

provided by the second phospho-relay protein, CheB. A methylation protein, CheR, is constantly methylating the transducer, but CheB can demethylate the transducer. The number of methyl groups attached to the transducers in the cell membrane influences the efficiency of sensor kinase phosphorylation. The competition to methylate (ChyR) or demethylate (CheB) the transducer adapts the cell's response to a stimulant.

F7 DNA repair

Key Notes

DNA damage limitation	The damage caused to the structure and sequence of a DNA helix must be repaired by the cellular machinery before efficient replication and transcription can begin again.
Repair by polymerase proofreading	The DNA polymerases themselves have the ability to proofread nucleotides during replication. However, this system cannot cope with bases that have been extensively modified and sometimes proofreading can cause mutation in the regions of DNA modification.
Mismatch repair	Newly replicated DNA spends some time in a hemimethylated form. The methyl-directed mismatch repair mechanism can correctly identify the nonmutated strand by means of this methylation. A series of proteins can then rectify the mutation using the parental strand as a template.
Nucleotide excision repair	The UvrABC system scans the genome and initiates the process in regions of distortion of the helix, such as those caused by thymine dimers. After the incorrect bases have been removed, the system recruits DNA polymerase I and DNA ligase to mend the break in the strand.
Base excision repair	Chemical base modification of intact DNA can be excised by specific DNA glycosylases, which break the ribose-base glycosyl bond. Apurinic or apyrimidinic sites are then removed by specific endonucleases. Together this is known as the base excision repair (BER) system.
Direct repair	The DNA photolyases are used for photoreactivation or light repair of pyrimidine dimers, and proteins such as O^6-methylguanine-DNA methyltransferase can repair the effects of alkylating agents such as methylmethane sulfonate.
The SOS response	Should a mutation still persist despite the action of the other systems, the cell can create a drastic system to resolve mismatches without reference to mother and daughter strands. The induction of the system depends on the balance between the proteins LexA and RecA. The cell can thus continue replication, but this response is extremely error-prone.
Related topics	(F4) Transcription (F10) Bacteriophages (F9) Recombination

DNA damage limitation

The damage caused to the structure and sequence of a DNA helix must be repaired by the cellular machinery before efficient replication and transcription can begin again. If the damage is not repaired, then the **mutation** will become inherited by subsequent generations. The existence of modified bases signals the induction of systems designed to remove specific types of damage. Some types of damage can be corrected by an enzyme-catalyzed reversal of the nucleotide modification, but more lasting damage means that stretches of one strand of DNA must be removed and replaced with new nucleotides. There are limitations to the capacity of the cell to repair its own DNA in that in most cases the repair machinery relies on a correct template against which to make corrections. If both sides of the helix, i.e. both base pairs, are damaged, then the cell is less likely to correct the error. Thus events that cause deletion of regions of the DNA are more likely to be inherited than those that only damage one strand, or a few bases of one strand, of the helix.

Although DNA repair systems are extremely efficient in organisms such as the Bacterium *Deinococcus radiodurans* (Section C6) and the Archaeon *Pyrodictium abyssi* (Section C6), there is a greater error in DNA repair mechanisms than in DNA replication. The systems described here are specific for the Bacteria, but analogous ones exist in the Archaea and the eukaryotes.

Repair by polymerase proofreading

The DNA polymerases themselves have the ability to proofread nucleotides during replication (Section F3). As such this provides a first line of defense in order to preserve the integrity of the genome. However, this system cannot cope with bases that have been extensively modified and sometimes proofreading can cause mutation in the regions of DNA modification.

Mismatch repair

Despite the high fidelity of DNA replication, and the backup proofreading system, it is still possible for the incorrect base to be opposite a template base. To fix such potential mutations, the repair mechanism must have some way of distinguishing between the template strand and the daughter strand synthesized against it. In many bacteria a mature chromosomal DNA has methylated bases at the sequence GATC. Daughter DNA strands experience some delay in methylation, so newly replicated DNA spends some time in a **hemimethylated** form (template strand methylated, daughter strand not). The **methyl-directed mismatch repair** mechanism can identify correctly the nonmutated strand by means of methylation (Figure 1). The system can only resolve single base-pair mismatches and needs the participation of Dam methylase, MutH, MutL, MutS, DNA helicase II, single-stranded binding protein, DNA polymerase III, exonuclease I, exonuclease VII, RecJ nuclease, exonuclease X, and DNA ligase.

Nucleotide excision repair

The removal of many nucleotides (by UvrABC exinuclease) and their replacement (by DNA polymerase I and DNA ligase) is ubiquitous in the Bacteria. In *E. coli* this mechanism has been shown to be largely error-free. It enables the removal of large areas of mutation, e.g. as a result of the presence of pyrimidine dimers, and is thus known as nucleotide excision repair (**NER**). The UvrABC system scans the genome and initiates the process in regions of distortion of the helix (Figure 2). Nucleotide excision repair differs from base excision repair in that the entire nucleotide is removed, causing a complete break in one strand of the helix.

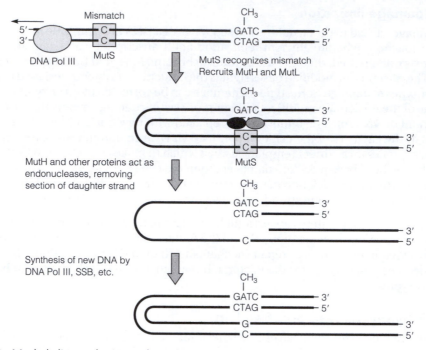

Figure 1. Methyl-directed mismatch repair.

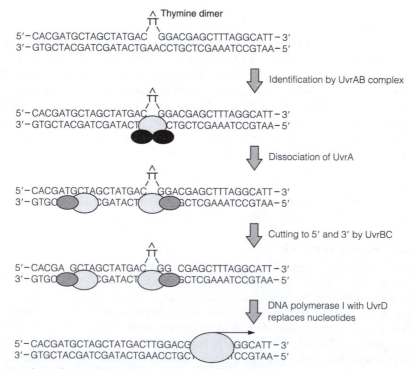

Figure 2. Nucleotide excision repair of thymine dimers.

Base excision repair

Some mutagenic events lead to the modification of the base part of the nucleotide. The action of the base modifiers can sometimes be corrected by very specific DNA glycosylases, which break the ribose-base glycosyl bond. This leaves the backbone of the DNA strand intact, with some sites depurinated or depyrimidinated. These **apurinic** or **apyrimidinic** sites (AP sites) are then removed by AP endonucleases. In contrast to NER, only one or two nucleotides are removed. The DNA strand is made entire again by the action of DNA polymerase I and DNA ligase, by the same mechanism as outlined in NER. The abnormal bases repaired by base excision repair (**BER**) include uracil, hypoxanthine, xanthine, and alkylated bases. In some organisms it is also the mechanism used to resolve pyrimidine dimers.

Direct repair

In addition to proofreading and the excision repair mechanisms, the Bacterial cell has a range of mechanisms for correcting chemical changes in DNA *in situ*. The **DNA photolyases** are used for **photoreactivation** or **light repair** of pyrimidine dimers, and proteins such as O^6-methylguanine-DNA methyltransferase (the Ada protein) can repair the effects of alkylating agents such as MMS (methylmethane sulfonate). This transferase is unusual in that it is not an enzyme in the strictest sense of the word. The methyl group is directly transferred to the Ada protein, which renders it completely inactive. Thus it cannot participate in another reaction, so does not play a catalytic role.

The SOS response

Should the base pairing between strands still remain incorrect, the Bacterial cell has a final mechanism to excise corrupted DNA. This mechanism may actually result in the persistence of mutations in the next generation, but prevents the accumulation of mismatches to the point where they become lethal. Normally a protein called **LexA** represses an overall response, but as mismatches accumulate, the concentration of another protein (**RecA**) also rises. A high concentration of RecA causes the cleavage of LexA. In the absence of LexA, 17 genes are turned on which constitute the **SOS response**. This system quickly resolves all DNA mismatches, but does not distinguish between parental and daughter strands. The cell can continue replication, but this response is extremely error-prone.

F8 Transfer of DNA between cells

Key Notes

DNA passes between microorganisms

Vertical transfer between members of the same species is well characterized, but more recently it has become apparent that microorganisms are capable of horizontal transfer between species, genera, phyla or even kingdoms. The best studied systems of transfer between donor and recipient include conjugation, transduction, and transformation.

Conjugation

A plasmid known as the F′ factor encodes genes that enable the plasmid to be passed from one individual to another. It has also been shown that F′ can integrate into the genome to form Hfr strains. When an Hfr strain conjugates with an F-recipient, parts of the donor's chromosome may be copied and transferred as well, but normally only sections of the genome close to the site of integration. If the acquisition of the donor's DNA causes the recipient to change its phenotype after recombination, the recipient is said to be a transconjugant.

Transformation

Transformation is the most widely used technique to introduce DNA into *Escherichia coli*, but the least well understood. Washing in ice-cold magnesium chloride solution renders the cells competent, a state in which they can take up DNA when heat shocked. After incubation in a recovery medium, a small proportion of the total cells maintain the plasmid and these are known as transformants. A similar protocol allows *Bacillus subtilis* cells to take up linear DNA.

Generalized and specialized transduction

DNA transfer can occur between cells by the action of a bacteriophage. If the phage is lysogenic and excises from the host's genome imperfectly, some host genes can be transferred to a new recipient on reinfection. This is the process of specialized transduction. In contrast, generalized transduction arises when parts of the host genome are packaged instead of phage DNA. In both cases the resulting recipient cells are known as transductants.

Related topics

(F9) Recombination (F11) Plasmids
(F10) Bacteriophages

DNA passes between microorganisms

While the Bacteria lack the sophisticated systems of sexual reproduction found in eukaryotes, they do not rely on mutation alone to generate diversity within their populations. Spontaneous mutagenesis alone is insufficient to respond to the varying environments that the majority of Bacterial species encounter. Bacteria can acquire variation in genes, complete new genes or even complete biosynthetic pathways via the acquisition of plasmids (Section F11), infection by bacteriophages (Section F10) or transposons (Section F9), and by direct, nonsexual interchange of genetic information (see below).

The phenomenon of Bacterial transfer of genetic information has long been studied in the context of same-species or **vertical gene transfer**. However, it has become increasingly apparent that transfer occurs between species too. **Horizontal gene transfer** can account for the rapid spread of antibiotic resistance in a variety of hospital-acquired infections, but also can be used to our advantage in the generation of consortia for the biodegradation of xenobiotics. In a mixed population, such as that found in a hospital or bioreactor, it is hard to pinpoint the exact mechanisms used to acquire DNA, but looking in the genomes of isolated organisms it can be easy to pinpoint the genomic regions that originated in other organisms by their unusual codon usage or other sequence signatures.

The best understood mechanism of gene transfer is **conjugation**, a method that gave the first genetic map of *Escherichia coli*. This is a plasmid-mediated effect, in contrast to **transduction** in which DNA moves from the **donor** to the **recipient** strains by means of a Bacterial virus or **bacteriophage**. The method used most commonly to introduce foreign genes into laboratory strains of Bacteria is **transformation**, a poorly understood process relying on a shock response to take up plasmids or other fragments of DNA.

Conjugation

Some strains of *E. coli* possess a plasmid (the **F′ factor**), which allows them to transfer DNA to a recipient of the same species. The transfer of DNA is unidirectional from the donor carrying the F′ factor (the strain designated F⁺) to a recipient (F⁻) that does not have the plasmid (Figure 1a). The copying of the F′ factor plasmid is accompanied by a physical joining of donor and recipient via a plasmid-encoded structure, the F-pilus. After conjugation is complete the recipient becomes F⁺.

Conjugation was first described in 1946 by Lederberg and Tatum, who noted that two Bacterial auxotrophs could be mixed together to make a new nonauxotrophic strain. They noted that the transfer of information could be stopped easily by shaking the cells, even though the shaking process did not kill the cells outright. Others separated the donor and recipient by a fine filter and demonstrated that physical contact between the cells was an absolute requirement for conjugation. The process was not fully understood until it became apparent that the strains that could act as donors (**high-frequency recombination** or **Hfr** strains) all possessed the large F′ plasmid. It was also shown that F′ could **integrate** into the genome of the Hfr strains. During the procedure, parts of the donor's chromosome may be copied and transferred as well, although rarely the entire complement (Figure 1b). If the acquisition of the donor's DNA causes the recipient to change its phenotype after recombination, the recipient is said to be a **transconjugant**.

Transformation

Although conjugation and transduction provide simple ways to introduce genes into microorganisms, both are time-consuming processes. The most commonly used way of

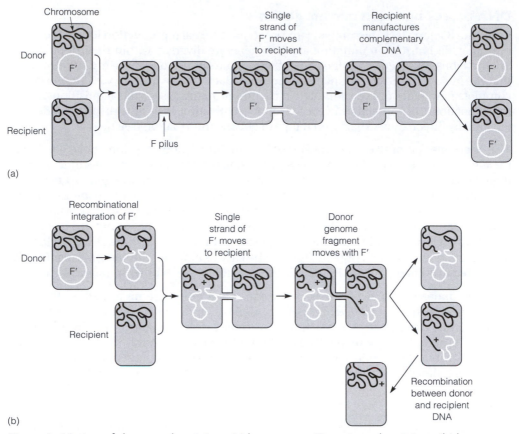

Figure 1. Mating of donor and recipient: (a) between an F⁺ strain and recipient; (b) between an Hfr strain and recipient.

introducing DNA into Bacteria in today's laboratory is via transformation. Most routine molecular biology experiments result in the generation of plasmids (Section F11), in the form of DNA libraries, cloning experiments, or preparation of DNA constructs for other experiments.

The technique of transformation is well defined, but the mechanism is poorly understood. In *E. coli*, in order to make the cells available for manipulation, plasmid-free cells are grown to mid log phase, and then washed in a cold magnesium chloride solution to remove traces of growth medium. A high concentration of washed cells is suspended in cold calcium chloride. At this stage the cells are **competent**, a state mimicking a physiological state in which the Bacteria are able to take up DNA from their environment. It is thought that this is a starvation response and perhaps the influx of external DNA may provide an extra gene that might help the cell survive. In the laboratory, the plasmid to be **transformed** is added to the competent cells. The plasmid remains in solution until the cold cells are suddenly heat shocked by raising the temperature to 42°C. Only a small percentage of the cells take up plasmid DNA as a result of the heat shock, and all the cells are in a weakened state. A rich recovery medium is added, and then the cells are plated out on a selective medium after incubation. The cells that have successfully taken up the plasmid are known as **transformants**. Not all Bacteria can be transformed in the laboratory, but a similar protocol is used to generate *Bacillus subtilis* transformants, although they can only take up linear DNA and must maintain the acquired gene(s) by recombination into their chromosomes. However, the way in which either plasmid DNA in *E. coli* or linear DNA in *B. subtilis* crosses the cell membrane intact, or in what form it does so, is currently the subject of speculation.

Generalized and specialized transduction

DNA transfer can occur between cells by the action of **bacteriophages** (Section F10). Occasionally, the excision of a lysogenic phage's genome from the host chromosome is imperfect. Instead of just phage genes, the excision includes one or two host genes from locations adjacent to the site of phage recombination. When the altered phage infects another cell, it is possible that the extra genes will complement mutated genes in the second host's genome. This process is known as **specialized transduction**. In contrast **generalized transduction** can lead to the transfer of any gene in a host genome, not just those adjacent to a site of lysogeny.

An example of a generalized **transducing phage** is the bacteriophage P1 of *E. coli* (Figure 2). Using this phage to transfer DNA between *E. coli* strains is a relatively quick and easy way of making chromosomal mutations, and is sometimes abbreviated to P1 transduction. The principle is outlined in Figure 2, and for the most part follows the normal life cycle of a lysogenic bacteriophage (Section F10). The difference occurs just as the pro-phage leaves the host genome on the induction of the lytic cycle. The phage genome leaves completely and induces the degradation of the host chromosomes. The phage replicates inside the cell, but during this process some random Bacterial DNA is packaged as if it were phage DNA.

When the donor cell is lysed, the P1 phage released can be used to infect a recipient. If the donor was a wild-type cell and the recipient was, for example, an auxotroph for histidine, some of the P1 phage may carry the DNA to correct the mutation. A small number

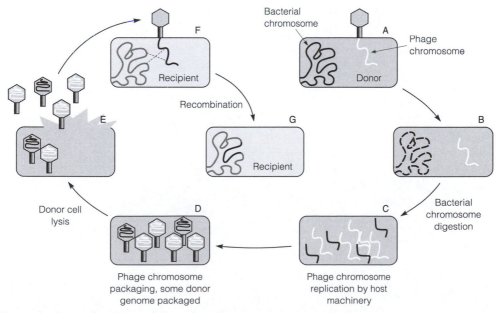

Figure 2. Generalized transduction by phage P1.

of infected recipients will receive a fragment of wild-type donor DNA instead of phage P1 DNA and this can recombine with a recipient chromosome. While a useful technique in *E. coli*, where P1 phage has been engineered to increase the frequency of generalized transduction, few similar systems exist. The number of genes that can be transferred is limited to the maximum amount of DNA that can be packaged into a P1 phage head.

F9 Recombination

Key Notes

Homologous recombination

This is sometimes called generalized recombination and involves the exchange of DNA between two regions of homology (i.e. similar but not necessarily identical DNA sequence) on two different DNA molecules. RecBCD digests double-stranded DNA until it reaches a chi sequence, where the activity of RecBCD then changes to a single strand-specific 5′–3′ exonuclease, so that a single DNA strand remains as a long 3′ overhang. The free single-stranded DNA becomes coated with RecA protein, and can migrate into the other DNA molecule in a process of strand invasion. Once the nicks in the DNA strands have been joined, the two DNA molecules adopt a cross-shaped formation known as the Holliday junction. The binding of RuvA to the Holliday junction leads to the binding of RuvB to adjacent double-stranded DNA. RuvB is a translocase, and promotes the movement of the Holliday junction along the DNA molecules (branch migration). The single-stranded parts of the junction are cleaved (strand exchange) and the two molecules resolve when RuvC binds to the RuvAB–DNA complex.

Site-specific recombination

Recombination between regions of DNA homology of only a few base pairs can occur via site-specific recombination mediated by XerC and XerD. A similar method is used to integrate bacteriophage λ DNA into its host's genome. λ Integrase makes cuts on both strands about 15 bp apart at a specific site. Integration host factor (IHF) catalyzes recombination between the homologous sites on the phage and host genome, resulting in the λ genome insertion into the *E. coli* DNA. The reverse of the process is catalyzed by the phage-encoded excisionase, stimulated by host IHF and the protein Xis.

Transposition

Transposition is the insertion of short DNA fragments into any position in the genome. This is sometimes called illegitimate recombination as it appears to require no homology between sequences. Insertion sequences (IS) use this mechanism to move between sites and can be detected in many chromosomes and plasmids. Transposable elements could be looked on as very basic viruses without the means to encode a protein coat for protection outside the cell. The IS encoded transposase makes cuts at another site in the host chromosome and is excised to be integrated into the new site. The IS is flanked by inverted terminal repeats, which aid in the transposition process. As the transposase cuts are staggered, the bases either side of the repeats are duplicated after integration. The transposable viruses do not excise themselves during transposition, but place a copy elsewhere on the genome instead (replicative transposition). Transposon mutagenesis can offer a convenient method of generating mutants via *Tn5* derivatives.

Related topics   (F4) Transcription   (F10) Bacteriophages

Homologous recombination

This process is sometimes called **generalized recombination** and involves the exchange of DNA between two regions of homology (i.e. similar but not necessarily identical DNA sequence) on two different DNA molecules. One of these regions might be on a plasmid, the other on the chromosome. In rarer instances these two regions could be on the same chromosome but many hundreds of base pairs apart or more. After recognition and alignment of the regions of homology by the enzyme RecA (Figure 1a), a protein complex, **RecBCD,** digests dsDNA until it reaches a *chi* sequence (crossover hot instigator, sequence GCTGGTGG). The activity of RecBCD then changes to a single strand-specific 5′–3′ exonuclease, so that a single DNA strand remains as a long 3′ overhang. This is represented by the arrowheads in Figure 1b. The free single-stranded DNA becomes coated with RecA protein, and can migrate into the other DNA molecule in a process of **strand invasion** (Figure 1c). Once the nicks in the DNA strands have been joined, the two

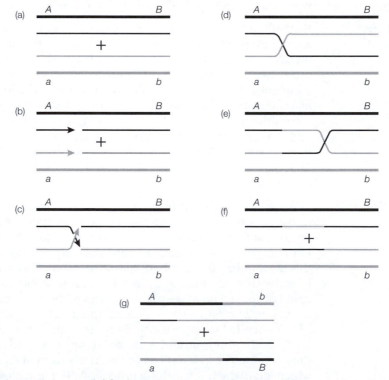

Figure 1. The Holliday model for homologous recombination. From Turner P, McLennan A, Bates A & White M (2005) *Instant Notes Molecular Biology*, 3rd ed. Garland Science.

DNA molecules adopt a cross-shaped formation known as the **Holliday junction**. This is shown in Figure 1d, and is named after the person to first propose it.

The binding of RuvA to the Holliday junction leads to the binding of RuvB to adjacent double-stranded DNA. RuvB is a translocase, and promotes the movement of the Holliday junction along the DNA molecules (Figure 1e). This is known as **branch migration**. The single-stranded parts of the junction are cleaved (**strand exchange**) and the two molecules become separate molecules again (**resolve**) when RuvC binds to the RuvAB–DNA complex. How the resolution proceeds is still the subject of some debate, but it has been postulated that equilibrium is established between a RuvA planar Holliday junction and tetrahedral junctions stabilized by another protein, RusA. Depending on the direction of a final cleavage, there are two possible outcomes (Figures 1f and 1g).

Site-specific recombination

Recombination can occur between regions of DNA with apparently no homology, where only a few base pairs are the same. This is an important mechanism in the XerC/XerD-mediated resolution of the two circular molecules of DNA in *E. coli* chromosomal replication (Section F3), and does not require RecA. A similar method is used to integrate bacteriophage λ DNA into its host's genome (Section F10). The integration site is a 15 bp sequence present in both the host and λ DNA. The λ **integrase** makes cuts on both strands about 15 bp apart. **Integration host factor** (IHF) catalyzes recombination between the homologous sites and the λ genome is inserted into that of *E. coli*. The reverse of the process is catalyzed by the phage-encoded **excisionase**, stimulated by host IHF and Xis.

Transposition

Transposition is the insertion of short DNA fragments into any position in the genome, a process carried out by **transposons**, **insertion sequences** (IS), and some specialized phages (Table 1). This is sometimes called **illegitimate recombination** as it appears to require no homology between sequences. However, analysis of transposon positions after transposition suggests that it is a pseudo-random event, and that the recombination is site-specific. However, the recognition sites are extremely degenerate and occur so frequently in most genomes that almost all genes can be interrupted by the entry of a transposon.

The presence of insertion sequences can be detected throughout the genomes of many Bacteria and Archaea as well as in their plasmids, and they are essential to some functions. The integration of the F′ plasmid into the *E. coli* chromosome is mediated by homologous recombination between identical insertion sequences. The ecology and microbiology of transposons is extremely complex, given that some of these elements can move around the genome in the absence of chromosomal replication and initiate a form of

Table 1. Transposable elements in *E. coli*

Type of element	Example	Notes
Insertion sequence	IS*21*	Carries only the information necessary for transposition
Transposon	Tn*5*	Carries transposition genes plus other genes, e.g. for antibiotic resistance
Transposable	Mu	Carries many genes and also has a protein-coated bacteriophage form for survival outside the host

conjugation to transfer themselves from one cell to another. In some senses, transposable elements could be looked on as very basic viruses especially given that transposition is the method of replication for bacteriophage such as Mu.

A simple transposition event is illustrated by the conservative transposition of an IS element. The IS encoded **transposase** makes cuts at another site in the host chromosome (in a pseudo-random manner) and is excised and integrated into the new site in much the same way as λ achieves lysogeny. The IS is flanked by **inverted terminal repeats**, which aid in the transposition process. As the transposase cuts are staggered, the bases either side of the repeats are duplicated after integration (Figure 2).

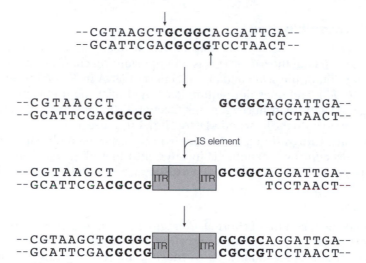

Figure 2. Transposition of an insertion sequence (IS) element into a host DNA with duplication of the target site (shown in bold). From Turner P, McLennan A, Bates A & White M (2005) *Instant Notes Molecular Biology*, 3rd ed. Garland Science.

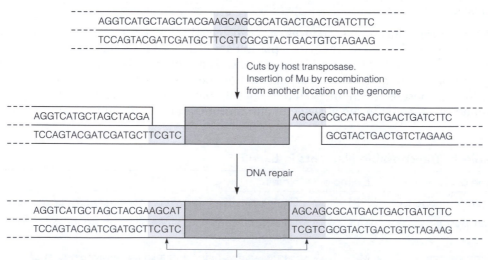

Figure 3. Insertion of bacteriophage Mu by transposition into a bacterial genome.

The transposable viruses do not excise themselves during transposition, but place a copy elsewhere on the genome instead. This process of **replicative transposition** (Figure 3) allows Mu to replicate its genome in the host in terms of copy number per chromosome as well as during cell division.

Transposons offer a convenient method of generating mutants, particularly when gene interruption is required. **Transposon mutagenesis** is most developed for Tn5 derivatives, in which additional genes such as those for kanamycin resistance or β-galactosidase activity have been added. A bacterial strain can be transformed with a plasmid bearing the modified transposons. The transposons can then jump from the plasmid to the chromosome and cause random interruption of genes. Auxotrophs or other mutants can then be identified as normal (Section F3).

F10 Bacteriophages

Key Notes

Viruses of bacteria   The viruses that infect the prokaryotes are the bacteriophages, sometimes abbreviated to phages. The replicative form (the virion) can be a variety of shapes including filamentous and icosahedral, or may have a distinct head and tail section. Most possible variations of nucleic acid content – double-stranded DNA (dsDNA), single-stranded DNA (ssDNA), dsRNA, and ssRNA – have been observed in bacteriophages. In common with other viruses, after entry to the cell they may either replicate and cause lysis of the cell, or enter into a dormant form by integration of viral DNA into the host genome.

Modes of replication   Virulent phages have a lytic mode of replication, where the bacteriophage genome is injected into a cell, replicates, and then kills its host on the formation of many new virions. The alternative used by lysogenic phages is where (post-infection) rather than triggering more virion production, the viral genome recombines with the host genome (known as lysogeny) and does no further damage. In response to an environmental stimulus, the viral DNA excises from the host genome and a lytic cycle ensues.

The lytic cycle   The lytic or virulent bacteriophages attack and lyse their hosts during the replication cycle. After injection of the viral genome, bacteriophage T4 first transcribes nucleases to cut up the Bacterial genome. T4 genomes are synthesized as a concatemer, with each copy linked end to end with another. T4-specific nucleases cut the concatemeric DNA into pieces. Phage head, tail, sheath, and other proteins necessary for the construction of the protein coat are then synthesized, assemble around the T4 chromosome, and are released on cell lysis.

The lysogenic cycle   The lysogenic phages are exemplified by bacteriophage λ. On entry to the host cytoplasm, the phage genomic cos sites at either end of the λ genome allow the molecule to circularize into a plasmid-like form. At this point the bacteriophage can replicate and lyse the cell much as T4 does, or enter into a lysogenic form where the entire λ genome integrates into the host genome by recombination. The phage genome (in this state known as a prophage) is replicated as a single copy during each round of host cell division. Induction of a further recombinational event excises the phage DNA from the host. The replication cycle then follows a lytic path. The replication of the λ genome is not accomplished using a bidirectional replication fork but by rolling circle replication, resulting in a long double-stranded, multi-copy molecule. This then has to be cleaved to yield individual genomes ready for packaging.

| **Related topics** | (F4) Transcription | (K1) Virus structure |
| | (F9) Recombination | (K2) Virus taxonomy |

Viruses of bacteria

In common with most living organisms, microorganisms can be infected by viruses. The virology of higher organisms is discussed further in Section K. The viruses that infect the Bacteria and Archaea are the **bacteriophages**, sometimes abbreviated to **phages**. Bacteriophages are not living entities and are incapable of reproduction outside of their host cell. In essence they are genetic elements that have acquired the ability to maintain themselves outside the host cell by coating their nucleic acids with proteins, but are no more alive than the protein-free DNA of plasmids, transposons or insertion sequences (Sections F9 and F11). As with the viruses of eukaryotes, the form and reproductive cycle of bacteriophages varies considerably. The extracellular **virion** form can be filamentous (e.g. bacteriophage M13), icosahedral (e.g. bacteriophage T4, but see Section K1 for a further discussion of icosahedral symmetry) or have a distinct head and tail section (Figure 1). Some bacteriophages are surrounded by a membrane, which is often derived from the host. The diversity of shapes of the protein part of the virus is accompanied by a diversity of nucleic acid complements. Most possible variations of nucleic acid content – dsDNA, ssDNA, dsRNA, ssRNA – have been observed in bacteriophages. Almost all species of Bacteria and Archaea have been found to have one or more bacteriophage that infects them, and some phages have an extremely broad host range.

All bacteriophages have common methods of replication. Attachment to the host cell is followed by entry to the cytoplasm by penetration or injection. Phage gene expression commences and the phage genome is copied. Genomes and phage proteins are then assembled into virions ready for release and to begin the infection cycle again. An archetypal bacteriophage has a head containing the genome, and a tail to attach to the

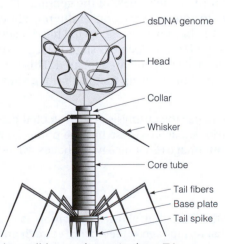

Figure 1. The structure of the well-known bacteriophageT4.

host. The tail fibers target specific receptors on the surface of the host cell membrane. A sheath surrounds a core tube, through which the viral genome is injected into the host cytoplasm. Contraction of the sheath is ATP-dependent, and in a manner analogous to muscle contraction. Transcription of the bacteriophage genome is tightly controlled, but very quickly takes over the host cell's functions, shutting down all activities and dedicating all transcription and translation towards the manufacture of new virions.

Modes of replication

Bacteriophages can have one of two reproductive strategies. In the first **lytic** mode of replication, the bacteriophage enters a suitable cell, replicates, and then kills its host. The alternative is a **lysogenic** mode. A whole virion infects the cell, but rather than triggering more virion production, the viral genome recombines with the host genome and does no further damage. The lysogenized virus's only function is to code for proteins that prevent reinfection with bacteriophages of the same type. The viral genome is passed on to daughter cells during normal host cell division. In response to an environmental stimulus the viral DNA excises from the host genome and a lytic cycle ensues. **Virulent bacteriophages**, such as T4, which only have a lytic replication cycle, run the risk of exterminating the host population. In the laboratory, these bacteriophages are characterized by the complete lysis of all the host cells in a liquid culture. Those with the ability to perform **lysogeny** (such as bacteriophage λ) can maintain a pool of host cells capable of releasing virions gradually. In liquid culture, Bacteria infected with lysogenic phages fail to reach the expected optical density, but still appear to be dividing normally.

The lytic cycle

The lytic or **virulent** bacteriophages attack and lyse their hosts during the replication cycle (Figure 2). This is exemplified by bacteriophage T4, whose relatively large genome contains about 200 genes. T4 attaches to the cell membrane and injects its genome into the cytoplasm. The first genes transcribed code for nucleases to cut up the bacterial genome, effectively stopping any further host cell activity. The remainder of the phage replication cycle relies on existing host ribosomes, RNA polymerases, and enzymes in the cell, in addition to the phage-encoded products. Phage genes are now expressed that enable the T4 DNA to be copied. The T4 genomes are synthesized as a **concatemer**, with each copy linked end to end with another. T4-specific nucleases cut the concatemeric DNA into pieces, each approximately 105% of the genome. Phage head, tail, sheath, and other proteins necessary for the construction of the protein coat are now synthesized, and coalesce around the T4 chromosome. Each viral head receives the same overall complement of genes, but as the DNA molecule is slightly larger than the minimum number of genes required, the genes at each end are repeated. This **terminal redundancy** allows a simple but reliable method of ensuring that each phage head always receives at least one copy of each gene.

Finally, phage genes that cause the autolysis of the infected cell are transcribed, with the release of around 300 virions. Each of these has the potential to infect more cells, but if left in a population without phage resistance lytic phages will eventually wipe out all their local hosts.

The lysogenic cycle

The problem of killing the host that allows replication is overcome by the **lysogenic phages**, exemplified by bacteriophage λ. Free virions attach and inject their genome into

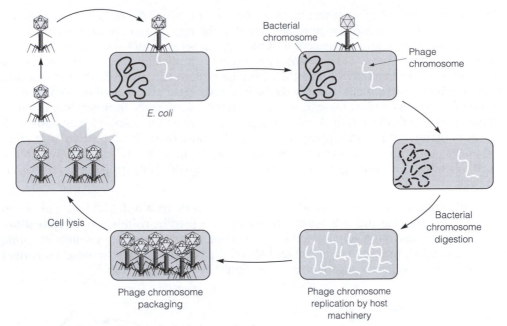

Figure 2. Replication cycle of a lytic bacteriophage.

the cytoplasm in the same way as the virulent phages, but then have the option of either establishing a carrier population among the host cells, or replicating, lysing, and reinfecting (Figure 3). On entry to the host cytoplasm, the *cos* sites at either end of the λ genome allow the molecule to circularize into a plasmid-like form. At this point the bacteriophage can replicate and lyse the cell much as T4 does, or enter into a form that perpetuates as the host divides.

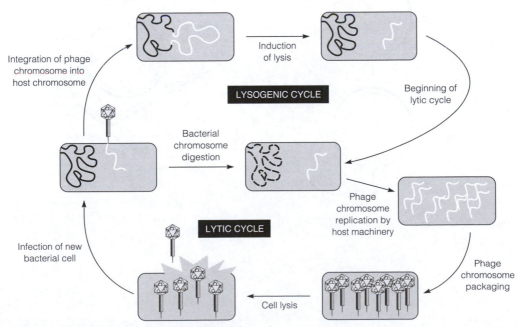

Figure 3. The lysogenic cycle of bacteriophage λ.

The carrier or lysogenic state results from the integration of the entire λ genome into the host genome by recombination (Section F9). The regions of homology allowing the recombination are found in the center of the λ genome, and are known as att sites. The phage genome (in this state known as a **prophage**) is replicated as a single copy during each round of host cell division so that λ is passed vertically through the generations. To ensure that new colonies of hosts can be infected, a stimulus related to DNA damage (ultraviolet light, chemical mutagens, etc.) induces a further recombinational event to excise the phage DNA from the host. The replication cycle then follows a lytic path, with digestion of the host genome, phage genome replication, phage head protein formation, and packaging until the host cell is lysed and phage progeny are released. However, the mechanism by which the λ genome is copied differs considerably from bacterial replication.

The replication of the λ genome is not accomplished using a bidirectional replication fork. Instead, the plasmid-like cytoplasmic form is copied by **rolling circle replication**. This allows copies of the genome to be made in one direction only, but results in a long double-stranded, multi-copy molecule. This then has to be cleaved to yield individual genomes ready for packaging, but has the advantage of being very rapid.

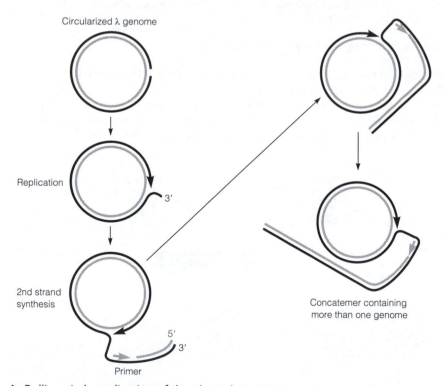

Figure 4. Rolling circle replication of the phage λ genome.

F11 Plasmids

Key Notes

Structure of plasmids	A plasmid is defined as an entire molecule of DNA that is replicated independently of the host chromosome. When over 100 kbp or more, they are known as megaplasmids. Removal of a plasmid is known as curing. There are more than 300 different types of naturally occurring plasmids for *Escherichia coli* alone, some of which have been adapted for applications in recombinant DNA technology.
Replication of plasmids	The plasmids of *E. coli* are complete, closed, circular (ccc) double-stranded DNA molecules. Plasmids are replicated by the host cell machinery. Episomal plasmids integrate themselves into the chromosome and are replicated along with it. The number of plasmids per cell is known as the copy number, and this can vary between very low (stringent plasmids) with only one or two copies per cell, and very high (relaxed plasmids) with hundreds of copies per cell. As maintenance of a plasmid in a cell is energetically expensive, the cell will try to delete the plasmid. Thus to maintain a plasmid in a strain, some form of selective pressure is applied to ensure that as many cells as possible keep the plasmid.
Plasmid incompatibility	In common with the bacteriophages and other viruses, some plasmids have evolved mechanisms to repel reinfection by different types of plasmid. When this occurs, the two plasmids are said to be incompatible. Plasmids can be grouped according to their compability.
Function of plasmids	The best known of the *E. coli* plasmids is the F′ plasmid. The genes that govern transfer are known as *tra* genes and code for proteins with a variety of functions including enablement of the movement of DNA and the rolling circle replication associated with it. Other plasmids carry genes that confer heavy metal resistance, entire metabolic pathways, toxins, virulence factors or antibiotics. There is transfer of plasmids between members of the same species, but promiscuous plasmids may be transferred between cells of different species.
Vectors for recombinant DNA technology	The plasmids used in molecular biology have been engineered to remove most of the unfavorable aspects of the wild type. Most are small (<5 kbp) so that they are easy to manipulate *in vitro* and are less likely to recombine with the host chromosome. The *tra* genes are absent so that the recombinant DNA will be contained in the designated host cell.

Recombinant DNA	Recombinant strains of *E. coli* have been developed that can harbor foreign DNA but cannot survive outside the laboratory. Adapted *E. coli* plasmid cloning vectors maintain foreign DNA in an easily accessible form, while expression vectors are plasmids used to generate large amounts of recombinant protein. Shuttle vectors allow the easy creation of specific arrangements of genes or gene fragments in *E. coli* before the construct is transferred to another organism. Part of the attraction of *E. coli* is the ease with which genomic and plasmid DNA can be isolated. Plasmids can be isolated from lysed *E. coli* by techniques such as cesium chloride density centrifugation or by the use of the many commercially available spin columns.	
Related topics	(F8) Transfer of DNA between cells	(F10) Bacteriophages

Structure of plasmids

A plasmid is defined as an entire molecule of DNA that is replicated independently of the host chromosome. This alone does not differentiate them from viruses, so in addition a virus is defined as having extracellular structures for protection outside the host. A plasmid should carry no genes that are essential for the existence of the cell. This helps us to distinguish between very large plasmids (over 100 kbp or more, known as **megaplasmids**) and the smaller chromosomes. Plasmids often carry genes that help the cell to survive environmental challenges, such as a sudden rise in the concentration of mercury, but do not carry any of the genes for the central metabolic pathways. This can be tested experimentally by **curing** the cell, which is removing the plasmid. If the cell can still grow when cured, the DNA is a plasmid and not a chromosome. As there are more than 300 known types of naturally occurring plasmids for *E. coli* alone, this section will be restricted to a discussion of Bacterial plasmids, specifically those from *E. coli*. These have also been adapted to be the most useful, with applications in **recombinant DNA technology**.

Replication of plasmids

The plasmids of *E. coli* are complete, closed, circular (**ccc**) double-stranded DNA molecules. This means that they are circular, joined at both ends, and are supercoiled within the cytoplasm. Typically the plasmids are about 5% of the genome size, and the minimum size is only around 2000 bp. The plasmid is replicated by the host cell machinery, using DNA polymerase III and all the other apparatus associated with it. Some plasmids (the **episomes**) integrate themselves into the chromosome and are replicated along with it, much like the lysogenic bacteriophages (Section F10). The number of plasmids per cell is known as the **copy number**, and this can vary between very low (**stringent plasmids**) with only one or two copies per cell, and very high (**relaxed plasmids**) with hundreds of copies per cell. The relaxed plasmids in particular place a considerable energetic burden on the cell, since replicating each base pair requires one molecule of ATP. The cell will try to relieve itself of the burden of carrying the plasmid if there is no reason to keep it, so to maintain a plasmid in a strain, some form of **selection** must be applied. In order to be replicated by the host machinery, a plasmid requires an origin of replication and a region of site-specific recombination much like a chromosome. However, some plasmids are replicated unidirectionally.

Plasmid incompatibility

There can be more than one different type of plasmid in a cell, as exemplified by wild types of *Borrelia burgdorferi*, which carry 17 circular and linear plasmids. In common with the bacteriophages and other viruses, some plasmids have evolved mechanisms to repel reinfection by different types of plasmid. When this occurs, the two plasmids are said to be incompatible. Testing which plasmids are compatible with which gives a table of **plasmid incompatibility groups**. This can be useful to the biotechnologist who may wish to introduce recombinant genes in successive experiments.

Function of plasmids

The best known of the *E. coli* plasmids is the F′ plasmid (Section F8), which confers the ability of a cell to transfer this plasmid by conjugation. The genes that govern transfer are known as *tra* genes, and take up about one-third of the F′ plasmid's 99 159 bp. These genes code for proteins with a variety of functions including enablement of the movement of DNA and the rolling circle replication associated with it. The F′ plasmid confers a very specific property on *E. coli*, but some plasmids carry genes that confer heavy metal resistance, while others can carry entire metabolic pathways to deal with unusual carbon sources. The plasmids of most concern in medical microbiology encode toxins, virulence factors or antibiotics. For example, enteropathogenic *E. coli* colonize the small intestine and produce toxins that cause diarrhea and hemolysis in humans. Both the protein that enables attachment to the intestine wall and the toxins themselves are plasmid encoded.

The current model for plasmid ecology is that a great diversity of genes is held extrachromosomally in any indigenous bacterial population. There is constant exchange of genetic material, but the majority of cells do not retain the plasmids acquired after a few cell divisions. When the population is stressed, only those bacteria carrying the plasmid-borne genes essential to deal with that stress survive and quickly divide to outnumber plasmid-free organisms. There is some transfer between members of the same species, but **promiscuous** plasmids may be transferred between cells of different species. This horizontal gene transfer has been noted in the spread of antibiotic and mercury resistance.

Vectors for recombinant DNA technology

The plasmids used in molecular biology have been engineered to remove most of the unfavorable aspects of the wild type. Most are small (<5 kbp) so that they are easy to manipulate *in vitro* and are less likely to recombine with the host chromosome. The *tra* genes are absent so that the recombinant DNA will be contained in the designated host cell.

Recombinant DNA

The Bacterium *E. coli* plays a pivotal role in what is popularly known as genetic engineering. Our detailed knowledge of its microbiology, physiology, and biochemistry has allowed us to develop mutants that can be used to clone pieces of DNA from most known organisms, safe in the knowledge that the strains harboring foreign DNA (**recombinants**) cannot survive outside the laboratory. Adapted *E. coli* plasmids are used to maintain foreign DNA in an easily accessible form (plasmids known as **cloning vectors**). **Expression vectors** are plasmids used to generate large amounts of recombinant protein (sometimes up to 10% of the total cell protein) so that enzymes can be purified for research, biotechnological, food, and medical uses. **Shuttle vectors** can also be constructed, which allow the easy creation of specific arrangements of genes or gene fragments in *E. coli* before the

construct is transferred to another organism via transformation (Section F8), transduction (Section F8), electroporation or transfection (the latter two techniques are not discussed in this text). As the demand for very specific proteins subject to post-translational modification becomes more sophisticated, other organisms have been used for cloning and expression, but *E. coli* is still the most widely used.

Part of the attraction of *E. coli* is the ease with which its genomic and plasmid DNA can be isolated. *E. coli*, in common with many other bacteria, will lyse on the addition of a sodium hydroxide solution. After neutralization, the genomic DNA can be purified by techniques such as **cesium chloride density centrifugation** or by the use of the many commercially available **spin columns**. The many plasmids used in biotechnology can also be separated from the rest of the cytoplasmic contents after alkaline lysis. If the lysate is treated with sodium dodecyl sulfate (SDS) and neutralized with sodium acetate, larger macromolecules (proteins, membrane debris, and genomic DNA) become entangled (both physically and chemically) with the white precipitate formed by SDS. The plasmids remain in solution and can be purified further with a few simple steps. This means that plasmid DNA can be isolated from *E. coli* in less than 15 minutes, rendering the organism suitable for building up libraries of DNA fragments in plasmids for screening projects or archiving.

In a text such as this it is not appropriate to discuss the full range of DNA and RNA techniques available to the microbiologist.

G1 Biotechnology

Key Notes	
Biotechnology	Biotechnology is the use of living organisms in technology and industry. Prokaryotes have been exploited for many years in the manufacture of food and other useful products. For large-scale applications, prokaryotes are grown in fermenters using batch, fed-batch or continuous-culture processes.
Gene technology and biotechnology	The efficient production of some proteins and chemicals can only be carried out by mutant or recombinant microorganisms. Gene technology provides a means of providing virus-free human proteins, as well as the overexpression of useful enzymes from extremophiles.
Bioremediation	The biodeterioration of man-made compounds by microbes can be a problem (e.g. in the case of paints and plastics) or a benefit when it comes to cleaning up the environment. Bioremediation is the use of microbes to treat waste water and soil where unacceptable levels of pollutants have accumulated. Even highly toxic compounds such as polychlorinated biphenyls (PCBs) can be metabolized by some Bacteria.
Related topics	(D3) Large-scale and continuous culture (F11) Plasmids

Biotechnology

Microorganisms have been used to make a wide variety of products for many thousands of years, but lately this has been called biotechnology. In order to produce products or intermediates, microbes are grown in large vats (fermenters) protected from contamination and changes in pH, temperature, and dissolved oxygen concentration. Industrial fermenters vary in size from a few liters to several thousand liters (see Section D3). Modern biotechnology relies heavily on developments in chemical engineering to regulate and monitor the processes occurring in these fermenters, and this regulation makes the growth of Bacteria or fungi under these conditions somewhat different from that found in ordinary shake flasks. The optimization of fermenters requires a thorough understanding of the kinetics of growth of the organism used in the bioprocess.

Industrial fermenters allow the growth of microorganisms by one of three processes: batch, fed-batch, and continuous fermentation (see Section D3). A batch fermenter is sterilized, inoculated, and allowed to grow before the entire culture volume is harvested. A fed-batch system follows a similar regimen, but instead only some of the culture is harvested after growth, and fresh medium is added to restart the process. Continuous culture uses a constant in-flow of medium offset by the exit of spent medium, product, and biomass, and is not used so much in the production of products for human consumption because of perceived problems with culture sterility. Most developments in fermentation systems focus on the provision or exclusion of oxygen from the vessel, depending on the process required. The amount of dissolved oxygen in a culture can be increased by a variety of methods, from the simple mechanical stirred vessel to the airlift fermenter (see Section D3).

Biotechnology now plays an important role in many industries, including the food-processing industry, providing both enzymes (e.g. invertase for the manufacture of glucose syrup) and biomolecules (e.g. sodium glutamate).

Gene technology and biotechnology

The exploitation of biological potential for industrial processes currently relies on metabolic diversity of microorganisms to provide biological solutions to common problems. Improvement of biotechnological processes would seem an attractive start for the modern molecular biologist, but public opinion and government regulation of the use of genetically modified organisms has restricted the widespread use of gene technology.

However, the production of many drugs and fine chemical products relies on **recombinant** or **mutant microorganisms** to produce enzymes (Section G3) and many human proteins have been cloned in *Escherichia coli* and other microorganisms.

Bioremediation

The metabolic diversity of microbes has allowed their commercial exploitation in the field of waste-water and soil clean-up. The **biodeterioration** of paints, plastics, and other man-made compounds by microbes is often seen as a problem, but where unacceptable levels of these compounds accumulate, this property can be a benefit. Although the technology is still in its infancy, compounds previously considered **recalcitrant** are now being subjected to treatment by a variety of microbes. Pilot studies have shown the efficacy of the use of bacteria against compounds such as trichloroethylene (**TCE**) and polychlorinated biphenyls (**PCBs**). The use of microbes is not without problems: the **biosurfactants** of *Pseudomonas* species have been used to allow endemic soil bacteria **bioavailability** to emulsified crude oil after tanker spillages. However, the emulsified oil proved much more mobile than the crude slick, and a bigger problem was created as the hydrocarbons moved deep into gravel and towards potable water sources. Experimental waste-water treatment plants, where polluted medium is much more contained, have been far more successful than *in situ* approaches.

G2 Food microbiology

Key Notes

Foods from Bacteria

Prokaryotes are used in the manufacture of a surprising diversity of processed food, e.g. cheese, yogurt, and soy sauce. The role of Bacteria is predominantly to convert carbohydrates and alcohols into organic acids, lowering the pH and increasing the storage life and palatability of the food. A large part of food microbiology is concerned with the elimination and detection of food-borne pathogens.

Flavorings and additives

Chemically complex flavorings and additives can be made cost-effectively by microorganisms. The nucleotides, lipids, amino acids, and vitamins added to processed foods can be obtained from microbial metabolites or as a result of microbial bioconversions.

Organic acids

Vinegar and citric acid are produced microbially in very large quantities. The most significant commercial product is the stationary phase production of citrate by *Aspergillus niger*.

Starter cultures

Starter cultures are used to reduce the duration of the fermentation process and reduce the chance of contamination by microorganisms endemic to the fermentation substrate.

Food hygiene

Bacteria, viruses, fungi, protozoa, algae, and prions are all responsible for causing diseases spread by food. Food poisoning can be caused either by the growth of the cells in the body or by the presence of microbial products in food.

Food spoilage

Food must be stored under conditions that do not favor the growth of microorganisms, particularly pathogens. Low temperature, low pH, high salt, and high osmolarity can prevent the spoilage of foods, as can complete sterilization using heat or radiation. Pasteurization provides a method for medium-term storage of mainly liquid foods while preserving palatability.

This section mostly deals with the role of Bacteria in food microbiology. The Archaea have little economic significance in the food industry, except rarely as spoilage organisms in salted foods. In contrast, the molds and fungi play a vital role in brewing, bread making, and directly as human food in the form of edible mushrooms and Quorn (Section I4).

Foods from Bacteria

Today's industrial fermented foods have their origins in much older processes. Foods that use bacteria in their production include soy sauce (*Pediococcus* species), cheese, yogurt (*Lactobacillus*, *Enterococcus*, and *Bifidobacterium* species as well as many others), sauerkraut (*Lactobacillus*, *Streptococcus*, *Leuconostoc* species), and the traditional manufacture of vinegar (*Acinetobacter* species). In general, prokaryotes are used to acidify foodstuff to allow storage, and a table of foods made by microorganisms is shown in Table 1.

Table 1. Microbes in the food industry

Foods	Flavorings
Fermented meat	Vinegar
Soy sauce	Nucleotides
Cheeses/milk/yogurt	Amino acids
Mushrooms and edible fungi	Vitamins
Baker's yeast	Lipids
Coffee	
Pickles, sauerkraut, olives	
Organic acids	Starter cultures
Citric acid	Dairy industry (cheese, etc.), silage
Itaconic acid	Leguminous crops

Bacterial food fermentations are generally complex and involve a succession of different organisms to render the final product palatable. The process rarely involves bacteria alone, but the production of fermented cabbage (sauerkraut) uses bacteria with progressively higher tolerance to acid. *Streptococcus faecalis* and *Leuconostoc mesenteroides* initiate the conversion of plant sugars into organic acids, and are succeeded by *Lactobacillus brevis* and finally *Lactobacillus plantarum*. A similar process is used to preserve cucumbers (gherkins) and olives.

Flavorings and additives

The worldwide market for food additives and supplements is worth tens of billions of dollars yearly and demand grows annually by 4%. The complex organic compounds that make up food flavors can be made using synthetic organic chemical approaches, but are made more cheaply as microbial products or as a result of microbial bioconversions. Flavors as a whole are a diverse group of compounds including aldehydes, esters, lactones, polyols, and terpenes and a range of microorganisms are used to produce them. Food regulations in the US and Europe allow the classification of flavors (and many other products) as 'natural' even if they are produced by bulk fermentation of a Bacterial or fungal culture. For example, citronellol produced by *Kluyveromyces lactis* can be used as a natural flavoring in fruit drinks.

Flavors are often targeted by biotechnologists as the equivalents in unprocessed foods are expensive or difficult to isolate from plant material. For example, cocoa butter from cocoa beans is high in value and is prone to developing off flavors during storage or processing as the fats within it become rancid. Tea seed oil is a byproduct of tea processing and can be converted by a microbial or fungal inter-esterase into a fatty acid strongly resembling true cocoa butter. The reaction uses 1,3-dipalmitoyl-2-oleyl glycerol mixed with stearic acid, and this becomes 1,3-distearoyl-2-oleyl glycerol (the product) and palmitic acid after inter-esterase treatment.

Some of the biggest value microbial products include the nucleosides. **Guanosine triphosphate** is a base used in the synthesis of RNA but can be overproduced by strains of *Bacillus subtilis*. It is added to food to give the '**umami**' taste. It is often used in combination with another product of *Bacillus*, disodium inosinate, to enhance the flavor of instant noodles, cured meats, tinned vegetables, and dried soups.

Food additives now include the '**neutraceuticals**,' compounds that are believed to be important for nutrition but are not classified as drugs in most countries. Ascorbic acid (vitamin C, produced from *Acetobacter suboxydans*), cyanocobalamin (vitamin B_{12}, produced from a number of bacteria including *Pseudomonas denitrificans* and *Streptomyces griseus*), and riboflavin (vitamin B_2, produced from recombinant *Bacillus subtilis*) are added to many foods, including cereals, or are sold as vitamin supplement pills.

Organic acids

Small quantities of amino acids such as glutamic acid, tryptophan, and lysine are produced by microbial fermentations and used in food flavoring. However, microbes are also used to produce bulk organic acids, the most important of which are acetic and citric acids. Acetic acid is the active ingredient of vinegar, and although vinegar itself is a complex mixture of many compounds and flavors, the key reaction in its production is the microbial oxidation of alcohol to acetate. Vinegar has been produced for thousands of years, but in modern controlled processes, *Gluconobacter* or *Acetobacter* are used to carefully convert any alcohol-containing liquid (e.g. wine, cider, fermented malt) into acetic acid. The fermentation occurs in vats or the Bacteria are immobilized onto wood shavings so that a simple trickle filter column is made. Alcoholic liquid is poured in at the top, trickles past the Bacteria, where the oxidation takes place, and then product can be collected at the bottom.

Citric acid is used very widely to give acidity to foods and to enhance fruit flavors and forms a major constituent of some carbonated drinks. It is produced by an aerobic fermentation of sucrose by the mold *Aspergillus niger* as a secondary metabolite. The organism overproduces citrate during stationary phase to sequester iron, and this has been exploited and enhanced by biotechnologists.

Starter cultures

Many fermented foods can be manufactured merely by allowing microorganisms already present in the environment to act on a foodstuff. An example of this is in the agricultural manufacture of silage (fermented cut grass used for winter feeding of cattle). In the absence of air the acidification of grass will occur as microorganisms on the grass slowly convert carbohydrates into organic acids. However, there is a significant danger that pathogenic Bacteria or toxins will build up during the fermentation, so **starter cultures** containing *Enterococcus*, *Pediococcus*, and *Lactobacillus* are grown up in the laboratory and mixed with the grass as it is cut. The presence of the large amounts of starter culture increases the speed of the fermentation and reduces the incidence of contamination as the starter culture out-competes endemic microorganisms.

The principle of using starter cultures is employed in the large-scale manufacture of beer and wine, where the brewer or vintner has a choice of *Saccharomyces* strains (yeast) that will give particular flavors distinct to different types of beer or wine.

Food hygiene

A large part of food microbiology is concerned with the elimination and detection of food-borne pathogens. Food poisoning can be caused by all the types of organism mentioned in this book, except the Archaea. The food microbiologist must make sure that food products do not spread diseases caused by Bacteria, viruses, fungi, protozoa, algae, and prions. The most serious diseases are generally those that result in microbial growth inside the human body, but food poisoning can also be caused by microbial products (Table 2; Sections I5 and J5). The mechanisms by which microbes cause human disease are dealt with in more detail in *Instant Notes in Medical Microbiology*.

Table 2. Examples of microbes causing food poisoning

Microbes	Foods associated with outbreaks	Disease
Whole cells		
Campylobacter jejuni	Milk, poultry, meat Contaminated salads	Mild to severe gastroenteritis
Salmonella species	Meat, eggs, foods made with under-cooked eggs such as ice cream, mayonnaise	Salmonellosis (diarrhea, vomiting, fever, headache) caused by *Salmonella* serovars. Progression to septicemia can be caused by *S. typhi* and *S. paratyphi*

Other Bacteria causing food poisoning include *Shigella* spp., *Listeria monocytogenes*, *Yersinia entercolitica*, *Escherichia coli*, *Vibrio* spp., and *Brucella* spp.

Microbes	Foods associated with outbreaks	Disease
Cell products		
Staphylococcus aureus enterotoxin	Meat and meat products, milk, fish, canned food	Nausea, diarrhea, vomiting
Clostridium botulinum neurotoxin	Low acid canned food	Botulism (diarrhea, vomiting and neurological effects, e.g. vision problems, paralysis)
Bacillus cereus enterotoxins	Rice, milk, cereals	Diarrhea or nausea and vomiting depending on enterotoxin type
Viruses		
Norovirus	Many, including sewage-polluted shellfish	Nausea, diarrhea, vomiting
Rotavirus	Many fecally contaminated foods	Nausea, diarrhea, vomiting (particularly the under 5s and over 70s)
Hepatitis A virus	Many fecally contaminated foods	Jaundice (liver necrosis)

Most viruses spread by food contaminated via the fecal–oral route. More than 100 DNA and RNA viruses have been associated with food-borne disease

Microbes	Foods associated with outbreaks	Disease
Eukaryotes		
Fungal mycotoxins	Mushrooms, toadstools, cereals	See Section I5
Parasitic chlorophyta and protista	Sewage-polluted shellfish	See Section J5

Food spoilage

The food microbiologist must balance the use of microbes to make foods with the need to keep food fresh. Many pathogenic and nonpathogenic microorganisms take advantage of improperly protected foods as carbon and energy sources, rendering the food dangerous to eat or inedible. Preservation of food can take the form of storing the food under conditions that do not favor the growth of most microorganisms. Most approaches aim to reduce the amount of water available for growth (the **water activity**) by altering the salt concentration, pH, temperature, osmotic potential or hydration of the foodstuff.

Few organisms that can use human foodstuffs for growth are capable of extremophilic growth, so food will be more suitable for long-term storage if it is stored at very low pH (e.g. sauerkraut) or very high salt (e.g. cured meats such as bacon, salt beef, salt cod). Many fermentation processes leave the foodstuff in such a state. Low temperature (0–4°C) inhibits the growth of organisms, but food spoilage can still occur by the action of psychrophiles such as *Pseudomonas* spp. Bacterial growth is very restricted below –5°C, but fungi, yeasts, and molds can continue slow growth at as low as –12°C (e.g. *Debaryomyces* spp.). Very high sugar concentrations prevent the growth of most Bacteria. Jams, preserves, and chutneys use sucrose and fructose to prevent Bacterial spoilage, although some fungi and molds can still grow under these conditions.

Many foods become unpalatable when these approaches are used. In this case, the food microbiologist will try to eliminate or greatly reduce the numbers of microorganisms in foods. **Sterilization** by heat can be used to preserve foods such as milk for long periods as long as the food remains protected from contamination. However, very high temperatures (100–121°C) must be used to accomplish complete sterilization, which can alter the taste of the food considerably. More recently exposure to radioactive material has also been used to sterilize food.

Pasteurization provides a method for medium-term storage of milk, liquid egg or fruit juices while preserving palatability. The temperature and time used for pasteurization are dependent on the food, with egg being heated for 2.5 minutes at 64.4°C, whereas milk is heated at 71.7°C for 15 seconds. The advent of pasteurization of milk was a major step forward in public health, reducing the incidence of brucellosis to virtually zero in the Western world. However, the process is designed to kill pathogens and has the side-effect of increasing shelf-life. Eventually pasteurized products will spoil.

G3 Recombinant microorganisms in biotechnology

Key Notes	
Overexpression of prokaryotic genes	Although some useful Bacterial proteins and metabolites can be harvested from wild-type cells or their media, most are at such low levels that strains need to be improved to enhance recovery. The main approaches to overexpressing a particular gene or pathway are: mutation of secondary genes to relieve repression, increasing gene dosage, control of gene-specific promoters, and fusion to another protein to increase stability or ease of purification.
Production of recombinant mammalian proteins	Proteins and metabolites purified from mammals can cause immune reactions or carry disease. Bacteria and other microorganisms can be used to produce proteins of the same sequence as their mammalian counterpart as long as introns within the gene are removed so that coding DNA alone remains. However, microorganisms lack the capability to add sugar molecules in the same way as the human body, so recombinant hormones can have a lesser activity than the wild type.
Metabolic engineering	As our understanding of biochemical and regulatory pathways increases, we can begin to use a combination of systems biology and genetic engineering to create organisms programmed to produce biotechnologically important proteins or metabolites. This approach has been used to overproduce polyketides or biofuels in *Escherichia coli*.
Related topics	(F4) Transcription (F11) Plasmids

Overexpression of prokaryotic genes

Although some useful Bacterial proteins and metabolites can be harvested from wild-type cells or their media, most are at such low levels that strains need to be improved to enhance recovery. There are four main approaches to overexpressing a particular gene or pathway:

- Mutation of other genes causing repression
- Increasing gene dosage
- Controlling gene-specific promoters
- Fusion to another protein

Frequently a combination of all four techniques is used. The cell is a complex network of repressors and activators (Section F4), some of which serve to repress the cellular levels of molecules or proteins that have biotechnological significance. Random or site-directed mutation can result in derepression and accumulation of bioproducts. Mutational approaches can also be used to alter cell membrane transporters, allowing the cell to use substrates not available to the wild type.

Mutational approaches can lead to significant increases in product yield at the research scale, but where the cell is being exploited as a 'microbial factory,' economics dictate that even higher yields are required. Cloning genes into plasmids known as expression vectors (Section F11) will result in more copies of the gene being present in the cell. This will lead to more mRNA and hopefully more protein. This protein can then be harvested, or may act to produce more metabolite if this is the goal. However, increasing the gene dosage in this way can result in too much protein being produced in the cytoplasm. This can aggregate to form **inclusion bodies**, which are hard to isolate active protein from, or the protein can block membrane transport processes. The biotechnologist thus has a range of expression vectors available, from low copy number plasmids for toxic proteins, to high copy plasmids for amenable proteins.

In addition to manipulating gene dosage, the time at which gene expression begins can be crucial. A protein expressed during early logarithmic growth may prove toxic, but can be expressed at high levels if the recombinant gene is only switched on during late log phase or stationary phase. Regulation of protein expression is achieved by using adapted promoters, placed just upstream of the gene of interest in the expression vector. The first regulated expression vectors used the *lac* promoter (Section F4) so that the gene of interest could be switched on using IPTG (isopropyl-β-D1-thiogalactopyranoside). However, better, more controllable expression could be induced by using a promoter that was a hybrid of the *lac* and *trp* promoters. This '*trc*' promoter retains the ability to be turned on by IPTG but is of greater strength than *lac* alone. Other tunable promoters have been designed that turn on genes in response to low levels of uncommon carbohydrates such as xylose and arabinose. This alleviates the need to use the relatively expensive IPTG. All these different promoters allow the biotechnologist to turn on a gene merely by adding an inducer at any point during a fermentation.

Even though all these techniques can result in up to 10% of a Bacterial cell's protein being composed of the recombinant, some medically and industrially important proteins can be hard to purify, cannot be expressed at high levels due to cell toxicity, are not stable at high concentration, or prove to be particularly susceptible to protease activity. In these cases the recombinant protein can be 'fused' to another small protein to aid both stability and purification. For example, the C-terminal of a fusion protein can be made of 220 amino acids normally found in Bacterial glutathione-S-transferase (GST). After expression, this will allow the rapid purification of the fusion protein by using a column containing agarose coated with glutathione. The glutathione binds strongly to the GST domain of the fusion protein and other cellular constituents can be washed away. Although the fusion protein approach is useful in many applications, the protein of interest is physically very close to GST and this can lead to reduction or loss of activity. Inclusion of a thrombin domain between the GST and the protein of interest allows cleavage of the more valuable polypeptide away from GST. Other commonly used fusion proteins are based around the Bacterial maltose binding protein or synthetic histidine tagging genes.

Production of recombinant mammalian proteins

Proteins and metabolic products from animal cells are an important target for biotechnologists, with important applications in medicine and, to a lesser extent, food. We are now more aware of the dangers of using mammals as direct sources of protein and fortunately the days of using human cadavers as sources of therapeutic hormones are over. An example of this is insulin, which before 1982 was extracted from cows or pigs for use in diabetes therapy. However, yields were low and a minority of patients developed an immune response to the injections. These patients could have been given insulin from human sources, but the risk of immune reaction from co-purified proteins remained, along with the possibility of contracting diseases such as hepatitis (or latterly AIDS). The development of Humulin (human insulin expressed in *E. coli*, developed from 1978 by Genentech) surmounted some of these problems, and this is generally held to be the first recombinant Bacterial bioproduct used in medicine.

For a human gene to be expressed in a bacterium such as *E. coli*, a number of precautions must be taken to ensure that the resulting protein is similar to the human equivalent. The gene must be free of introns, as Bacteria use different signals for post-transcriptional processing (Section F5). The most efficient way of doing this is to use human cDNA, made by the action of reverse transcriptase on a messenger RNA sample. For more details of cDNA synthesis, see *Instant Notes in Molecular Biology*. Once the introns are removed from the gene, it may also be necessary to use site-directed mutagenesis to change which codons are used by the bacterium during translation – the set of codons used by humans is not identical to that used by *E. coli*, and absence of a codon-specific tRNA can lead to lower yields, random insertion of amino acids or a complete cessation of translation.

Once the gene has been adapted to function in a Bacterial cell, it can be placed under the control of a suitable promoter (see above) and production of the recombinant protein can start. A list of some of the proteins that have been engineered in this way appears in Table 1. However, even though the recombinant protein has the identical amino acid sequence to its human counterpart, the product may not have the same activity during therapy as the native drug target. There may be a number of reasons for this, as follows:

- It is not possible to make an absolutely pure preparation of any protein – a contaminating bacterial protein may be causing inhibition or immune reaction.

- Bacterial bioproducts may be contaminated with bacterial lipopolysaccharides (Section C8), which are toxic.

- The protein may only function properly when a certain number of sugar molecules are attached (when the protein is glycosylated). Bacteria use different signals for glycosylation than human cells.

Table 1. A selection of human proteins cloned in *E. coli* and their therapeutic use

Protein	Function	Therapeutic use
Urokinase	Plasminogen activator	Anticoagulant
Serum albumin	Major blood protein	Synthetic plasma constituent
Factor VIII, factor X	Blood clotting	Prevention of bleeding in hemophiliacs
Interferons	Can cause cells to become resistant to some viruses	Antiviral therapy
Growth hormone releasing factor (HGH)	Permits the action of growth hormone in the body	Growth promotion, recovery from physical stress
Erythropoietin (EPO)	Stimulates production of red blood cells	Replacement of cells after chemotherapy; treatment of anemia

Glycosylation problems mean that Bacteria are not always suitable for recombinant expression. Yeast or fungi can be used, and although the number of sugar molecules increases over Bacteria, glycosylation still will not be identical. An example of this is recombinant erythropoietin (EPO), which is used to boost red blood cell counts in leukemia and other cancer patients. Bacterial EPO needs to be administered in much higher doses than that produced from human cadavers, and the difference in glycosylation is used as a test for EPO's illegal nontherapeutic use as a blood booster in endurance sports.

Few mammalian recombinant proteins are used outside of medicine, but a few commonly used enzymes are produced in this way. Recombinant bovine rennin provides a way of making curd for cheese that is suitable for vegetarians and is also consistent in quality, avoiding its collection from calves' stomachs.

Metabolic engineering

We understand more about how the Bacterial cell works than any other organism. Our understanding of the **systems biology** of Bacteria such as *E. coli* and *Bacillus subtilis* has allowed us to build fairly accurate predictive computer models of how the biosynthetic and regulatory networks of these cells are influenced by some external factors. Although we do not yet fully understand all the subtleties of all the regulatory systems, we are in a position to engineer Bacteria to overproduce complex metabolites of our choice.

Metabolic engineering has had considerable success in engineering pathways for the production of fine chemicals (such as polyketide antibiotics) and biofuels (e.g. the production of biobutanol by *E. coli* using adapted genes from *Clostridium beijerinckii*). The approach relies on combining predictive modeling of pathways with a modular approach to genetic engineering. Whole new pathways can be added to genetically amenable Bacteria, and then some of the approaches to increasing yield are used to fine-tune production.

G4 Microbial bioproducts

Key Notes	
Biocatalysis	Microorganisms are a convenient source of enzymes. The regiospecific and stereospecific properties of enzymes can be exploited to produce enantiomerically pure preparations of some chemicals. These bacterial enzymes may either be used directly or held in immobilized cells for ease of use. The use of prokaryote enzymes can significantly reduce the cost and increase the yield of reactions that an organic chemist would find difficult to perform.
Biofuels	As the cost of recovering oil and gas from underground reserves increases along with our awareness of the environmental impact of using these fuels, biofuel derived from living plant material becomes more cost-efficient. Two main approaches have been taken to biofuels: generation of biomass from sunlight for combustion and direct generation of organic solvents for use in combustion engines.
Biopharmaceuticals	Microorganisms can be used to produce chemicals and enzymes for therapeutic use. The high regioselectivity and stereoselectivity of some reactions mean that ultrapure chemicals can be generated. Metabolic engineering can be used to further enhance production.

Biocatalysis

Microorganisms can be used as a convenient source of enzymes, since these same proteins provide them with the means to grow on a variety of complex organic compounds. The enzymes can be used in their purified form (either immobilized or in solution) but are sometimes more stable when whole cells are used. Bacterial cells may be rendered nonviable by **immobilization** or **permeabilization** but still retain catalytic activity. This can alleviate the need to provide cofactors such as ATP, which can add significantly to costs. The reactions the enzymes catalyze are often **regiospecific** (attacking a single group on a molecule but leaving others of the same chemical composition) and **stereospecific** (attacking one enantiomer such as D-glucose, but not the corresponding stereoisomer). The specificity of enzymes has allowed industrial chemists to perform reactions that would be impossible by normal synthetic routes, but mostly **biotransformations** are cheaper to perform and have a higher yield.

Examples of biotransformations are provided in Table 1 but one of the most economically significant biotransformations is the production of acrylamide (a polymer used in many chemical processes as well as in the cosmetics industry). It is possible to use a copper catalyst to convert acrylonitrile into acrylamide, but this reaction must be performed at 100°C, after which the catalyst must be regenerated and the unreacted highly toxic acrylonitrile must also be rigorously separated from the product. However, these problems are avoided by the use of immobilized *Pseudomonas chlororaphis* performing

Table 1. Examples of industrial production of organic compounds by prokaryotes

Prokaryote source	Chemical	Major application
Acetobacter	Acetic acid	Solvent, starting compound for many synthetic reactions
Clostridium	Isopropanol	Solvent, antifreeze
Clostridium	Acetone	Solvent, starting compound for many synthetic reactions
Bacillus	Acrylic acid	Precursor for acrylonitrile and other polymers
Bacillus	Propylene glycol	Solvent, antifreeze, antifungal compound

the biotransformation in a bioreactor. The reaction can be run at 10°C, so heating costs are reduced and the bacterial enzyme responsible (nitrile hydratase) converts over 99.9% of the acrylonitrile into acrylamide. Around half a million tonnes of acrylamide are produced annually by this process.

Although biotransformations have been used in the bulk chemical industry, the main application is in the production of **fine chemicals** such as antibiotic derivatives. The cost savings can be dramatic: cortisone was first synthesized as a 31-step organic synthesis starting from 615 kg of deoxycholic acid. This yielded 1 kg of cortisone, which was sold as an anti-inflammatory drug in the 1940s at around $200 per gram. Use of enzymes from the fungus *Aspergillus niger* in some of the steps reduced the cost to $6 per gram in 1952. The use of mycobacterial enzymes allowed plant sterols to be used as much cheaper starting compounds, so that by 1980 the price of cortisone had dropped to 46 cents in the United States, about a quarter of 1% of the original cost.

Biofuels

All our fossil fuels can be considered to be biofuels, since both methane and crude oil arise from the action of microorganisms on ancient carbon sources. However, as the cost of recovering oil and gas from underground reserves increases along with our awareness of the environmental impact of using these fuels, biofuel derived from living plant material becomes more cost-efficient. Two main approaches have been taken to biofuels: generation of biomass from sunlight for combustion and direct generation of organic solvents for use in combustion engines.

The sun provides us with more energy than we can harness, and many phototrophic organisms have been examined as possible biofuels. The idea that algal farms could be constructed that would use atmospheric carbon dioxide and sunlight to generate biomass seems to provide a zero carbon emission solution to energy production. Although many research groups and some start-up companies have designed realistic systems using algae or phototrophic Bacteria, a cost-effective solution has yet to be devised – it is currently cheaper to use fossil fuels. Algae do have the advantage that as well as acting as a source of combustible biomass, during phototrophic growth they can also produce algal oils that may act as biodiesel, or produce excess carbohydrate that can be fermented further to produce biosolvents such as ethanol or butanol. Species of *Botryococcus*, *Chlorella*, and *Sargassum* have all been used in algal biomass pilot projects.

More controversially, waste agricultural products and even food products can be converted into ethanol or butanol by Bacteria. The rise in food prices in 2009 was partly attributed to the use of wheat and rice as substrates for biofuel production. Brazil has long championed the use of sugar cane as a substrate for *Zymomonas mobilis*, providing this fossil-fuel poor state with a ready source of ethanol for adapted automobiles. The use of bioethanol is also rising in the United States, using food-grade wheat as substrate.

Although fermentations to produce bioethanol are well understood from the long human history of brewing, ethanol is not an ideal fuel for automobiles. Longer chain alcohols such as butanol or propanol can be used more efficiently and are less prone to evaporation. However, the pathways for the production of butanol are normally found in anaerobic Bacteria such as *Clostridium acetobutyliticum*, which in biotechnology are less well understood. However, recent exciting research has begun to combine the ability of some better known Bacteria to use waste products (such as corn steep liquor, olive oil waste or even potato washings) with metabolic engineering to produce *E. coli* that can produce butanol aerobically.

Biopharmaceuticals

The regiospecific and stereospecific nature of enzymes, coupled to the possibility of genetic engineering, mean that microorganisms can be used to make significant quantities of proteins and metabolites for therapeutic use. As more enzymes are studied with a view to industrial application, those from the extremophiles seem more attractive. Although bacteria growing at high temperature or extremes of pH are often easier to keep sterile on a large scale, protein yields are often low. Cloning into well studied prokaryotes such as *E. coli* or *Bacillus subtilis* allows much more controlled and efficient heterologous overexpression. The production of biopharmaceuticals is discussed in Section G3 in the context of overexpression and metabolic engineering.

H1 Taxonomy

Key Notes

Current taxonomic status of the eukaryotic microbes

The objective of current taxonomic schemes is to create monophyletic groups of microorganisms which are assumed to have a single ancestor. This is achieved by studying characters of an individual and comparing them with those of other, closely related, organisms. Using such information phylogenetic trees can be constructed that can indicate probable evolutionary sequences of the eukaryotic microbes.

The protista

Previously, members of this group were grouped into two form groups, the algae and protozoa. A more modern approach is to group both colorless and pigmented species together in a monophyletic taxonomy which currently has five supergroups.

Opisthokonts: the fungi

The true fungi, animals, and close relatives are now within the Opisthokonts. These all have flattened mitochondrial christae, and where they have flagella they are single and inserted into the posterior end of the cell.

Archaeplastida

This group contains the blue green, red, and green algae, and gave rise to the land plants. Within this group mitochondria have lamellar christae.

Excavata

The excavata contain the euglenozoa, heterolobosea, kinetoplastids, and fornicata. The group includes many significant human and animal parasites, and also the photosynthetic euglenids.

Chromalveolata

This supergroup contains the apicomplexa, dinoflagellates, ciliates, stramenopiles, cryptomonads, and haptophytes. The Rhizaria, containing amebae, flagellates, and ameboflagellates may be a monophyletic group within the chromalveolata.

Amoebozoa

The many different types of amebae are now grouped within this supergroup.

Sequence of evolutionary events

A number of phenomena prevent us from knowing the exact events of evolution. Both primitive and derived features can be found in some protists, and loss of features, parallel evolution, and transfer of genetic material can all contribute to the difficulty in creating a definitive tree.

Related topics

(B1) Prokaryotic systematics
(B3) Inference of phylogeny from rRNA gene sequence
(I1) Fungal structure and growth

(J1) Archaeplastida, Excavata, Chromalveolata, and Amoebozoa: taxonomy and structure
(K2) Virus taxonomy

Current taxonomic status of the eukaryotic microbes

Establishing relationships within the different members of the fungi, chlorophyta, and protistan microbes relies on studying **characters**, which are features or attributes of an individual organism that can be used to compare it with another organism. These features can be morphological, anatomical, ultrastructural, biochemical or based on sequences of nucleic acids. The objective of such a study is to create **monophyletic** groups that are assumed to have a single ancestor, usually extinct; a similar approach is used in the creation of classification systems for prokaryotes (Section B1). A **cladistic** approach would not assume features from an ancestor, but would merely define a monophyletic group on the basis of shared characters. New information based on the presence and type of mitochondria, and the DNA sequencing of ribosomal RNA, place the fungal, chlorophytan, and protistan members of the eukaryotic microbes into a complex **phylogenetic** *tree* (Figure 1). For more details of these trees, see Section B3.

The protista

The protista are a paraphyletic group of organisms that are not animals, true fungi or green plants. There are approximately 200 000 named species. Traditionally they were classified into nonmonophyletic, adaptive groups called the flagellates, algae, and protozoa. There are about 80 different patterns of organization, clustered into 60 lineages.

In this book we have followed a classification based on current information published on the Tree of Life website (www.tolweb.org).

The major groups at the root of the Tree of Life are under review. Our previous understanding of the monophyly of the Archaea, and the place of the amitochondrial protists at the base of the eukaryotic branch, has been shown to be untenable. The supergroup view

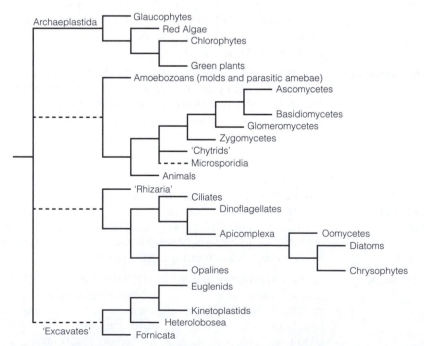

Figure 1. Tentative phylogenetic tree of the probable evolutionary sequence of protistan microbes.

is favored, where there are currently proposed up to five large, structurally and meta-bolically diverse, groups of microorganisms. Some supergroups have good supporting evidence, others less so. Again the Tree Of Life website provides an excellent discussion of current views.

Opisthokonts: the fungi

The true fungi, animals, and close relatives are now within the Opisthokonts. These all have flattened mitochondrial christae, and where they have flagella they are single and inserted into the posterior end of the cell. There is strong molecular support for this supergroup (Section I1).

Archaeplastida

This group contains the blue green, red, and green algae, as well as the land plants, and within this group mitochondria have lamellar christae. This group is well supported by molecular evidence (Sections I1 and J1).

Excavata

Until recently this assemblage was less well supported than the archaeplastida, but recent analysis that excludes rapidly evolving taxa places them as a monophyletic super-group containing the euglenozoa, heterolobosea, kinetoplastids, and a recently erected phylum, the fornicata. The group includes many significant human and animal parasites, many of which have anaerobic or microaerophyllic lifestyles, and also the photosynthetic euglenids (Section J1).

Chromalveolata

This supergroup contains the alveolates, including the apicomplexa, dinoflagellates, and ciliates, as well as the stramenopiles, cryptomonads, and haptophytes. Much of the supporting evidence for this grouping is molecular, plus when photosynthetic, the common origin of the plastid from a red algal endosymbiont. It appears that the Rhizaria, containing amebae, flagellates, and ameboflagellates, is a monophyletic group within the chromalveolata (Section J1).

Amoebozoa

The many different types of amebae are now grouped within this supergroup, and includes the lobose, lobose testate, and slime molds, which are ameboid for part of the life cycle. It also includes amebae lacking in mitochondria (Section J1).

Sequence of evolutionary events

Exact orders of evolution are difficult to determine, because of morphological inconsis-tencies within groups. For instance, dinoflagellates have a number of less specialized fea-tures, but they also have more derived features including being biflagellate and having scales, characteristics that bring them close to the alveolata (Section J1).

Even sequencing of nucleic acid to provide phylogenetic **gene trees** may not answer all the questions. For instance, sequencing the small subunit nuclear-encoded rRNA reveals that one can follow the evolution of an individual gene, but not necessarily the whole organism, as different regions of rRNA genes evolve at different rates. Phylogenetic pat-terns may also be confused because of transfer of genetic information between lineages as a consequence of **endosymbiosis** and other mechanisms. **Parallel** evolution within these organisms is also likely to occur, and superficial similarities within distinct lineages may arise because of loss of features, which is likely to be a significant factor within these apparently simple organisms.

Even the current proposal for five supergroups seems likely to change; recent analysis of opisthokonts and amoebozoa shows that they are closely related, and possibly are still a polyphyletic group. It seems probable that they are close to the root of the tree of life.

H2 Eukaryotic cell structure

Key Notes

Eukaryotes

Eukaryote cells have complex, membrane-bound, subcellular organelles which compartmentalize cell functions. The distinguishing feature of a eukaryotic cell is the nucleus.

Plasma membrane

The plasma membrane is a semi-permeable barrier between the outside and inside of the cell, and it is involved in cell–cell recognition, endo- and exocytosis, and adhesion to surfaces. Transport systems in the membrane allow it to import materials selectively into the cell.

Cytoplasm

The cytoplasm is 70–85% water, but also contains proteins, sugars, and salts in solution. The organelles of the cell are suspended in the cytoplasm. Both fungal and photosynthetic protista have single membrane-bound vacuoles in their cells.

Cytoskeleton

The cytoskeleton of the cell is made up of microtubules, intermediate filaments, and microfilaments, which maintain the shape of the cell and carry out numerous functions such as motility and the transport of organelles.

Nucleus and ribosomes

The nucleus is a double-membrane-bound organelle that contains the chromosomal DNA of the cell. Inside the nucleus is the nucleolus, which is the site of ribosomal RNA synthesis. Ribosomes are made of two subunits of RNA plus proteins, and they are the site of DNA translation and protein synthesis.

Endoplasmic reticulum

The endoplasmic reticulum (ER) is a complex of membrane tubes and plates, which is continuous in places with the nuclear membrane. The ER can be smooth or it may be termed rough where ribosomes are attached to it. The main function of this organelle is the synthesis and transport of proteins and lipids.

Golgi body

The Golgi body is a series of flattened, membrane-bound fenestrated sacs and vesicles. Vesicles secreted from the ER fuse with the *cis*-Golgi, and their contents are then further processed by resident biochemical processes. Processed materials are then secreted from the *trans*-Golgi in vesicles that fuse with other organelles or with the plasma membrane.

Lysosomes and peroxisomes

Lysosomes and peroxisomes are membrane-bound sacs secreted from the Golgi body. Lysosomes contain acid hydrolases involved in intracellular digestion. Peroxisomes contain amino and fatty acid-degrading enzymes and the enzyme catalase, which detoxifies hydrogen peroxide released by degradative processes.

Mitochondria	Mitochondria are the site of respiration and oxidative phosphorylation in aerobic organisms. They are bound by a double membrane, the inner one being infolded to form plates or tubes called cristae. ATP production is located in particles attached to the cristae.
Hydrogenosomes	Hydrogenosomes are organelles found in some amitochondrial, anaerobic groups. Their function is energy production. They contain enzymes of electron transport which use terminal electron acceptors that generate hydrogen.
Glycosomes	Glycosomes contain the enzymes of glycolysis and are found only in the apicomplexa.
Chloroplasts	Chloroplasts are double-membrane-bound organelles that contain the photosynthetic pigment chlorophyll. Within the chloroplast are stacks of flattened sacs called thylakoids where the photosynthetic systems are located.
Cell walls	Cell walls are found in the photosynthetic protista (cellulose-based) and the fungi (chitin-based). They delimit the outside of the cell from the environment and are important in maintaining cell rigidity and controlling excess of water influx due to osmosis.
Flagella	Flagella are microtubule-containing extensions of the cell membrane. They provide the cell with motility by their flexuous bending, which is controlled by the microtubule motor protein dynein.
Cilia	Cilia have the same internal structure as flagella but they are smaller and more numerous. Co-ordinated movement, which can be seen as wave-like beating, is required for motility.
Contractile vacuoles	Contractile vacuoles are found in free-living freshwater protozoa. They expel water absorbed into the cell by osmosis through pores in the cell surface.
Related topics	(A1) The microbial world (C7) Composition of a typical prokaryotic cell (C8) The bacterial cell wall (E1) Heterotrophic pathways (E2) Electron transport, oxidative phosphorylation, and β-oxidation of fatty acids (E3) Autotrophic reactions (F3) DNA replication (F4) Transcription (F5) Messenger RNA and translation (H1) Taxonomy (H3) Cell division and ploidy (J1) Archaeplastida, Excavata, Chromalveolata, and Amoebozoa: taxonomy and structure

Eukaryotes

Eukaryotic cells are compartmentalized by **membranes**. The cell contains several different types of membrane-bound organelles in which different biochemical and physiological processes can occur in a regulated way (Figure 1). Membranes also transport

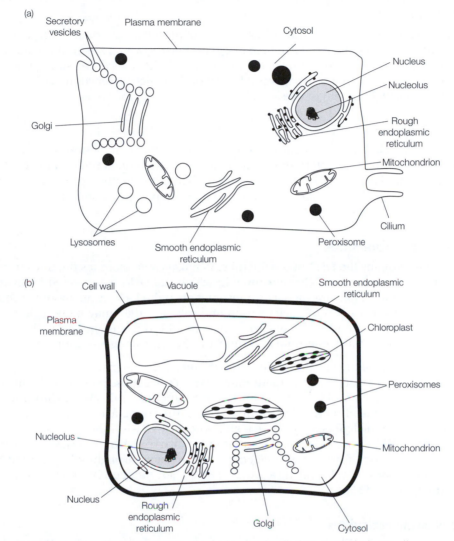

Figure 1. Eukaryotic cell structure: (a) a typical animal cell, (b) a typical plant cell.

information, metabolic intermediates and end products from the site of biosynthesis to the site of use.

Plasma membrane

The plasma membrane of eukaryotes is a **semi-permeable** barrier that forms the boundary between the outside and the inside of the cell. It is similar to that of the prokaryotes (Sections C7 and C8), except that it contains **sterols**, flat molecules that give the membrane a greater rigidity and that stabilize the eukaryotic cell. There are transport systems in the membrane that selectively import materials into the cell, and it is also involved in **endocytosis** and **exocytosis**, where food particles are engulfed and waste products are expelled from the cell in membrane-bound vesicles. The plasma membrane is involved in key interactive processes between cells, like **cell–cell recognition** systems, as well as adhesion of the cell to solid surfaces.

Cytoplasm

The cytoplasm is a dilute solution (70–85% water) of proteins, sugars, and salts in which all other organelles are suspended. It has **sol–gel** properties, which means it can be liquid or semi-solid depending on its molecular organization. Vacuoles act as storage sites for nutrients and waste products in a very weak solution. The high water content of the vacuole maintains a high cell turgor pressure.

Cytoskeleton

The eukaryotic cell is further stabilized by a cytoskeleton made up of **tubulin** containing **microtubules** (25 nm diameter), **microfilaments** (4–7 nm diameter), and **intermediate fibers** (8–10 nm diameter). The cytoskeleton is a dynamic structure, providing support for the cell but also the machinery for ameboid movement, cytoplasmic streaming, and nuclear and cell division.

Nucleus and ribosomes

The nucleus contains the DNA of the microbe. In eukaryotic microbes the nucleus usually contains more than one **chromosome**, in which the DNA is protected by **histone** proteins. In diploid organisms these chromosomes are paired. However, in some protista there are two distinct types of nuclei within one cell, and they may be polyploid. The larger of the two nuclei is termed a **macronucleus** and this nucleus is associated with cellular function. The smaller **micronucleus** functions in controlling reproduction.

The nucleus is surrounded by the **nuclear membrane**, a double membrane that is perforated by many pores where the two membranes fuse. It is via these pores that the nucleus remains in constant control of the rest of the cell machinery via **mRNA** and **ribosomes**. In places the nuclear membrane is also continuous with the endoplasmic reticulum. Within the nucleus is the **nucleolus**, an RNA-rich area where rRNA is synthesized (see Sections F4 and F5 for DNA transcription and translation). Eukaryotic ribosomes are essentially very similar to those of prokaryotes (Section F5) but they are slightly larger, their two subunits are of 60S and 40S, making a dimer of 80S. Their function is the same as that of prokaryotes (Section F5).

Endoplasmic reticulum

The outer membrane of the nucleus is in places continuous with a complex, three-dimensional array of membrane tubes and sheets, the endoplasmic reticulum (ER). Tubular ER can be studded with ribosomes, and described as **rough ER** (RER), where **ribosomal translation** and **protein modification** take place (Section F5). These proteins can either be secreted into the lumen of the ER or inserted into the membrane. Plates of **smooth ER** are associated with **lipid synthesis** and **protein** and **lipid transport** across cells.

Golgi body

The Golgi body is composed of stacks of a flattened series of membrane-bound sacs or **cisternae**, surrounded by a complex of tubes and vesicles. There is a definite **polarity** across the stack, the *cis* or forming face receiving vesicles from the ER, the contents of these vesicles then being processed by the Golgi, to be budded from the sides or the *trans* (maturing) face of the organelle. The Golgi apparatus processes and packages materials for secretion into other subcellular organelles or from the cell membrane. Golgi bodies in fungi are less well developed than in algae, and tend to have fewer or single cisternae. They are sometimes termed **dictyosomes**.

Lysosomes and peroxisomes

The Golgi body generates these single-membrane-bound organelles which contain enzymes (**acid hydrolases** in the lysosome, **aminases**, **amidases**, and **lipases** plus **catalase** in the peroxisomes) needed in the digestion of many different macromolecules. The internal pH of the lysosome is **acidic** (pH 3.5–5) to enable the enzymes to work at the optimum pH, and this pH is maintained by proton pumps present on the membrane. The breakdown of amino and fatty acids by the peroxisomes generates **hydrogen peroxide**, a potentially cytotoxic by-product. The enzyme catalase, also present in the peroxisome, degrades the peroxide into water and oxygen, protecting the cell.

Mitochondria

Mitochondria are double-membrane-bound organelles where the processes of **respiration** and **oxidative phosphorylation** occur (Sections E1 and E2). They are approximately 2–3 μm long and 1 μm in diameter. Their numbers in a cell vary. They contain a small, circular DNA molecule, which encodes some of the mitochondrial proteins, and 70S ribosomes (Section F5). The inner membranes of the mitochondria contain an **ATP/ADP transporter** that moves the ATP, which is synthesized in the organelle, outwards into the cytoplasm. ATP production is located on particles attached to the cristae, the inner infolded mitochondrial membrane (Figure 2). Their structures differ slightly between the protistan groups. Not all protista have mitochondria (Section J1) and in these cells metabolism is essentially anaerobic (Section J2). In the aerobic, very primitive eukaryotes the mitochondrial **cristae** are discoid. Mitochondria of the fungi are large and highly lobed with flat, plate-like cristae, while those of the chlorophyta have much more inflated cristae.

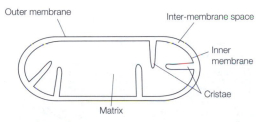

Figure 2. Structure of a mitochondrion.

Hydrogenosomes

Hydrogenosomes are unique organelles found in anaerobic protista that lack mitochondria. They are membrane-bound organelles containing **electron transport pathways** in which **hydrogenase** enzymes transfer electrons to terminal electron acceptors, which generate molecular hydrogen (Section J2) and ATP.

Glycosomes

Glycosomes are unique to the protistan group Apicomplexa. This organelle is surrounded by a single unit membrane and contains the **enzymes of glycolysis** (Section E1).

Chloroplasts

Chloroplasts are **chlorophyll**-containing organelles that can use light energy to fix carbon dioxide into carbohydrates (**photosynthesis**) (Sections E3 and J2). They are bound by double membranes and contain flattened membrane sacs called **thylakoids** where the light reaction of photosynthesis takes place (Figure 3). In photosynthetic protista and the chlorophyta these organelles are large, almost filling the cell. The **pyrenoid** is a proteinaceous region within the chloroplast where polysaccharide biosynthesis takes place.

Cell walls

The protoplasts of fungal and photosynthetic protistan cells are in most cases surrounded by rigid cell walls. In the fungi the cell wall is composed of a microcrystalline polymer of **chitin** (repeating units of β1–4-linked NAG) and amorphous β-glucans, while in the chlorophyta and some other photosynthetic protista, cell walls are composed of **cellulose** (repeating units of β1–4-linked glucose) and **hemicelluloses**.

Flagella

Flagella are membrane-bound extensions of the cell, which contain microtubules (Section J1). The microtubules are arranged as a bundle of nine doublets around the periphery of the flagellum, with a pair of single microtubules running within them. This structure is called the **axoneme**. Flagella provide cells with motility, because they flex and bend when supplied with ATP. Each outer pair of microtubules has arms projecting towards a neighboring doublet (Figure 4) and a spoke extending to the inner pair of microtubules. Microtubules are formed from **tubulin**, a self-assembling protein. Tubulin is composed of two subunits, α and β, arranged in a helical fashion. The projecting arms between outer subunits are made up of the protein **dynein**. This protein is involved in converting the energy released from ATP hydrolysis into mechanical energy for flagellar movement. Movement is produced by the interaction of the dynein arms with one of the microtubules of adjacent doublets. A basal body (**kinetosome**) anchors the flagellum within the cytoplasm.

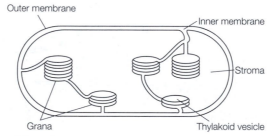

Figure 3. Structure of a chloroplast.

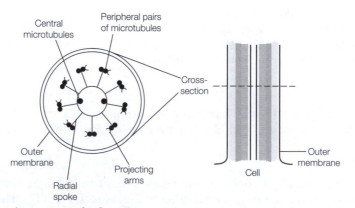

Figure 4. General structure of a flagellum or cilium.

Cilia

Cilia have the same internal structure as flagella but they are typically shorter in length. They are usually present on a cell in great numbers and propel the cell by co-ordinated beating seen as waves over the surface of the organism.

Contractile vacuoles

Contractile vacuoles are found in free-living freshwater nonphotosynthetic protistans. Their function is to regulate osmotic pressure within the cell by expelling water from a central vacuole through a pore in the outer surface. The simplest contractile vacuoles consist of a vacuole that can form anywhere in the cell, to a fixed structure that is surrounded by bands of microtubules and surrounded by collecting canals that collect fluid from the cytoplasm (Figure 5).

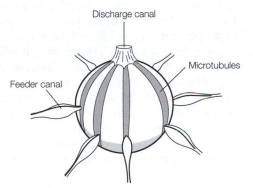

Figure 5. Diagram of a contractile vacuole.

H3 Cell division and ploidy

Key Notes

Cell cycle

The cell cycle describes all the events that occur in a cell from the end of one cell division to the end of the next. There are four phases in the eukaryotic cell cycle, which include cell growth (G_1), DNA synthesis (S), a second gap or growth phase (G_2), and finally nuclear division (M phase).

Mitosis and asexual cell division

Mitosis is nuclear division that results in progeny nuclei that are identical to the parent. It is usually followed by cytokinesis, cell division, which produces cells that have the same phenotype as the parent. In some eukaryotic microbes multiple nuclear divisions may occur without cytokinesis, giving rise to large, multinucleate cells. There are four stages of mitosis, prophase, metaphase, anaphase, and telophase. During prophase chromosomes duplicate to form chromatids, joined at the centromeres. Centromeres are attached to spindle microtubules. During metaphase, chromatids are arranged across the center of the cell, and during anaphase the microtubules pull the sister chromatids apart to the poles of the dividing cell. During telophase the microtubules disappear and the nuclear envelopes fuse.

Meiosis and sexual cell division

Meiosis is a nuclear division where there is a halving of chromosome numbers from a diploid to a haploid state. In organisms with an extended diploid phase of their life cycle, meiosis produces haploid gametes, and it is immediately followed by gamete fusion and formation of a new diploid organism. In organisms with an extended haploid stage, diploid formation is immediately followed by meiosis which produces the new haploid organism. There are eight stages to meiosis; prophase 1, metaphase 1, anaphase 1, and telophase 1, and prophase 2, metaphase 2, anaphase 2, and telophase 2. During prophase 1, homologous chromosomes associate and duplicate. At this point there may be recombination. During metaphase 1, homologous chromosomes assemble on the spindle across the cell, and they are separated during anaphase 1 and telophase 1. Prophase and metaphase 2 are transient and the chromatids assemble across the cell to be separated during anaphase 2. During telophase 2 cell division usually occurs.

Chromosomes

Eukaryotic microorganisms package the large amount of DNA they contain into chromosomes. Chromosomes contain a single, linear double strand of DNA tightly bound with histone proteins. There are usually between four and eight chromosomes per cell.

Histones	Histones are basic proteins that bind to DNA to condense and fold it. They are vital to the structure of DNA and have been highly conserved during evolution.	
Related topics	(C9) Cell division	(I3) Reproduction in fungi
	(F2) Genomes	(J3) Archaeplastida, Excavata,
	(F3) DNA replication	Chromalveolata, and
	(F9) Recombination	Amoebozoa: life cycles

Cell cycle

The cell cycle consists of an **interphase** during which growth of the cell (G_1) occurs, the cell increases in volume to maximum size and there is synthesis of cytoplasmic constituents and RNA (Figure 1). Synthesis of DNA occurs next, during the **S** phase, as chromosomes are duplicated in preparation for nuclear division. During the final part of the cell cycle a second gap or growth phase occurs (G_2), when specific cell division-related proteins are synthesized. Nuclear division (**M**) then follows to complete the cell cycle.

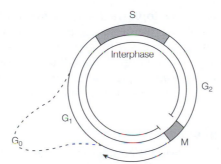

Figure 1. The eukaryotic cell cycle. The S phase is typically 6–8 hours long, G_2 is a phase in which the cell prepares for mitosis and lasts for 2–6 hours, mitosis itself (M) is short and takes only about 1 hour. The length of G_1 is very variable and depends on the cell type. Cells can enter G_0, a quiescent phase, instead of continuing with the cell cycle. From Hames D & Hooper N (2011) *Instant Notes Biochemistry*, 4th ed. Garland Science.

Mitosis and asexual cell division

Asexual cell division in unicellular eukaryotic organisms is synonymous with growth. Division is usually by **binary fission** after a single nuclear division. The parent cell divides, usually longitudinally, into two even-sized identical progeny cells. Division can be by **multiple fission** (Section J1), where many nuclear divisions occur, producing either a large multinucleate **coenocyte** or many uninucleate progeny cells. All cells produced from mitosis are genetically identical to their parent.

In both cases **somatic** cell division is preceded by the mitotic division of the cell nucleus. In mitosis the replicated DNA from the S phase of the cell cycle is separated equally into two progeny cells. The events of mitosis can be separated into four stages for convenience, but each flows into the other as a continuous process.

The first phase in mitosis is the **prophase**. In this phase microtubules form from the **microtubule organizing centers** (**MTOCs**). In fungi these structures are known as

spindle pole bodies (**SPBs**), located close to the nuclear envelope. In the motile species of protista the MTOC is called a **centriole** and it becomes surrounded by microtubules in a process termed **aster formation** (Figure 2a). The MTOCs begin to move towards opposite poles of the nucleus, and spindle microtubules appear between them. Single chromosomes, which have been duplicated to form **chromatids**, are joined together at their **centromeres**. These centromeres are also attached to spindle microtubules. By late prophase–early **metaphase**, MTOCs are opposite each other and the spindle is complete, with chromosomes aligned across the center in a **metaphase plate**. In many species of fungi, chromosomes remain extremely indistinct during mitosis, and they do not appear to assemble across a metaphase plate.

The nuclear envelope may disappear between the prophase and the beginning of the metaphase in some eukaryotic microbes, but in some protista the nuclear envelope remains intact throughout the process, the microtubules of the spindle penetrating through it.

In the next phase of mitosis, the **anaphase**, pairs of chromatids that were held together at the centromere begin to separate simultaneously, and spindle microtubules begin to pull them towards the two poles of the cell. By the end of the anaphase the chromatids have been pulled close to the MTOCs. In many species of the fungi there is an asynchronous chromosome separation during the anaphase.

In the final phase of mitosis, the **telophase**, the aster microtubules disappear, and the nuclear envelope re-forms if it has disintegrated. In the two progeny nuclei the MTOC duplicates, and **cell division** commences with the division of the cytoplasm by an invaginating plasma membrane or the formation of a **cell plate** by Golgi-derived vesicles across the midline between the two nuclei (Figure 2b). In some protista separation can be by **budding**, producing a progeny cell that is much smaller than the parent.

Meiosis and sexual cell division

Most protista are **haploid** for most of their life cycle. They have only one set of chromosomes. The **diploid** stage (two sets of chromosomes) is often very transient and found only in resting structures such as spores. The life cycle ends with meiosis, which returns the new cell to its haploid state. In organisms with a dominant diploid vegetative phase, meiosis occurs just before cell division, producing haploid gametes which then fuse to re-form the diploid.

At the end of the interphase and before meiosis begins, the duplication of chromosomes to chromatids occurs just as it does in mitosis. However, as the cell is diploid, **homologous** pairs of chromosomes associate during **prophase 1** (Figure 3), and at this point it is possible for genetic recombination to occur (Section F9).

During metaphase 1 chromosomes assemble across the metaphase plate, and in anaphase 1 homologous chromosomes are separated. Telophase 1 is very transient, and the chromosomes rapidly move into the prophase and metaphase 2, where pairs of chromatids assemble across the metaphase plate. The chromatids separate from each other at anaphase 2, and in telophase 2 cytoplasm begins to separate around the four progeny nuclei, each containing a haploid complement of chromosomes.

In some circumstances **multiple sets** of chromosomes can exist in a cell and this is termed **polyploidy**. Nuclear division in polyploid organisms is complex and often results in the loss of single chromosomes, leading to odd numbers of chromosomes in some progeny cells. This is termed **aneuploidy**. Polyploid and aneuploid cells are usually unable to participate in meiosis because of their odd chromosome numbers.

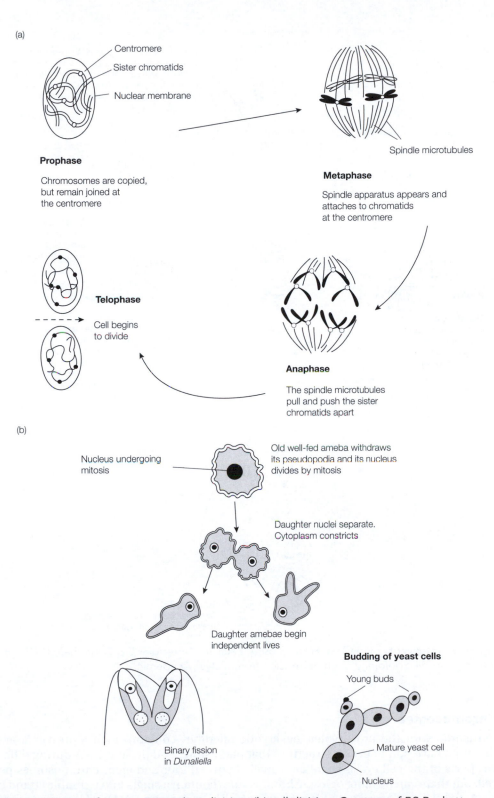

(a)

Centromere

Sister chromatids

Nuclear membrane

Spindle microtubules

Prophase

Chromosomes are copied,
but remain joined at
the centromere

Metaphase

Spindle apparatus appears and
attaches to chromatids
at the centromere

Telophase

Cell begins
to divide

Anaphase

The spindle microtubules
pull and push the sister
chromatids apart

(b)

Nucleus undergoing
mitosis

Old well-fed ameba withdraws
its pseudopodia and its nucleus
divides by mitosis

Daughter nuclei separate.
Cytoplasm constricts

Daughter amebae begin
independent lives

Budding of yeast cells

Young buds

Mature yeast cell

Nucleus

Binary fission
in *Dunaliella*

Figure 2. Events of mitosis: (a) nuclear division, (b) cell division. Courtesy of BS Beckett.

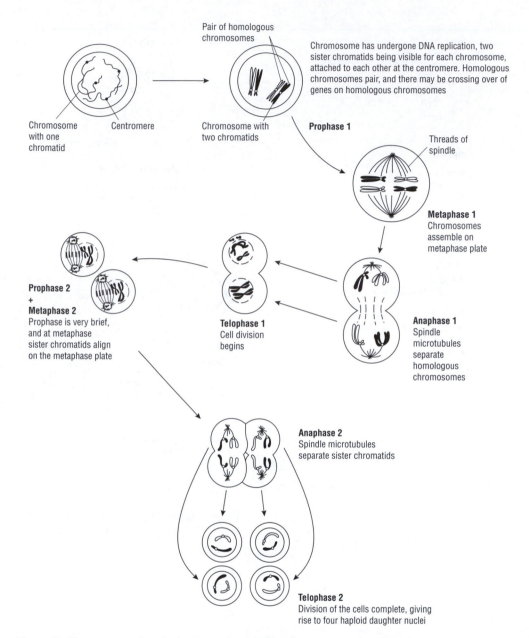

Pair of homologous chromosomes

Chromosome has undergone DNA replication, two sister chromatids being visible for each chromosome, attached to each other at the centromere. Homologous chromosomes pair, and there may be crossing over of genes on homologous chromosomes

Chromosome with one chromatid

Centromere

Chromosome with two chromatids

Prophase 1

Threads of spindle

Metaphase 1
Chromosomes assemble on metaphase plate

**Prophase 2
+
Metaphase 2**
Prophase is very brief, and at metaphase sister chromatids align on the metaphase plate

Telophase 1
Cell division begins

Anaphase 1
Spindle microtubules separate homologous chromosomes

Anaphase 2
Spindle microtubules separate sister chromatids

Telophase 2
Division of the cells complete, giving rise to four haploid daughter nuclei

Figure 3. The events of meiosis. From Gross T, Faull JL, Ketteridge S & Springham D (1995) *Introductory Microbiology*. With permission from Kluwer Academic.

Chromosomes

Compared with the prokaryotes, eukaryotic microbes contain much more DNA, and therefore have had to evolve structures that pack, store, and present DNA during different parts of the cell cycle. There are usually between four and eight chromosomes per cell, but there can be more or less. Chromosomes contain a single linear double strand of DNA, tightly condensed with **histone** proteins. The combined DNA and histone is often termed **chromatin**. The histone proteins bind to the DNA and create three levels of folding. During the prophase, chromatin is very condensed but, in the interphase, chromatin is described as dispersed.

Histones

Histones are very basic proteins because they contain many basic amino acids (lysine and arginine), with positively charged side-chains. These side-chains associate with the negatively charged phosphate groups on the DNA molecule. The histones are vital to the structure of DNA and have been **highly conserved** during evolution.

I1 Fungal structure and growth

Key Notes

Fungal taxonomy

Fungal taxonomy is undergoing radical revision based on molecular data and phylogenetic analysis. Currently there is one subclass, the dikarya, with up to seven further phyla and a number of unassigned taxa.

Fungal structure

Fungi are heterotrophic, eukaryotic organisms with a filamentous, tubular structure, a single branch of which is called a hypha. A network of hyphae is called a mycelium. Hyphae are bound by firm, chitin-containing walls and contain most eukaryotic organelles. Not all fungi are multicellular, some are single-celled and are termed yeasts.

Fungal wall structure and growth

Fungal walls are formed from semi-crystalline chitin microfibrils embedded in an amorphous matrix of β-glucan. In the Ascomycota and Basidiomycota, hyphae grow by tip growth followed by septation. In the Chytridiomycota and Zygomycota fungal hyphae grow by tip growth but remain aseptate.

Colonial growth

Colonial growth is characterized by the radial extension of mycelium over and through a substrate, creating a circular or spherical fungal colony.

Kinetics of growth

Fungal growth can be measured by measuring mycelial mass changes with time under excess of nutrient conditions. From this information the specific growth rate can be calculated. After a lag phase, a brief period of exponential growth follows as hyphal tips are initiated. As the new hypha extends, it grows at a linear rate until nutrient depletion causes a retardation phase, followed by a stationary phase.

Hyphal growth unit

Hyphal growth may also be measured by microscopy, and by counting the total numbers of hyphal tips, and dividing that number by the total length of mycelium in the colony, the average length of hypha required to support a growing tip can be calculated. This is termed the hyphal growth unit.

Peripheral growth zone

The peripheral growth zone is the region of mycelium behind the tip, which permits radial extension at a rate equal to the specific growth rate.

Related topics

(C7) Composition of a typical prokaryotic cell
(C9) Cell division
(D1) Measurement of microbial growth
(D2) Batch culture in the laboratory

(H1) Taxonomy
(H2) Eukaryotic cell structure
(I3) Reproduction in fungi
(J1) Archaeplastida, Excavata, Chromalveolata, and Amoebozoa: taxonomy and structure

Fungal taxonomy

In the past the fungi were a **polyphyletic** group that contained microorganisms that had very different ancestors. Current thinking now prefers a **monophyletic** classification where all groups within a phylum are descendants of one ancestor (Section H1).

Fungal taxonomy is currently undergoing a comprehensive review based on molecular data and phylogenetic analysis. The fungi currently contain around 70 000 described species, with an estimated 1.5 to 4 million fungi as yet undescribed. They are now considered to be a member of the eukaryotic crown, evolving about a billion years ago as a sister clade to the Animals.

The original classification of the fungi into four major phyla based largely on form groups is now being replaced with a classification that includes one sub-kingdom, the Dikarya, which contains the phyla the **Ascomycota** and **Basidiomycota**. These are fungi with complex mycelium with elaborate, perforate septa and a complex life cycle that includes a phase where a pair of unfused nuclei exist together in one cell (the dikaryon).

Up to seven other phyla are proposed, including the **Glomeromycota**, and the **Zygomycota**, fungi with nonseptate mycelium, asexual sporangiospores, and sexual zygospores (grouped together in the Zygomycetes in the old system) and three phyla containing the fungi with motile spores, the **Blastocladiomycota**, **Chytridiomycota**, and **Neocallimastigomycota**. This group is loosely termed the chytrids.

However, it seems that neither the Zygomycota nor the Chytridiomycota are monophyletic and the phyla may yet be deconstructed as data accumulate. Two chytrids in particular illustrate the difficulty in developing a classification for this group. *Rozella* is an intracellular parasite of other chytrid fungi with an extremely reduced thallus, which appears to have diverged very early from the main fungal branch. From recent molecular data it appears to be a sister group to the **Microsporidia**, until very recently considered to be a member of the animal kingdom. *Olpidium*, a fungus with a growth habit very like that of *Rozella* as an intracellular parasite, appears to have evolved a lot later than most of the chytrids and may be closer to the Zygomycetes.

A more startling change is the inclusion of the microsporidia in the fungi. These are obligate, highly specialized intracellular parasites, about 1300 described species, infecting animals, humans, insects, and fish. Originally described as schizomycete fungi, then placed at the base of the eukaryote tree, as amitochondrial primitive species, molecular data now find them within the fungal branch of the eukaryotic tree.

Within each of the major phyla are several classes of fungi, indicated by names ending with -*etes*, and within classes can be subclasses or orders, indicated by names ending in -*ales*, within which there are genera and then species (Section B1). The important differences between fungi used to distinguish taxonomic groups are summarized in Table 1.

Fungal structure

Fungi are filamentous, nonphotosynthetic, eukaryotic microorganisms that have a heterotrophic nutrition (Section E1). Their basic cellular unit is described as a hypha (Figure 1a). This is a tubular compartment that is surrounded by a rigid, chitin-containing wall. The hypha extends by tip growth, and multiplies by branching, creating a fine network called a mycelium. Hyphae contain nuclei, mitochondria, ribosomes, Golgi, and membrane-bound vesicles within a plasma membrane-bound cytoplasm (Section H2). The subcellular structures are supported and organized by microtubules and endoplasmic reticulum (ER). The cytoplasmic contents of the hypha tend to be concentrated towards

Table 1. Features of the main groups of fungi

	Perforate septae present or absent	Asexual sporulation	Sexual Group sporulation
Dikarya			
Ascomycota	Present	Conidiospores	Ascospores
Basidiomycota	Present	Rare	Basidiospores
Glomeromycota	Absent	Nonmotile spores	None found
Zygomycota	Absent	Nonmotile sporangiospores	Zygospore
Chytrids			
Blastocladiomycota	Absent	Motile zoospores	Oospore
Chytridiomycota	Absent	Motile zoospores	Oospore
Neocallimastigomycota	Absent	Motile zoospores	Oospore
(Microsporidia)	Absent		

the growing tip. Older parts of the hypha are heavily vacuolated and may be separated from the younger areas by cross walls called septae. Not all fungi are multicellular, some are unicellular and are termed yeasts (Figure 1b). These grow by binary fission or budding.

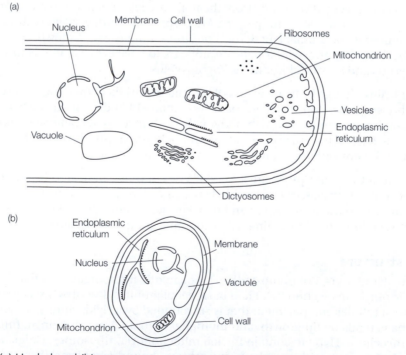

Figure 1. (a) Hyphal and (b) yeast structures. From Grove SM, Bracker CE & Morré BJ (1970) *American Journal of Botany* 57, 245–266. With permission from The Botanical Society of America.

Fungal wall structure and growth

Fungal walls are rigid structures formed from layers of semi-crystalline chitin **microfibrils** that are embedded in an amorphous matrix of β 1–3 and β 1–6 **glucan**. Some protein may also be present. Growth occurs at the hyphal tip by the fusion of membrane-bound vesicles containing wall-softening enzymes, cell wall monomers, and cell wall polymerizing enzymes derived from the Golgi with the hyphal tip membrane (Figure 2). The fungal wall is softened, extended by turgor pressure, and then rigidified.

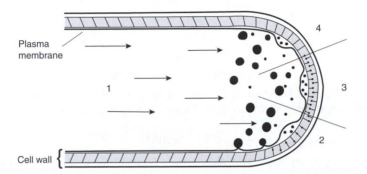

Figure 2. Hyphal tip growth. Step 1, vesicles migrate to the apical regions of the hyphae. Step 2, wall-lysing enzymes break fibrils in the existing wall, and turgor pressure causes the wall to expand. Step 3, amorphous wall polymers and precursors pass through the fibrillar layer. Step 4, wall-synthesizing enzymes rebuild the wall fibrils. From Isaac S (1991) *Fungal–plant interactions*. With permission from Kluwer Academic.

Septae are cross walls that form within the mycelium. Growth in the Zygomycota and Chytridiomycota is not accompanied by septum formation, and the mycelium is coenocytic. Septae only occur in these groups to delimit reproductive structures (Section H3) from the parent mycelium, and they are complete. In the dikarya growth of the mycelium is accompanied by the formation of incomplete septae. Septae in the Ascomycota are perforate, and covered by ER membranes to limit movement of large organelles such as nuclei from compartment to compartment. This structure is called the **dolipore** septum. In dikaryotic Basidiomycota, septum formation is co-ordinated with divisions of the two mating-type nuclei (Section H3), maintaining the dikaryotic state by the formation of **clamp connections**. These septae resemble crozier formation in the formation of asci and the structures may be homologous.

Colonial growth

Hyphal tip growth allows fungi to extend into new regions from a point source or **inoculum**. Older parts of the hyphae are often emptied of contents as the cytoplasm is taken forwards with the growing tip. This creates the radiating colonial pattern seen on agar plates (Figure 3), in ringworm infections of skin, and in fairy rings on grass lawns.

Kinetics of growth

When fungi are filamentous their growth rate cannot be established by cell counting using a hemocytometer or by turbidometric measurements (which can be used to measure bacterial and yeast growth, Section D1). However, by measuring mass (M) changes with time (t) under excess nutrient conditions the **specific growth rate** (m) for the culture can be calculated using the formula:

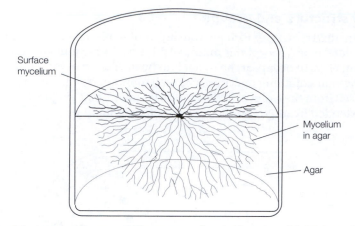

Figure 3. Colonial growth patterns of filamentous fungi. From Ingold CT & Hudson HJ (1993) *The Biology of Fungi*. With permission from Kluwer Academic.

$$dM/dt = mM$$

Fungal growth in a given medium follows the growth phases of lag, acceleration, exponential, linear, retardation, stationary, and decline (Section D1, Figure 3).

Exponential growth occurs only for a brief period as hyphae branches are initiated, and then the new hypha extends at a linear rate into uncolonized regions of substrate. Only hyphal tips contribute to extension growth. However, older hyphae can grow aerially or differentiate to produce sporing structures.

Hyphal growth unit

It is also possible to observe hyphal growth by microscopy, measuring tip growth and branching rates of mycelium. From these data the hyphal growth unit and the **peripheral growth zone** can be calculated. The hyphal growth unit (G), which is the average length of hypha that is required to support tip growth, is defined as the **ratio** between the total length of mycelium and the total number of tips:

total number of tips/total length of the mycelium = hyphal growth unit

In most fungi the hyphal growth unit includes the tip compartment plus two or three subapical compartments. The ratio increases exponentially from germination, but stabilizes to give a constant figure for a particular strain under any given set of environmental conditions.

Peripheral growth zone

The peripheral growth zone is the region of mycelium behind the tip, needed to support maximum growth of the hyphal tip. This zone permits radial extension at a rate equal to that of the **specific growth rate** of unicells in liquid culture. The mean rate of hyphal extension (E) is a function of the hyphal growth unit and the specific growth rate (μ):

$$E = G\mu$$

The **radial extension rate** (K_r) is a function of the peripheral growth zone (ω) and the specific growth rate (μ):

$$K_r = \omega\mu$$

I2 Fungal nutrition

Key Notes	
Carbon nutrition	Fungi require organic carbon compounds to satisfy their carbon and energy requirements. They obtain this carbon by saprotrophy, symbiosis or parasitism. The carbon must be available in a soluble form in order to cross the rigid cell wall, or must be broken down by enzymes secreted by the fungal cells.
Carbon metabolism	Fungi normally utilize glycolysis and aerobic metabolism of carbohydrates. Some can use fermentative pathways under reduced oxygen levels. A few fungi are truly anaerobic.
Nitrogen nutrition	Fungi cannot fix gaseous nitrogen but can utilize nitrate, ammonia, and some amino acids as nitrogen sources.
Macronutrients, micronutrients, and growth factors	Most macronutrients and micronutrients that fungi require are present in excess in their environments. Phosphorus and iron may be in short supply, and fungi have specific mechanisms to obtain these nutrients. Some fungi may require external supplies of some vitamins, sterols, and growth factors.
Water, pH, and temperature	Fungi require water for nutrient uptake and are therefore restricted to damp environments. They occupy acidic environments between pH 4 and 6, and by their activity further acidify them. Most fungi are mesophilic, growing at between 5° and 40°C, but some can tolerate high or low temperatures.
Secondary metabolism	Secondary metabolites, derived from many different metabolic pathways, are produced by fungi when vegetative growth becomes restricted by nutrient depletion or stress. Such compounds may offer a competitive advantage to the producer.
Related topics	(D1) Measurement of microbial growth (E1) Heterotrophic pathways (E3) Autotrophic reactions (E4) Other unique microbial biochemical pathways (H2) Eukaryotic cell structure	(J1) Archaeplastida, Excavata, Chromalveolata, and Amoebozoa: taxonomy and structure (J2) Archaeplastida, Excavata, Chromalveolata, and Amoebozoa: nutrition and metabolism

Carbon nutrition

Fungi are heterotrophic for carbon (Section E1). They need organic compounds to satisfy energy and carbon requirements. There are three main modes of nutrition: **saprotrophy**, where fungi utilize dead plant, animal or microbial remains; **parasitism**, where fungi utilize living tissues of plants and animals to the detriment of the host; and **symbiosis**, where fungi live with living tissues to the benefit of the host.

Carbohydrates must enter hyphae in a soluble form because the rigid cell wall prevents endocytosis. Soluble sugars cross the fungal wall by diffusion, followed by active uptake across the fungal membrane. This type of nutrition is seen in the symbiotic and some parasitic fungi. For the saprophytic fungi most carbon in the environment is not in a soluble form but is present as a complex polymer like cellulose, chitin or lignin. These polymers have to be broken down enzymically before they can be utilized. Fungi release **degradative enzymes** into their environments. Different classes of enzyme can be produced, including the **cellulases**, **chitinases**, **proteases**, and multi-component **lignin-degrading enzymes**, depending on the type of substrate the fungus is growing on. Regulation of these enzymes is by **substrate induction** and **end-product inhibition**.

Carbon metabolism

Once within the hypha, carbon and energy metabolism is by the processes of **glycolysis** and the **citric acid cycle** (Section E1). Fungi are usually aerobic, but some species, for example the yeasts, are capable of living in low oxygen tension environments and utilizing **fermentative** pathways of metabolism (Section E1). Recently, truly **anaerobic** fungi have been discovered within animal rumen and in anaerobic sewage-sludge digesters.

Nitrogen nutrition

Fungi are heterotrophic for nitrogen. They cannot fix gaseous nitrogen, but they can utilize nitrate, ammonia, and some amino acids by direct uptake across the hyphal membrane. Complex nitrogen sources, such as peptides and proteins, can be utilized after extracellular proteases have degraded them into amino acids.

Macronutrients, micronutrients, and growth factors

Phosphorus, potassium, magnesium, calcium, and sulfur are all macronutrients required by fungi. All but phosphorus are usually available to excess in the fungal environment. Phosphorus can sometimes be in short supply, particularly in soils, and fungi have the ability to produce extracellular **phosphatase** enzymes which allow them to access otherwise unavailable phosphate stores.

Micronutrients include copper, manganese, sodium, zinc, and molybdenum, all of which are usually available to excess in the environment. Iron is relatively insoluble and therefore not easily assimilated, but fungi can produce **siderophores** or organic acids, which can **chelate** or alter iron solubility and improve its availability.

Some fungi may require preformed vitamins, for example, thiamin and biotin. Other requirements can be for sterols, riboflavin, nicotinic acid, and folic acid.

Water, pH, and temperature

Fungi require water for nutrient uptake and they are therefore restricted to fairly moist environments such as host tissue if they are parasites or symbionts, or soils and damp substrates if they are saprophytes. Desiccation causes death unless the fungus is specialized, as they are in the lichens (Section I4). Some fungi are wholly aquatic.

Fungi tend to occupy acidic environments, and by their metabolic activity (respiration and organic acid secretion) tend to further acidify them. They grow optimally at pH 4–6.

Most fungi are **mesophilic**, growing at between 5° and 40°C. Some are **psychrophilic** and are able to grow at under 5°C, others are **thermotolerant** or **thermophilic** and can grow at over 50°C.

Secondary metabolism

Nutrient depletion, competition or other types of metabolic stress that limit fungal growth promote the formation and secretion of secondary metabolites. These compounds can be produced by many different metabolic pathways and include compounds termed **antibiotics** (active against bacteria, protista, and other fungi), **plant hormones** (gibberellic acid and indoleacetic acid – IAA), and **cytotoxic** and **cytostimulatory** compounds.

13 Reproduction in fungi

Key Notes

Life cycles	All fungi undergo a period of vegetative growth during which their mycelium exploits a substrate. This stage is followed by asexual and/or sexual reproduction, which differs in each of the phyla.
Reproduction in the Chytridiomycota and related taxa	Asexual reproduction in the Chytridiomycota is usually by the formation of motile, uniflagellate zoospores within spore-containing structures called sporangia. Sexual reproduction is by the formation of diploid oospores following the fusion of two haploid cells. These may undergo meiosis, or there may be a period of vegetative growth in the diploid state.
Reproduction in the Zygomycota	Fungi in the Zygomycota reproduce asexually by the formation of nonmotile sporangiospores within sporangia elevated on aerial hyphae. Sexual reproduction is by the formation of diploid zygospores, within which meiosis occurs.
Reproduction in the Glomeromycota	Sexual reproduction appears not to occur in this group. All members of this group are symbiotic and the life cycle is completed within plant roots producing only asexual spores.
Reproduction in the Dikarya: Ascomycota	Fungi in the Ascomycota reproduce asexually by the formation of conidiospores from hyphal tips. Sexual reproduction is by the fusion of hyphae, rapidly followed by meiosis and the production of ascospores.
Reproduction in the Dikarya: Basidiomycota	Basidiomycete fungi rarely reproduce asexually. Sexual reproduction is by the formation of basidiospores on the gills or pores of large fruit bodies we know as mushrooms and toadstools.
Fungal spores	Spores allow fungi to spread, to maintain genetic diversity, and to survive adverse conditions. Spores produced asexually are generally adapted for dispersal while those produced sexually are often adapted for survival.
Spore discharge	Spores may be discharged from parent mycelium by passive or active means. Passive mechanisms include using wind and water as dispersants; active mechanisms use explosive principles.
Air spora	Spores in the atmosphere can affect human, animal, and plant health. They can cause allergies and spread plant disease.

Related topics	(H3) Cell division and ploidy	(I4) The fungi and related
	(I1) Fungal structure and growth	phyla: beneficial effects
	(I2) Fungal nutrition	(I5) The fungi and related phyla: detrimental effects

Life cycles

Each of the fungal groups is characterized by differences in their life cycles. All fungi are characterized by having a period of vegetative growth where their biomass increases. The length of time and the amount of biomass needed before sporulation can occur vary. Almost all fungi reproduce by the production of **spores**, but a few have lost all sporing structures and are referred to as *mycelia sterilia*. Different types of spore are produced in different parts of the life cycle.

Reproduction in the Chytridiomycota and related taxa

Fungi in the Chytridiomycota and related groups are quite distinct from other fungi as they have extremely simple thalli and motile zoospores. Some species within this group can be so simple that they consist of a single vegetative cell within (**endobiotic**) or upon (**epibiotic**) a host cell, the whole of which is converted into a **sporangium**, a structure containing spores. These types are termed **holocarpic** forms.

Other members of this group have a more complex morphology, and have **rhizoids** and a simple mycelium. Asexual reproduction in the Chytridiomycota is by the production of motile **zoospores** in sporangia that are delimited from the vegetative mycelium by complete septae. The zoospores have a single, posterior flagellum. Sexual reproduction occurs in some members of the Chytridiomycota by the production of **diploid** spores after either somatic fusion of haploid cells, or fusion of two different mating-type mycelia, or fusion of two motile **gametes**, or fusion of one motile gamete with a **nonmotile egg** (Figure 1). The resulting spore may undergo meiosis to produce a haploid mycelium or it may germinate to produce a diploid vegetative mycelium, which can undergo asexual reproduction by the production of diploid zoospores. The diploid mycelium can also produce resting sporangia in which meiosis occurs, generating haploid zoospores that germinate to produce haploid vegetative mycelium.

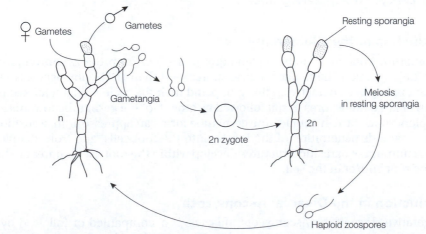

Figure 1. Chytrid life cycle.

Reproduction in the Zygomycota

In the Zygomycota, asexual reproduction begins with the production of aerial hyphae. The tip of an aerial hypha, now called a **sporangiophore**, is separated from the vegetative hyphae by a complete septum called a **columella** (Figure 2). The cytoplasmic contents of the tip differentiate into a sporangium containing many asexual spores. The spores contain haploid nuclei derived from repeated mitotic divisions of a nucleus from the vegetative mycelium (Section H3). Dispersal of the spores is by wind or water.

In sexual reproduction, two nuclei of different mating types fuse together within a specialized cell called a **zygospore** (Figure 2). In some species the different mating-type nuclei may be within one mycelium (**homothallism**). In other species, two mycelia with different mating-type nuclei must fuse (**heterothallism**). In both cases, fusion occurs between modified hyphal tips called **progametangia**, which once fused are termed the zygospore. Within the developing zygospore meiosis occurs; usually three of the nuclear products degenerate, leaving only one nuclear type present in the germinating mycelium (Section H3).

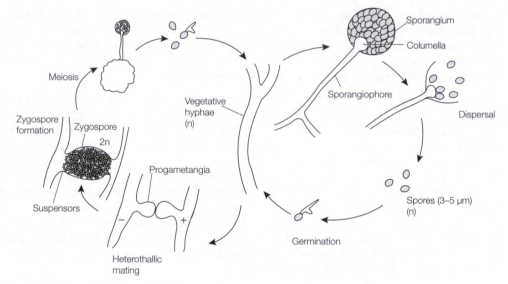

Figure 2. Life cycle of a typical zygomycete.

Reproduction in the Glomeromycota

The Glomeromycota are a monophyletic group, distinct from the Zygomycota on the basis of their symbiotic habit, lack of zygospores, and their rDNA phylogeny. No sexual reproduction has been detected in this group and their life cycle is characterized by germination of large mitospores in soil followed by growth of the germ tube towards a compatible plant root. Once in contact, the germ tube forms an appressorium, a cushion-like structure that aids penetration of the fungus into the root cells, and from the penetration site a number of parasitic structures develop within the root. New spores are formed inside the root and/or in the soil.

Reproduction in the Dikarya: Ascomycota

The vegetative stage of the Ascomycota life cycle is accompanied or followed by asexual sporulation by the production of single spores called conidia from the tips of aerial hyphae called conidiophores (Figure 3). The spores can be delimited by a complete transverse wall formation followed by spore differentiation (Figure 3a) termed **thallic** spore formation, or more usually by the extrusion of the wall from the hyphal tip, termed **blastic** spore formation (Figure 3b). These spores can be single-celled and contain one haploid nucleus, or they can be multicellular and contain several haploid nuclei produced by mitosis (Section H3).

Spores can be produced from single, unprotected conidiophores or they can be produced from **aggregations** that are large enough to be seen with the naked eye (Figure 3c). The conidiophores can aggregate into stalked structures where the spores produced are exposed at the top (**synnema** or **coremia**). Alternatively, varying amounts of sterile fungal tissue can protect the conidia, as in the flask-shaped **pycnidia**. Some species produce conidia in plant tissue, and the conidial aggregations erupt through the plant epidermis as a cup-shaped **acervulus** or a cushion-shaped **sporodochium**.

Sexual reproduction in this group occurs after **somatic** fusion of different mating-type mycelia. A transient diploid phase is rapidly followed by the formation of **ascospores** within sac-shaped **asci** differentiated from modified hyphal tips. In the initial stages of ascal development hooked hyphal tips form, called **croziers** or **shepherds' crooks** because of their shape. They have distinctive septae at their base, which insure that two different mating-type nuclei are maintained in the terminal cell. Formation of the septae is co-ordinated with nuclear division (Figure 4). In yeasts all these events occur within one cell, after fusion of two mating-type cells, the whole cell being converted into an ascus.

In more complex Ascomycota many asci form together, creating a fertile tissue called a **hymenium**. In some groups the hymenium can be supported or even enclosed by large amounts of vegetative mycelium. The whole structure is called a **fruit body** or **sporocarp** and is used as a major taxonomic feature (Figure 5). They can become large enough to be seen with the naked eye. Flask-shaped sexual reproductive bodies are called **perithecia**, cup-shaped bodies are called **apothecia**, and closed bodies are called **cleistothecia**.

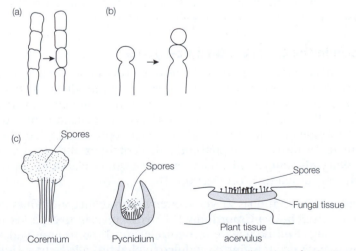

Figure 3. Asexual reproduction in the Ascomycetes. (a) Thallic spore formation; (b) blastic spore formation; (c) aggregations of conidiophores. From Ingold CT & Hudson HJ (1993) *The Biology of Fungi*. With permission from Kluwer Academic.

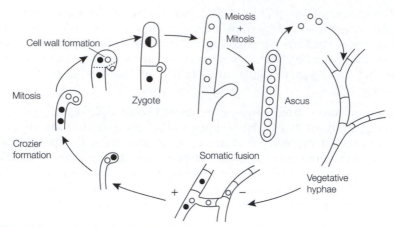

Figure 4. Sexual reproduction in the Ascomycetes.

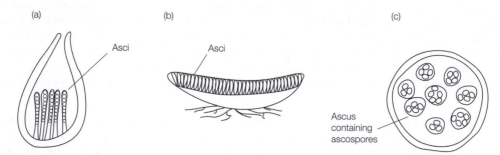

Figure 5. Structures of sexual sporocarps in the Ascomycetes. (a) Perithecium; (b) apothecium; (c) cleistothecium. From Ingold CT & Hudson HJ (1993) *The Biology of Fungi*. With permission from Kluwer Academic.

These structures have evolved to protect the asci and assist in spore dispersal, but the hymenium itself is unaffected by the presence of water.

Reproduction in the Dikarya: Basidiomycota

This group of fungi is characterized by the most complex and large structures found in the fungi. They are also distinctive in that they very rarely produce asexual spores. Much of the life cycle is spent as vegetative mycelium, exploiting complex substrates. A preliminary requisite for the onset of sexual reproduction is the acquisition of two mating types of nuclei by the fusion of compatible hyphae. Single representatives of the two mating-type nuclei are held within every hyphal compartment for extended periods of time. This is termed a **dikaryotic** state, and its maintenance requires elaborate septum formation during growth and nuclear division (Section I1).

Onset of sexual spore formation is triggered by environmental conditions and begins with the formation of a **fruit body primordium**. Dikaryotic mycelium expands and differentiates to form the large fruit bodies we recognize as mushrooms and toadstools. Diploid formation and meiosis occur within a modified hyphal tip called a **basidium** (Figure 6).

Four spores are budded from the basidium. Basidia form together to create a hymenium, which is highly sensitive to the presence of free water. The hymenium is distributed

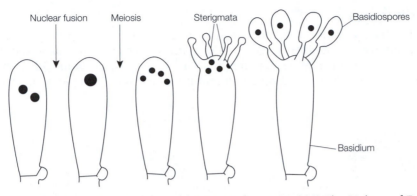

Figure 6. Basidium formation. From Ingold CT & Hudson HJ (1993) *The Biology of Fungi*. With permission from Kluwer Academic.

over sterile, dikaryotic-supporting tissues that protect it from rain. The hymenium can be exposed on **gills** or **pores** beneath the fruit body, seen in the **toadstools** and **bracket fungi**, or enclosed within chambers as in the **puffballs** and **truffles** (Figure 7).

Fungal spores

Fungi have two conflicting requirements for their spores. Spores must allow fungi to spread, but they must also allow them to survive adverse conditions. These requirements are met by different types of spores. Small, light spores are carried furthest from parent mycelium in air and these are the dispersal spores. They are usually the products of asexual sporulation, the sporangiospores and the conidiospores, and so spread genetically identical individuals as widely as possible. Genetic diversity is maintained by sexual reproduction, and the spore products are often large **resting spores** that withstand adverse conditions but remain close to their site of formation. Spores therefore vary greatly in size, shape, and ornamentation, and this variation reflects specialization of purpose.

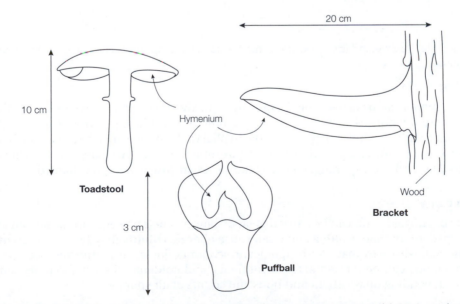

Figure 7. Structure of sexual sporocarps in the Basidiomycetes. From Ingold CT & Hudson HJ (1993) *The Biology of Fungi*. With permission from Kluwer Academic.

Spore discharge

Spores that have a dispersal function can be released from their parent mycelium by **active** or **passive** mechanisms (Figure 8). As many spores are wind-dispersed, they are produced in dry friable masses that are passively discharged by wind. Other spores are passively discharged by water droplets splashing spores away from parent mycelium.

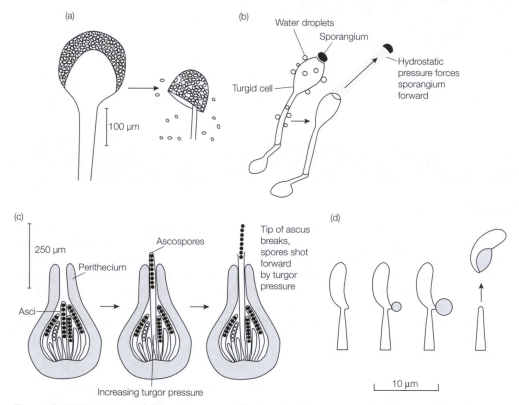

Figure 8. (a) Passive and (b) active spore discharge; (c) ascospore discharge; (d) basidiospore discharge.

Fungal spores can be actively discharged by **explosive** mechanisms. These mechanisms use a combination of an increasing turgor pressure within the spore-bearing hypha, combined with an inbuilt weak zone of the hyphal wall. This ensures that when the hypha bursts the spore discharge is directed for maximum distance. Asci are usually dispersed in this way, and a few sporangia too. Basidiospores are also actively discharged.

Air spora

Airborne fungal spores can be carried great distances. Their presence in the air can have an impact on human health as they can cause **allergic rhinitis** (hay fever) and **asthma**. Many plant diseases that cause significant economic losses are airborne (Section I5). Spore clouds can be tracked across continents, and epidemic disease forecasts can be made, depending on weather conditions and counts of air spora.

I4 Beneficial effects

Key Notes

Bread and brewing	The products of yeast fermentation (CO_2 and alcohol) are exploited in bread making and alcohol brewing. Both processes enhance the value of the substrate but contribute little to its nutritional value.
Symbioses	Fungi can enter into specialized and intimate, mutually beneficial associations with higher plants, other microbes and animals. The associations can be external to the host cell, as in the ectomycorrhizae and lichens, or inside the cell as in the endomycorrhizae and endophytic fungi.
Decomposers	Fungi are the main agents of decay of plant wastes in the environment, decomposing substrates to CO_2, H_2O, and fungal biomass, and releasing other nutrients back to the biosphere.
Biological control	Fungi can be used to control insect pests, weed plants, and plant diseases by exploiting their natural antagonistic, competitive, and pathological attributes.
Bioremediation	The degradative abilities of fungi can be exploited to decompose man-made pollutants such as hydrocarbons, pesticides, and explosives. They may decompose substrates into CO_2 and H_2O by respiratory pathways, or they may reduce toxicity by co-metabolic activity. The toxicity of some compounds can also be increased in this manner.
Industrially important natural products of fungi	Fungi naturally produce antibiotics, immunosuppressants, acids, enzymes, and several other classes of useful natural products. They also can be used to produce large quantities of protein, including the popular meat substitute Quorn.
Related topics	(E1) Heterotrophic pathways (I3) Reproduction in fungi (I1) Fungal structure and growth (I5) The fungi and related phyla: detrimental effects (I2) Fungal nutrition

---

Bread and brewing

The metabolic products of yeast metabolism are exploited by humans for **bread making** and **alcohol brewing**. Yeast metabolism of flour starch by respiration generates CO_2, which is trapped within the gluten-rich dough and forces bread to rise (Section E1). In the brewing of alcohol, yeasts are forced into fermentative metabolism in the sugar-rich, low-oxygen environments of beer wort or crushed grape juice. The fermentative metabolism is inefficient, and only partially metabolizes the available substrates, yielding CO_2 and **ethanol**. Both processes considerably enhance the value of the original substrate while contributing little to its nutritional status!

Symbioses

Fungi can enter into close associations with other microbes and with higher plants and animals. These beneficial associations are termed **symbioses**, and in most cases the symbiotic fungus gains carbohydrates from its associate, while the associating organism gains nutrients and possibly protection from predation and herbivory or plant pathogens.

Fungal symbioses are common on plant roots, and the symbiotic roots are termed **mycorrhizae** (Figure 1). Their presence enhances nutrient uptake by plant roots and plant performance. The fungal association can be predominantly external to the root tissue. These associations are called **ecto**mycorrhizae and can be seen on beech and pine trees, for example, and the fungal partners are members of the Dikarya. Other associations are predominantly within the plant root, and are termed **endo**mycorrhizae. These associations are seen on the roots of herbaceous species like grasses, but are also found in the roots of tropical species of tree and shrub. These partnerships are with fungi in the Glomeromycota.

Other associations between fungi and plants can occur within leaves or stems, and these are called **endophytic** fungi. They live almost all their life cycle within the host, grow very slowly, and do not cause any signs of infection. They appear to protect their host from herbivory and fungal infection by the production of metabolites. However, these products can have dramatic effects on herbivorous animals, causing symptoms of fungal toxicosis similar to St Anthony's Fire (Section I5).

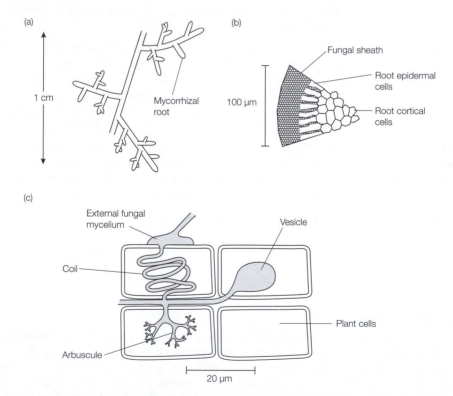

Figure 1. Structure of ectomycorrhiza and endomycorrhiza. (a) Macromorphology of ectomycorrhiza; (b) micromorphology of ectomycorrhizae; (c) micromorphology of endomycorrhizae. From Isaac S (1991) *Fungal–plant interactions*. With permission from Kluwer Academic.

Some fungi can form very intimate associations with algal species. These associations are termed **lichens**, and they have a form quite distinct from that of either component species, with **crustose**, **foliose** or **fruticose** thalli (Figure 2). They are slow-growing, and are adapted to occupy extreme or marginal environments, like bare rock faces, walls, and house roofs. As they are under quite extreme stress they are highly sensitive to pollution, for instance, from acid rain or heavy metals, and their presence or absence in an environment has become a useful indicator of urban and industrial pollution.

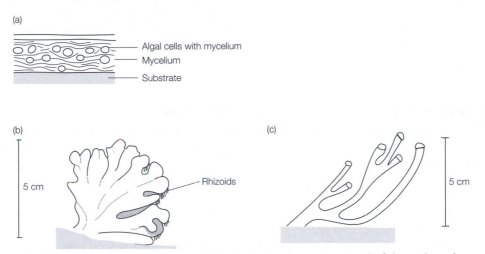

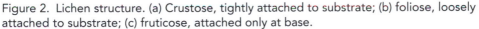

Figure 2. Lichen structure. (a) Crustose, tightly attached to substrate; (b) foliose, loosely attached to substrate; (c) fruticose, attached only at base.

Fungi can also associate with insect species in symbioses of varying intimacy. Some species of ants culture specific fungi on cut plant remains within their nests, and then browse on the fungal mycelium that develops. **Termites** have symbiotic fungi within their guts, which – in association with a consortium of other microbes, including protista and bacteria – help the termite digest its woody gut contents (Section J4).

Decomposers

The degradative processes that fungi perform with their extracellular enzymes are essential to the terrestrial biosphere (Section I2). They are the main agents of decay of cellulosic wastes produced by plants, which in the tropical rain forest can amount to 12 000 kg hectare^{-1} year^{-1}. They decompose this material into CO_2, H_2O, and fungal biomass, which in turn is decomposed by other microbes, returning mineral nutrients like phosphorus, nitrogen, and potassium to the biosphere. This process is termed mineralization.

Biological control

The natural attributes of fungi as disease-causing organisms can be exploited by humans to control weed plant populations and insect pests. They are even capable of parasitizing plant disease-causing fungi. This process is termed **biological control** and can be an alternative to the application of chemical pesticides. Applications of fungal propagules to targeted problem populations can cause epidemic disease or overwhelm and monopolize a niche.

Bioremediation

The degradative processes that fungal enzymes catalyze on their natural substrates can be used on other, man-made substrates to provide **biological clean-up**. Hydrocarbons like oils can be degraded by fungi and other microbes to CO_2 and H_2O by aerobic respiration. These activities are termed bioremediation, and contaminated areas of land can be actively bioremediated by the addition of fungal propagules.

Pesticides, explosives, and other recalcitrant molecules can be changed by **co-metabolic** activities of fungi, where enzymes normally used for one metabolic process within the fungus coincidentally catalyze another reaction. The products of co-metabolic reactions are not utilized any further by the fungus, but can sometimes be used by other microbes in the ecosystem. Reactions like these can lead to reduced toxicity of some contaminants, but in other cases can lead to **activation** of the pollutant, leading to an increase in the toxicity of the compound.

Industrially important natural products of fungi

Fungi are able to produce many different types of metabolite that are of commercial importance. These include antibiotics (e.g. penicillin and cephalosporin) and **immuno-suppressants** (cyclosporins) important in medicine, **enzymes** that are used in the food industry (e.g. α-amylase, rennin), other enzymes (e.g. cellulases, catalase), **acids** (e.g. lactic, citric), and several other products.

Protein extracted from fungi is also an important commercial product. In the 1960s protein from yeasts (single-cell protein) was developed. Currently, fungal protein extracted from *Fusarium graminearum*, known as Quorn, is a useful meat alternative.

I5 Detrimental effects

Key Notes

Biodeterioration	The degradative activities of fungi in unwanted situations cause significant economic losses. Materials containing large quantities of cellulose, leather, and hydrocarbons can be used as substrates by fungi, provided that there is an adequate water supply.
Plant disease	Fungi are capable of causing significant losses to crops both before and after harvest. However, this can be countered by the use of fungicides, storage conditions that do not favor fungal growth, and the development of resistant plant varieties.
Animal and human disease	Fungi can cause both superficial and deep, life-threatening infections of both man and animals. The latter are particularly dangerous in immunocompromised individuals.
Fungal toxicosis	Ingestion of fungi or their secondary metabolic products, accidentally or deliberately, can cause intoxication and occasionally death in both humans and animals.
Related topics	(I1) Fungal structure and growth (I2) Fungal nutrition (I3) Reproduction in fungi

---

Biodeterioration

The same extracellular enzymes that are important in the degradation of leaf litter and the recycling of nutrients in the biosphere can cause massive economic losses when they occur in circumstances where they are not wanted. Fungi can attack and utilize as substrates paper, cloth, leather, and hydrocarbons, but also can cause degradative change in other materials, for instance, glass and metal, because of their ability to produce acid as they grow. The supply of water is a key control point in these processes, and keeping substrates dry is an effective way of avoiding these changes.

Plant disease

Fungi are capable of attacking all plant species, causing serious damage and in some circumstances even death. In crop production over half of potential crop yield is lost to plant pathogens, most of it to **fungal disease**. In storage, up to one-third of the harvested product can be lost to post-harvest disease, again mostly as a result of the activities of fungi. Use of **fungicides** can reduce both pre- and post-harvest disease, and **plant breeding** programs can introduce disease-resistant strains of crop plants. Post-harvest losses can be reduced by storage of products at low temperatures and low moisture levels.

Animal and human disease

Human and animal epidermis can be attacked by fungi, causing superficial damage and discomfort like **ringworm** infections, **athlete's foot**, and **thrush**. Other deeper, systemic fungal infections of the lung and central nervous and lymphatic systems cause much more serious diseases, for example, **aspergillosis**, **coccidiomycosis**, **blastomycosis**, **histoplasmosis**, and **pneumocystis** pneumonia are all caused by fungi. Although most humans experience superficial fungal infections and survive, these deeper diseases are especially dangerous for the immunocompromised patient after transplantation, and the HIV-positive population.

Fungal toxicosis

Accidental or deliberate consumption of wild fungi or fungally contaminated food can lead to poisoning or **toxicosis** of the consumer because some fungi naturally contain toxic metabolites called **mycotoxins**. Deliberate toxicosis can arise from consumption of mushrooms and toadstools that are known to contain naturally hallucinatory drugs like **psilocybins**, which lead to euphoric states followed by extreme gastrointestinal distress. Accidental consumption of mis-identified fungal fruit bodies can lead to fatal mushroom poisoning from fungal toxins, causing total liver failure between 8 and 10 hours later. Consumption of food accidentally contaminated by fungal metabolites also leads to human and animal death. For instance, rye flour contaminated by the ergots of the fungus *Claviceps purpurea* leads to the symptoms of **St Anthony's fire**, where peripheral nerve damage is caused by the presence of **ergometrine** in the fungal tissue. This can be followed by gangrene of the limbs and death. Detection of fungal mycotoxins such as **ochratoxin** in apple juice and **aflatoxin** in peanuts has also caused problems for food producers and consumers and has forced improvements in product processing.

J1 Taxonomy and structure

Key Notes

Taxonomy of the Archaeplastida, Excavata, Chromalveolata, and Amoebozoa

Until fairly recently, the photosynthetic and nonphotosynthetic protistan genera were divided into two polyphyletic form groups, the algae and protozoa, based on the presence or absence of chloroplasts. Currently, a monophyletic taxonomy of the protista is being developed, based on molecular data. Three supergroups of eukaryotic microorganisms are currently proposed, the Archaeplastida, Excavata, and Chromalveolata. The Amoebozoa are currently under revision.

Structure of the Archaeplastida: chlorophytes

Members of the Archaeplastida considered as microorganisms are unicellular, photosynthetic organisms belonging to the chlorophytes (green algae). Some have a filamentous or membranous morphology. They have a cellulose cell wall. Many species are flagellate, and they contain chloroplasts, which vary in structure and pigment content.

Structure of the Excavata, Chromalveolata, and Amoebozoa

Members of the Excavata and Chromalveolata are heterotrophic or photosynthetic, unicellular eukaryotes. They vary greatly in shape and size, and contain most eukaryotic cell organelles. They also have some unique organelles. There are three groups within the Excavata, the euglenids, kinetoplastids, and fornicata, and within the Chromalveolata there are the alveolates (ciliates, dinoflagellates, and apicomplexans) and the stramenopiles (diatoms, chrysophytes, oomycetes, and opalines). Species in the Amoebozoa are naked, heterotrophic cells with an absorptive nutrition.

Growth in the Archaeplastida, Excavata, Chromalveolata, and Amoebozoa

Growth in the unicellular species is synonymous with longitudinal binary fission, but budding and multiple fissions occur in some groups. Coenocytic, tubular or filamentous species grow by tip growth like the fungi. Other filamentous or membranous species grow by intussusception of new cells into the filament. The kinetics of growth of unicellular species are similar to those of bacteria, but in addition to estimations of growth by mass measurement, cell counts and chlorophyll content can be assessed. Rapid cell division can lead to very high cell populations, only limited by nitrogen, phosphate or silicon availability.

Related topics	(A1) The microbial world	(J2) Archaeplastida,
	(C7) Composition of a typical prokaryotic cell	Excavata, Chromalveolata, and Amoebozoa: nutrition and metabolism
	(C9) Cell division	
	(H2) Eukaryotic cell structure	(J4) Archaeplastida,
	(H3) Cell division and ploidy	Excavata, Chromalveolata,
	(I1) Fungal structure and growth	and Amoebozoa: beneficial effects

Taxonomy of the Archaeplastida, Excavata, Chromalveolata, and Amoebozoa

In the past the eukaryotic photosynthetic and nonphotosynthetic microorganisms were divided into **form groups**, the algae and the protozoa, based on the presence or absence of chloroplasts.

The algae were further subdivided into groups based on pigmentation, the number and type of flagella, and other structural characteristics. Protozoa were divided on similar structural characteristics into four **polyphyletic** form groups, the ciliates, flagellates, sporozoans, and amebas.

Advances in molecular biology now allow us to begin to create a **monophyletic taxonomy** of the eukaryotic microorganisms and such a scheme includes many former members of the algae, fungi, and protista. **Monophyletic groups** are also called **clades** and are considered to be the only 'natural' kind of group. A suggested taxonomic scheme is shown in Figure 1 and characteristics of the major monophyletic groups are shown in Table 1.

Only those organisms included in the microbial world will be discussed here. Three supergroups, the **Archaeplastida**, **Excavata**, and **Chromalveolata**, have been proposed, which include most of the previous chlorophytan and protistan organisms. The **Amoebozoa** contain the amebae, and current work indicates that this branch of the eukaryotic tree is a sister group of the fungi within the opisthokonts. The organisms once considered as basal branches to the eukaryotic tree have now been shown not to be primitive, amitochondrial species (i.e. lacking mitochondria), but to have affinities with flagellate species in the Excavates.

Structure of the Archaeplastida: chlorophytes

The Archaeplastida contain the blue green algae, green algae, red algae, and higher plants. Only the green algae, the chlorophytes, are considered to be eukaryotic microorganisms. The chlorophytes are a monophyletic group and they range in complexity from unicellular motile or nonmotile organisms to **sheets**, **filaments**, and **coenocytes** (Figure 2). They are found in fresh and salt water, in soil, and on and in plants and animals. Most chlorophytan cell walls are formed from cellulose and they may be fibrillar, similar to those of the fungi, and sometimes impregnated with silica or calcium carbonate.

Chlorophytan cells contain nuclei, mitochondria, ribosomes, Golgi, and chloroplasts (Section H2). The internal cell structure is supported by a network of microtubules and endoplasmic reticulum. Chloroplasts in this group are very variable structures; they can be large and single, multiple, ribbon-like or stellate chloroplasts with chlorophylls

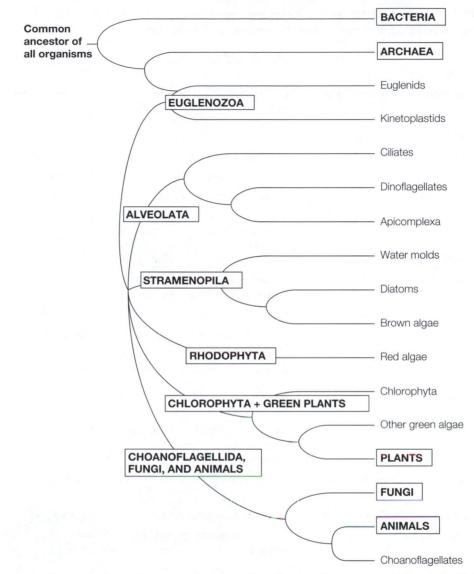

Figure 1. Current taxonomic scheme for the Archaeplastida, Excavata, and Chromalveolata

a and *b* and carotenoids and they store **starch**. Chlorophytan cells have a vegetative phase that is haploid, and sexual reproduction occurs when cells are stimulated to produce gametes instead of normal vegetative cells at binary fission.

They often possess flagella that have a 9 + 2 microtubule arrangement within them (Section H2). There may be one or two flagella per cell, which may be inserted **apically**, **laterally** or **posteriorly** and trail or girdle the cell. The flagellum can be a single whiplash or it can have hairs and scales. The presence of **eyespots** near the flagellar insertion point allows the cell to swim towards the light. Movement may be by lateral strokes or by a spiral movement that can push or pull the cell through the water.

Table 1. Characteristics of the major monophyletic groups of the Archaeplastida, Excavata, Chromalveolata, and Amoebozoa

Group	Common name	Characteristics	Examples
Archaeplastida			
Chlorophyta	Green algae	Chlorophyll *a* and *b*, cellulose cell walls	*Chlamydomonas*
Excavata			
Euglenozoa		**Flagellate unicells**	
Euglenids		Mostly photosynthetic	*Euglena*
Kinetoplastids		Single large mitochondrion	*Trypanosoma*
Fornicata		Contain mitosome, axially symmetric cells	*Giardia*
Chromalveolata			
Alveolata		**Unicellular; sacs (alveolata) below cell surface**	
Pyrrophyta	Dinoflagellates	Golden brown algae	*Peridinium*
Apicomplexa		Apical complex aids host cell penetration	*Plasmodium*
Ciliophora	Ciliates	Cilia; macro- and micronucleus	*Paramecium*
Stramenopila		**Motile stages with two unequal flagella, one with hairs**	
Bacillariophyta Diatoms		Unicellular, photosynthetic, two silicon- containing frustules	*Navicula*
Chrysophytes	Golden brown algae	Photosynthetic marine species	*Ceratium*
Oomycetes	Water molds	Heterotrophic coenocytes	*Saprolegnia*
Opalines		Endosymbiotic in amphibian bowels	
Amoebozoa			
Dictyostelids and Myxogastrids	Slime molds	Free-living soil amebae	*Physarum*
Naked and testate amebae		Free-living and pathogenic amebae	*Acanthamoeba*

Structure of the Excavata, Chromalveolata, and Amoebozoa

These members of the microbial eukaryotes are unicellular, heterotrophic, parasitic, symbiotic or photosynthetic organisms. They may be motile or nonmotile. They contain many of the organelles found in other eukaryotes, including nuclei, mitochondria (absent in some groups), chloroplasts (absent in some groups), ribosomes, endoplasmic reticulum (ER), Golgi vesicles, microtubules, and microfibrils.

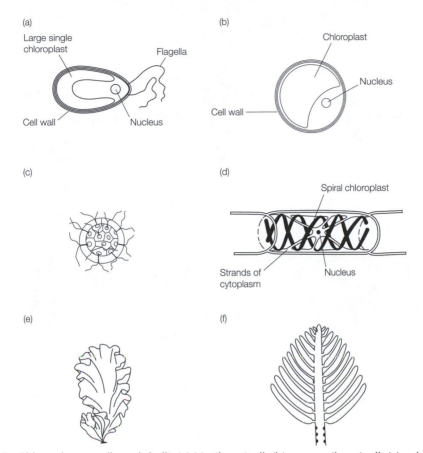

Figure 2. Chlorophytan cells and thalli. (a) Motile unicell; (b) nonmotile unicell; (c) colonial forms; (d) filamentous forms; (e) membranous (only two cells thick); and (f) tubular. Courtesy of CJ Clegg.

Chloroplasts can be very variable in this group, and their shape and pigment content are useful distinguishing taxonomic features. The number of ER membranes that surround the chloroplast, and the number of thylakoid stacks within them, are also important indicators of its symbiotic origin. Unique organelles found in the protista include **contractile vacuoles** that control the influx of water to the cells due to osmosis. Some of the protista have **hydrogenosomes** where fermentative reactions take place (Sections E2 and H2). Some excavates and chromalveolates have a cell covering of scales or plates formed of pseudochitin, and this may be impregnated with calcium or silica scales. These scales are secreted from the Golgi bodies within alveoli. Some species have the ability to form thick-walled cysts during a resting phase in the life cycle. These cysts can be very resistant to adverse conditions.

This group of organisms is characterized by a considerable variation in morphology (Figure 3). Free-floating species are radially or laterally symmetrical and streamlined into bullet or kidney shapes. At its simplest, the outer surface of the cells may be simply the plasma membrane. This is seen in the gametes of some species, in amebae, and in some of the intracellular parasites. Such a 'naked' cell has advantages in nutrient uptake that

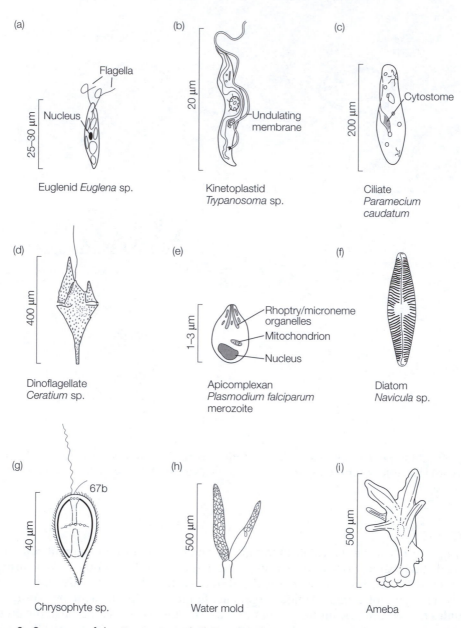

Figure 3. Structure of the Excavata and Chromalveolata.

outweigh its vulnerability. The naked membrane can be extended with lobes of cytoplasm called pseudopodia that are important in locomotion and feeding. This type of movement requires that the cell is in contact with a solid surface.

The cell is partitioned into two regions, the viscous ectoplasm and the more liquid endoplasm. Within the ectoplasm are myosin and actin filaments, while in the endoplasm only unpolymerized actin is present. To produce ameboid movement the microfilaments in the ectoplasm produce pressure on the endoplasm (Figure 4) moving the endoplasm to one part of the cell, and away from another.

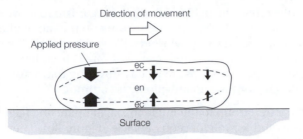

Figure 4. A proposed mechanism for ameboid movement. ec, ectoplasm; en, endoplasm.

Motility can also be via flagella or cilia that propel the cells through water by rhythmic beating. Flagella typically exhibit a sinusoidal motion in propelling water parallel to their axis (Figure 5a). The undulating action of the flagellum either propels water away from the surface of the cell body or draws water towards and over the cell body. Cilia exhibit an oar-like motion, propelling water parallel to the cell surface (Figure 5b).

Excavata

The Excavata supergroup contains the **Euglenozoa** within which are the **flagellate** forms the **euglenids**, the **kinetoplastids**, and a newly erected phylum the **Fornicata**, which contains some recently reclassified members of the **diplomonads** and **trichomonads**, previously considered to be basal members of the eukaryotic tree. Mitochondria are discoid throughout the group. There are two other unique features in the excavates, the **kinetoplast**, a very structured inclusion of DNA in the mitochondrion, and the **euglenid pellicle**, a series of interlocking protein strips beneath the cell membrane.

Euglenids (from the phylum Euglenozoa) are unicellular, motile organisms, with two flagella that originate from a pocket at the anterior end of the cell (Figure 3a). They may have a heterotrophic or photosynthetic nutrition. When photosynthetic, they contain chloroplasts with chlorophylls a and b and they store **paramylon** starch. These features suggest that the chloroplasts in this group originated from a chlorophytan ancestor. They inhabit nutrient-rich (**eutrophicated**) fresh and brackish water. Many species are capable of phagotrophy, and some are wholly saprophytic and devoid of chloroplasts. A few species are parasitic. They have a proteinaceous flexible pellicle that lies underneath the cell membrane very like that of the closely related alveolate species. The flexibility of this pellicle allows euglenids a characteristic flexibility termed **euglenoid movement** when

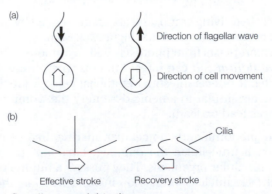

Figure 5. Motility in (a) flagellates and (b) ciliates.

cells are on solid substrates. The pellicle allows the organism to move through muds and sands. The structure of the pellicle varies depending on the mode of nutrition; bacteria-eating phagotrophic species have small rigid cells with around 12 strips in the pellicle, while the eukaryote-eating species are bigger and have 20–60 helical strips. In some of the parasitic species the pellicle strips are fused into a single structure. The nuclei appear to contain many chromosomes, and polyploidy is common.

Kinetoplastids, the sister group to the Euglenozoa, are parasitic flagellates with a single large mitochondrion and a proteinaceous pellicle (Figure 3b). They share many of the characteristics with the euglenids. The group is named after a large structure within the mitochondria called the **kinetoplast**, which contains DNA and associated proteins. The DNA is in the form of **maxicircles** and **minicircles**. As well as other proteins, maxicircles encode oxidative metabolism enzymes and minicircles encode unusual RNA-editing enzymes. Within this group are a number of human pathogens including trypanosomes *T. brucei gambiense*, *T. brucei rhodesiense* (spread by tsetse flies), and *T. cruzi* (transmitted by the bite of a triatomid bug), and the leishmaniasis pathogens, *Leishmania tropica*, *L. major*, and *L. donovani*, which are spread by sand flies.

Fornicata is a very recently devised clade that contains what were previously thought to be basal amitochondrial eukaryotes, the diplomonads and related species, but the recognition of a mitochondrion-like structure, the mitosome, and recent sequencing data place them within the Excavates. They have axially symmetric cell structures and the most notable member of the group is *Giardia intestinalis*, a human intestinal parasite. Parasite cysts are ingested in contaminated food and water, and germinate to form trophozoites, which stick to the intestinal epithelium and cause inflammation, cramps, and nonbloody diarrhea.

Chromalveolata

The second supergroup is the Chromalveolata, including the alveolates (ciliates, dinoflagellates, apicomplexa) and the stramenopiles (diatoms, chrysophytes, and oomycetes). They have a unique cell surface where the cell plasma membrane is underlain by a layer of vesicles called **alveoli**, which can be empty or contain cellulose plates or scales generated from the Golgi. These vesicles fuse with the plasma membrane and provide the cell with an effective cell wall. The plates can be impregnated with silica or calcium carbonate. Other unique characters include the presence of **micropores** through the cell surface and the possession of **extrusive organelles**. Mitochondria have tubular christae. Nutrition in this group is very varied, and includes predation, parasitism, and photoautotrophy.

Alveolata includes three groups; the ciliates, dinoflagellates, and apicomplexans:

- Ciliates are a largely free-living group, and are characterized by their dense covering of cilia and complex nuclear arrangements (Figure 3c). Many species are photosynthetic, having acquired chloroplasts from photosynthetic symbionts. Others are pathogenic, for example, **Balantidium coli** can infect humans when present in fecally contaminated food or water. It infects intestinal epithelial cells producing ulcers. Advanced cases show symptoms similar to amebic dysentery (i.e. vomiting, diarrhea, nausea) and the infection can lead to death.

- Dinoflagellates are predominantly unicellular, marine, free-living, motile organisms (Figure 3d). Although unicellular, the dinoflagellates are structurally an extremely diverse group of unicellular organisms. They possess a unique nuclear structure, the dinokaryon, which lacks nucleosomes, is low in basic proteins, and is high in DNA content. The chromosomes are very condensed and there is a prominent nucleolus. Most dinoflagellates have two flagella inserted into the cell at right angles to each other, around the midline of the cell. One is wrapped around the waist of the cell in a groove, the other extends from the posterior of the cell. There are also a number of extrusomes (organelles that can be extruded from the cell), including trichocysts, mucocysts, and nematocysts. Dinoflagellate chloroplasts contain chlorophylls a, $c1$, and $c2$ and the carotenoid **fucoxanthin**, and store **chrysolaminarin**, a $\beta1$–3-linked glucose. Some are capable of phagotrophy; others live within marine invertebrates and are termed **zooxanthellae** (Section J4).

- Apicomplexa are a wholly parasitic group and have a body form much like an ameba, but they have an apical complex that aids in the attachment of the parasite to the host cell membrane and assists in cell invasion. The complex is formed of a tubulin-containing conoid and rhoptries, which can be extruded from the apicomplexan cell (Figure 3e). The apicomplexa are all intracellular parasites, and the most significant in terms of human deaths is *Plasmodium falciparum*, the causative agent of malaria. There are other species of apicomplexans that are human pathogens but these seem to be distantly related and they likely diverged 5–8 million years ago at the point where humans and chimpanzees diverged from each other.

Stramenopiles include the photosynthetic **diatoms** and **chrysophytes**, and the heterotrophic or parasitic **oomycetes**, and **opalines**. Photosynthetic species have arisen after the symbiotic association between a nonphotosynthetic protistan and red or brown photosynthetic species some time ago in evolutionary history. The pigment content, the numbers of ER layers around the chloroplast, and the numbers of thylakoid lamellae confirm this hypothesis. Those species that contain red pigments from the cryptophytes can occupy the very deepest layers of the photic zone (Section J2). Most species in this group are unicellular, but some are colonial or filamentous. Life cycles are similar to those of the chlorophytes but the dominant vegetative stage is diploid:

- Diatoms contain chorophylls a and c and various accessory pigments, which give them a golden brown color. The chloroplasts have four membranes around them, demonstrating their origin as secondary symbionts from red algae. Diatoms differ from other members of the golden brown stramenopiles because their vegetative stage often lacks flagella, but they have a **gliding motility** on solid surfaces. They have a geometric shape supported by a silica-containing cell wall composed of a pair of 'nested' shells called **frustules** with a **girdle band** around them. The large half of the shell is termed the **epitheca**, the smaller is the **hypotheca** (Figure 3f). Nuclei are always diploid.

- Chrysophytes contain chlorophylls a, $c1$, and $c2$ and have two flagella inserted into the cell at near right angles to each other. Some species are covered in radially or bilaterally symmetrical scales (Figure 3g). They are also characterized by the formation of spores.

- Oomycetes and related groups. Members of the Oomycota are common aquatic species with a fungal-like thallus, predominantly saprophytes or parasites of animals and plants. They share many morphological characters of the chytrids (Section I1), but differences include cellulose-containing cell walls, tubular mitochondrial cristae, and a life cycle where the dominant somatic phase is diploid rather than haploid or dikaryotic (Figure 3h). They also possess biflagellate zoospores instead of the single flagellate zoospores found in the chytrids (Section J3). These organisms are capable of causing devastating disease in both plants and fish.

Members of the Hyphochytrids also have cellulose-containing cell walls, and these organisms have only recently been separated from the chytrids. A haploid vegetative stage is dominant in the life cycle. Members of this phylum of organisms are destructive

intracellular parasites of plants with an absorptive nutrition. They cause enlargement and multiplication of host cells, creating large and unsightly clubbed roots and often death of the plant. The best known disease is **club root** of crucifers. Like the Oomycota they share morphological and nutritional similarities with the chytrid fungi. They have some cellulose in their cell walls, and DNA sequence data also indicate a closer relationship with the dinoflagellate algae than that with the fungi.

Often grouped together, both the cellular and acellular slime molds exist for most of their life cycle as soil-dwelling amebae, feeding on soil bacteria. In both groups they are haploid until the onset of reproduction.

● Opalines are a group of specialized endosymbiotic species living in the bowels of amphibians. In this microaerophyllic environment they obtain nutrients by pinocytosis. They are multinucleate, and have a pellicle-like cell covering with closely spaced flagella that beat together.

Amoebozoa, amebae, and slime molds

Amebae are naked protistan cells that have an absorptive nutrition (Figure 3i) and within the Amoebozoa are the slime molds, the lobose amebae, testate amebae, and some pathogenic species. They have a wide distribution, living free in soil and water, and as parasites of animals and man.

Often grouped together, **Dictyostelids** and the **Myxogastrids** are phyla that contain the slime molds. The **Dictyostelids**, cellular slime molds, exist for most of their life cycle as haploid amebae that feed within soil by engulfing bacteria. They are uninucleate for most of their life cycle but form a plasmodium at sporulation. **Myxogastrids** are acellular slime molds. They exist as haploid amebae for their vegetative stage, but fuse in pairs to form a diploid cell that undergoes repeated mitotic nuclear divisions without cell division, forming a plasmodium. The life cycle is then very similar to the cellular slime molds.

Pathogenic amebae include *Entamoeba*, *Acanthamoeba*, and *Naegleria*. These pathogens have a number of different organ specificities. *Entamoeba* causes lesions in the gut causing mucosal destruction with abdominal pain, bloody diarrhea, and vomiting. *Acanthamoeba* infects the cornea, causing blinding keratitis, usually associated with contact lens use, but in immunocompromised patients it can also cause granulomatous encephalitis involving the central nervous system. *Naegleri* amebae enter the nasal passage (during swimming/diving into the water) and migrate along the olfactory nerves, through the cribriform plate into the brain. Infection is almost always fatal.

Growth in the Archaeplastida, Excavata, Chromalveolata, and Amoebozoa

Growth in the unicellular chlorophytes is synonymous with binary fission, budding or multiple fission. In most unicells, haploid or diploid nuclei undergo mitosis, and the cell then divides longitudinally to form two daughter cells. In some species there are two haploid divisions within the parent cell, followed by the formation of four motile daughter cells (Section H3). Multiple fission occurs where the nucleus divides several times without cell division. Cell division occurs after the formation of multiple nuclei. Some species produce buds, where the mitotic nucleus migrates to a bud forming on the parent cell surface. Some coenocytic filamentous species grow from the tip of the filament in a way very similar to that of hyphal growth (Section I1). Others grow by division of vegetative cells within filaments or sheets.

Accurate estimates of their growth rates can be made by cell counting or by estimating chlorophyll content of a culture. The kinetics of growth are similar to those seen in the bacteria (Section D1). However, for photosynthetic species, it is the depletion of nutrients other than carbon that leads to growth limitations and the onset of the stationary phase and cell death. Nitrogen, phosphates or silicon are frequently limiting.

J2 Nutrition and metabolism

Key Notes

Carbon and energy metabolism

The chlorophytes and photosynthetic species within the excavates and chromalveolates are photosynthetic organisms and obtain their carbon and energy requirements by the fixation of CO_2, using photosynthesis. In a terrestrial habitat, light levels are usually adequate to support photosynthesis, but in the aquatic habitat light energy is rapidly absorbed in the top 0.5 meter of the water column. Aquatic species have evolved three chlorophylls, *a*, *b*, and *c*, and a large number of accessory pigments to allow them to extend the depth to which they can grow. Photosynthetic species use aerobic respiration via glycolysis and the citric acid cycle. Nonphotosynthetic protista can obtain their carbon and energy requirements from the environment by diffusion, pinocytosis, and phagocytosis. In aerobic species, glycolysis, the citric acid cycle, and mitochondrial respiration provide the cell with energy and metabolites. Anaerobic protista may utilize fermentative pathways within the hydrogenosomes.

Oxygen and carbon dioxide

In the terrestrial environment, photosynthetic CO_2 and O_2 requirements are almost always satisfied by atmospheric gases. In the aquatic environment, the solubility of O_2 decreases with temperature and increasing dissolved CO_2 levels, and availability of O_2 therefore becomes limiting in warm waters.

Nitrogen nutrition

No eukaryotic microorganisms can fix nitrogen and therefore they all must obtain it in a fixed, inorganic or organic form. Most can utilize nitrate or ammonia; some require organic compounds. Nitrogen levels may be limiting to growth in marine environments.

Macronutrients, micronutrients, and growth factors

Most nutrients are available to excess in the aquatic environment, but phosphates and silicon are only poorly soluble in water and are often limiting to growth in fresh water. Some species are predominantly autotrophic, but many require an external supply of amino acids, vitamins, nucleic acids, and other growth factors; such species are described as auxotrophic.

Water, pH, and temperature optima

The members of the chlorophytes, excavates, and chromalveolates do not survive severe desiccation and are therefore found mostly in damp terrestrial habitats or in water. Phagocytic species have an absolute requirement for liquid water. Most species can tolerate a wide range of pH and temperature. Some are specialized and can inhabit

	extremely acidic, hot springs while others can complete their entire life cycle below 0°C.
Osmolarity	Freshwater species have a large difference between internal and osmotic pressure, and they must either have a rigid cell wall or a contractile vacuole to compensate for water uptake. Conversely, marine species are roughly isotonic with sea water.
Related topics	(D1) Measurement of microbial growth (E1) Heterotrophic pathways (E3) Autotrophic reactions (E4) Other unique microbial biochemical pathways (H2) Eukaryotic cell structure (I4) The fungi and related phyla: beneficial effects — (J1) Archaeplastida, Excavata, Chromalveolata, and Amoebozoa: taxonomy and structure (J4) Archaeplastida, Excavata, Chromalveolata, and Amoebozoa: beneficial effects

Carbon and energy metabolism

All species of the Archaeplastida, and many species of the excavates and chromalveolates are photosynthetic and therefore gain carbon and energy from the fixation of atmospheric or dissolved carbon dioxide using photosynthesis.

The photosynthetic reactions take place in the chloroplasts (Section H2), the light reactions occurring in the chloroplast thylakoids, and the light-independent reactions occurring in the stroma (Section E3). The chlorophyll pigments are membrane-bound within the thylakoids, and their properties include the ability to be excited by light. Different chlorophylls accept light energy of different wavelengths, depending on their structure. Accessory pigments, the **carotenoids**, **phycobilins**, and **xanthophylls**, also absorb light energy of different wavelengths and pass their excitation to chlorophyll, maximizing the breadth of wavelength over which light energy can be absorbed. The role of accessory pigments is particularly important as they allow photosynthetic organisms to occupy different parts of the **photic zone**, the shallow layer of water where sufficient light penetrates to support photosynthesis. The actual depth to which light will penetrate varies with turbidity and dissolved organic matter content (Section J4). Furthermore, light of different wavelengths penetrates water to different degrees (Figure 1a). Red light is absorbed rapidly by water, blue light least. The members of the green algae occupy only the shallowest of waters.

Chlorophylls *a*, *b*, and *c* are the photosynthetic pigments, with chlorophyll *a* being the primary photosynthetic pigment. The absorption **maxima** of chlorophyll *a* are at 430 and 663 nm (Figure 1b). Other chlorophylls are accessory pigments and have slightly different absorption maxima: chlorophyll *b* at 435 and 645 nm, chlorophyll *c*1 at 440, 583, and 634 nm, and *c*2 at 452, 586, and 635 nm. Phycobilins absorb light energy at 565 nm. There are also other accessory pigments found in the different algal groups, the carotenes and xanthophylls. Depending on the different accessory pigments present in individual species, light of different wavelengths can be absorbed and utilized in photosynthesis.

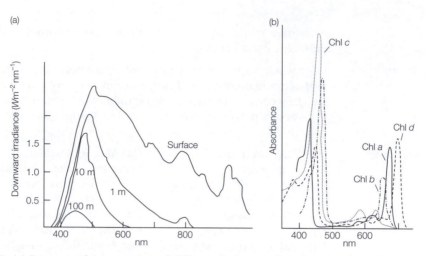

Figure 1. (a) Downward penetration of different light wavelengths. (b) Absorption spectra of chlorophylls.

Nutrient uptake in nonphotosynthetic excavates and chromalveolates may be by **diffusion**, **pinocytosis**, or **phagocytosis**. Within nutrient-rich tissues, such as those that intracellular parasites or blood-dwelling parasites live in, nutrient uptake is by diffusion and pinocytosis, the formation and digestion of small **vesicles** from the plasma membrane. These vesicles capture soluble nutrients from the environment.

In most larger, free-living species, particulate organic matter and bacteria are engulfed by phagocytosis. In this process large vacuoles can form anywhere over the cell surface in the amebae, or from specialized sites such as the cytostome in ciliates and flagellates. Digestion of the vesicle content occurs by the fusion of lysozyme-containing enzyme vesicles with the phagocytic vesicles. Once digestion is complete enzymes are recycled by the cell and sequestered into small vesicles, while cell debris is expelled from the cell by **reverse phagocytosis**, where the food vacuole re-fuses with the plasma membrane.

A large number of symbiotic species can be found in the excavates and chromalveolates in association with many different photosynthetic symbionts. Species of Alveolates and Euglenoids with these symbionts have an autotrophic nutrition, relying on photosynthetic products for carbon and energy sources (Section E3). In some cases the cytostome seen in nonphotosynthetic protistans is not present. Further evidence of the close taxonomic relationships between the photosynthetic and nonphotosynthetic groups is provided by the fact that it is possible to 'cure' some species of their symbionts, and once cured, these protista return to their holozoic nutrition and form food vesicles from a re-formed cytostome.

Almost all members of the green algae are aerobic and use mitochondrial respiration with oxygen as the terminal electron acceptor. In most of the archaeplastids, excavates, and chromalveolates, glycolysis and the citric acid cycle provide energy and intermediates for cellular metabolism (Section E1). Aerobic species use mitochondrial respiration, with oxygen as the terminal electron acceptor, generating ATP.

Anaerobic species have a fermentative metabolism (Section E1), which results in the incomplete oxidation of substrates. A few species that live in anaerobic lake sediments can use alternative electron acceptors such as nitrate.

There are several modifications to the usual glycolytic pathway (Section E1) found in some members of the excavates and chromalveolates. Some can utilize **inorganic pyrophosphate** rather than ATP, replacing enzymes like pyruvate kinase with pyrophosphate kinase. The advantage of this is that pyrophosphate kinase activity is reversible and can be used to synthesize glucose from other substrates.

In parasitic kinetoplastids glycolysis does not occur in the cytoplasm but in its own organelle, the **glycosome** (Section H2). Within this organelle a modified form of glycolysis can give rise to glycerol, which is the substrate of respiration in this group (Figure 2). In anaerobic species fermentative reactions occur within the **hydrogenosome**. Substrates like pyruvate and malate from glycolysis are taken up by this organelle and incompletely oxidized to end products like acetate, in the process forming ATP. Hydrogen is another characteristic end product of hydrogenosomal metabolism (Figure 3).

In many excavates and chromalveolates, mitochondrial metabolism resembles that of bacterial respiration, having a branched pathway, part of which is sensitive to cyanide (mammalian-like) and part of which has an alternative terminal oxidase that allows the mitochondrion to operate at low oxygen levels. These modifications are most developed in the specialized parasites, where both cellular and mitochondrial structure change as the parasite passes through different stages of the life cycle.

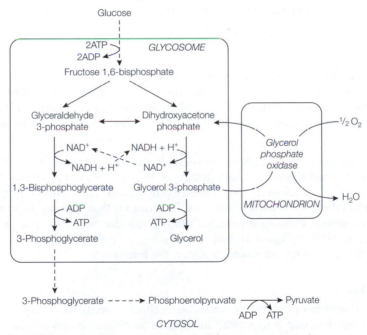

Figure 2. Aerobic glycolysis in bloodstream forms of trypanosomes, showing compartmentalization of much of the pathway in the glycosome. Dotted lines indicate several reaction steps.

Oxygen and carbon dioxide

In an aquatic environment temperature influences levels of dissolved oxygen. Oxygen utilization by living organisms increases with a rise in temperature; thus, availability of oxygen is likely to limit growth in warmer waters.

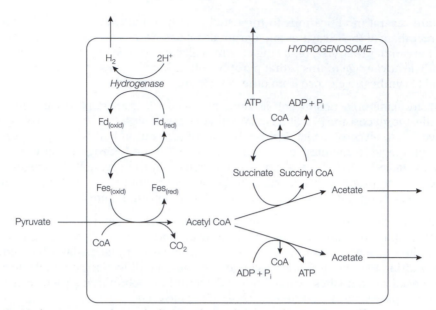

Figure 3. Hydrogenosomal metabolism in the trichomonads. Fes, iron-sulfur proteins; Fd, ferredoxin; oxid, oxidized state; red, reduced state.

Carbon dioxide, the levels of which in water vary inversely with dissolved oxygen, provides carbon to autotrophic species for photosynthesis. Most chlorophytes and photosynthetic excavates and chromalveolates utilize carbonic acid. Anaerobic photosynthesis occurs in a few species, using hydrogen sulfide or carbon dioxide as terminal acceptors (Section E3).

The absolute requirement for light in photosynthetic species means that they must have adaptations that allow them to remain in the photic zone of their environment. On land this is not problematic, but in the aquatic ecosystem there is a natural tendency for cells to sink, and thus there are cell modifications that counter this. Many zoospores and gametes are flagellate and can sense light, swimming towards it. Other species have structural modifications like spines that increase the resistance of the cell to sinking, and catch water uplift. Some have **gas vacuoles** that increase buoyancy.

Nitrogen nutrition

Members of the chlorophytes, excavates, and chromalveolates are heterotrophic for nitrogen and must obtain it in a fixed form such as nitrate, ammonia or amino acids. Amino acid requirement is common among the algae (see below). Poor availability of nitrogen is a common limiting factor to growth in the marine environment. Complex nitrogen sources such as peptides and proteins can be utilized after either extracellular or intracellular enzyme secretion has digested the polymers. Amino acids can also be stored as a nitrogen store.

Macronutrients, micronutrients, and growth factors

Carbon is rarely limiting for growth of photosynthetic species, but nitrogen and phosphorus often are. Freshwater productivity is often limited by phosphate availability. Silicon is often limiting to diatom growth in nutrient-poor, **oligotrophic** lakes. Most other nutrients are present to excess in the aquatic environment.

Although some photosynthetic species are completely **autotrophic**, many require an external supply of **vitamins**, frequently thiamin, biotin, B_{12}, and riboflavin, purines, pyrimidines, and other classes of growth factors. This requirement is termed **auxotrophy**, and it reflects the abundance of dissolved organic matter to be found in their environment, which allows the selection of populations that do not synthesize all their own metabolic requirements (Section D1).

Water, pH, and temperature optima

Almost all chlorophytes, excavates and chromalveolates are limited to damp environments. A few photosynthetic species that are in lichen symbiosis (Section I4) are protected from desiccation and can survive extremely dry conditions. Most members are tolerant of a wide range of pH; some are specialized and can inhabit highly acidic environments like those of hot sulfur springs. In highly illuminated, oligotrophic lakes, large changes in pH can be caused by variations in dissolved CO_2 concentrations and these changes can have detrimental effects on the populations of photosynthetic species.

Some dormant stages of chlorophytes, excavates, and chromalveolates are capable of withstanding 100°C for several hours. Other species can grow and divide at −2°C in sea water, and specialized 'snow algae' have growth optima between 1 and 5°C. Most species have a temperature optimum between 5 and 50°C.

Osmolarity

Sea water and fresh water have very different osmolarities. Freshwater species have an internal osmotic pressure of 50–150 mOs ml^{-1} while fresh water osmolarity is <10 mOs ml^{-1}. Those species that do not have a rigid cell wall to prevent excessive water uptake have contractile vacuoles that collect and expel excess water from the cell. Marine organisms have cytoplasm that is roughly isotonic with sea water and therefore usually do not require contractile vacuoles.

SECTION J – ARCHAEPLASTIDA, EXCAVATA,
CHROMALVEOLATA, AND AMOEBOZOA

J3 Life cycles

Key Notes

Life cycles in the Archaeplastida: chlorophytes

Many members of the Archaeplastida have a haploid vegetative phase, and gametes (motile or nonmotile) are formed by the differentiation of a vegetative cell. Zygote formation is followed by meiosis, producing haploid progeny cells. Other species have a diploid vegetative phase and produce haploid gametes by meiosis.

Life cycles in the Excavata, Chromalveolata, and Amoebozoa

Vegetative growth is associated with mitosis and cell division. Cell division can be by budding, binary or multiple divisions. Haploid life cycles are characterized by meiosis occurring at germination of the diploid zygote. Haploid cells form haploid gametes that fuse to form the diploid zygote. Species with a diploid life history are haploid only at gamete formation.

Life cycles in excavates

Life cycles in the excavates are characterized by asexual reproduction by longitudinal binary fission. Sexual reproduction is seen in the parasitic flagellates.

Life cycles in Chromalveolata

Asexual reproduction in ciliates is by homothetogenic cell division. Sexual reproduction occurs and the dominant phase of the life cycle is diploid. Dinoflagellates are haploid during most of their life cycle and reproduce by binary fission. Sexual reproduction is by the formation of macrogametes and microgametes. Asexual reproduction in the apicomplexa is characterized by multiple fissions. Sexual reproduction occurs in alternative hosts. In diatoms, diploid vegetative cells of diatoms reproduce asexually by mitosis. Centric diatoms form one or several macrogametes in an oogonium, fertilized by motile microgametes to generate a zygote within an auxospore. Chrysophyte asexual reproduction is by binary fission, and sexual reproduction results in a statospore. Asexual reproduction in oomycetes and related species is usually via flagellate motile diploid zoospores released from terminal sporangia. Sexual reproduction involves haploid gametes produced by terminal or subterminal antheridia and oogonia. Hyphochytrids reproduce via a multinucleate protoplast that is formed in the host cell, then releasing cysts into the soil that germinate into biflagellate zoospores able to invade new host plant roots.

Life cycles in Amoebozoa

Cellular slime mold amebae aggregate upon starvation to yield a multicellular pseudoplasmodium that differentiates into a fruiting body and generates haploid spores. Amebae of the acellular slime molds fuse to form a diploid cell (or this is produced via haploid flagellated cells), which grows into a large multinucleate plasmodium. Sporangia arise and meiosis within them generates haploid spores that germinate to release amebae. Pathogenic amebae have a simple life cycle existing as cysts outside the host and as amebae inside the host.

| Related topics | (H3) Cell division and ploidy | (I1) Fungal structure and growth |

Life cycles in the Archaeplastida: chlorophytes

The dominant vegetative phase of members of the chlorophytes can be haploid or diploid. Vegetative growth is associated with mitotic cell division. In species with a haploid life history, meiosis occurs at zygote germination and the cells remain haploid for the whole of their vegetative life. Diploid formation occurs only in the zygote. Species with a diploid life history are only haploid at gamete formation and after zygote formation continue their life cycles as diploids.

Motile compatible gametes first entangle flagella, cells then **conjugate** and nuclei fuse to form a zygote. The zygote may remain motile or it may form a thick-walled resting cyst. Meiosis occurs within the zygote and haploid, flagellate cells are released (Figure 1). Similar events are seen in the colonial forms; all cells of a colony may develop into free-swimming gametes after breakdown of the colony structure.

Filamentous species may reproduce sexually by conjugation. In this process, two vegetative cells form a **conjugation tube** between them and fuse. The cellular content from

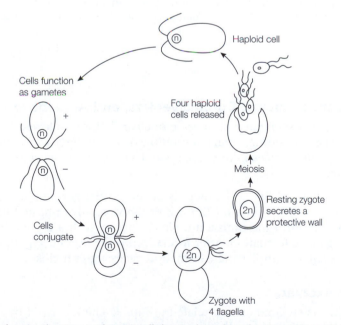

Figure 1. Sexual reproduction in the unicellular Archaeplastida. From Gross T, Faull JL, Ketteridge S & Springham D (1995) *Introductory Microbiology*. With permission from Kluwer Academic.

one cell then moves into the other, where nuclear fusion and zygote formation occur. The zygote encysts and meiosis occurs before the emergence of a haploid new filament (Figure 2). Other filamentous species produce motile gametes of two mating types from different vegetative cells. Often one gamete is considerably larger (**macrogamete**) than the other (**microgamete**). In some species, only one motile microgamete is produced and it fuses with a nonmotile gamete cell called an **oogonium** to form the zygote.

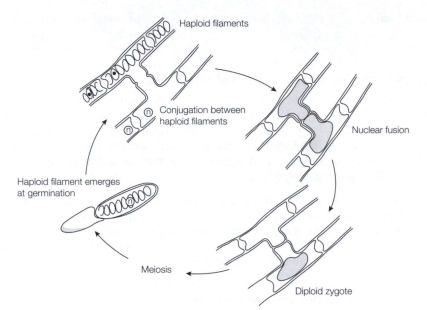

Figure 2. Conjugation between two filamentous green algal cells. From Gross T, Faull JL, Ketteridge S & Springham D (1995) *Introductory Microbiology*. With permission from Kluwer Academic.

Life cycles in the Excavata, Chromalveolata, and Amoebozoa

Vegetative growth is associated with mitotic nuclear division followed by cell division. Cell division can be by budding, binary or multiple divisions. Each of the different groups is characterized by differences in their life cycle. The dominant vegetative phase can be haploid or diploid.

In species with a haploid life cycle, meiosis occurs at the germination of the diploid zygote. The cells remain haploid for the rest of the vegetative phase, forming haploid gametes that fuse to form the transient diploid. Species with a diploid life history are haploid only at gamete formation. After formation of the diploid zygote they continue their life cycle as diploids until gamete formation necessitates meiosis.

Life cycles in excavates

Reproduction in the euglenoids and related groups is characterized by mitosis of the nucleus followed by cell division, which occurs along the longitudinal axis of the cell. This produces mirror image progeny (Section H3). Sexual reproduction is seen in some species of this group. The group includes many of the parasitic species including trypanosomes.

Life cycles in Chromalveolata

Alveolates

Ciliate asexual reproduction is also by binary fission after mitosis of the nucleus. Cell division is described as **homothetogenic**, across the narrow part of the cell. Ciliates have two types of nuclei, a **micronucleus** that is diploid, contains little RNA and a lot of histone, and a **macronucleus** that is polyploid and controls day-to-day cellular activities. There may be hundreds of macronuclei in one ciliate, and up to 80 micronuclei. During asexual mitotic cell division, both of these nuclei divide to provide progeny cells with at least one copy of both nuclei. During sexual reproduction the macronucleus degenerates and the micronucleus undergoes meiosis to form gametes. Gametes fuse to form a new diploid nucleus. This new nucleus divides, one copy of the nucleus remaining as the new micronucleus, and the others differentiate to provide a new polyploid macronucleus (Figure 3).

This unusual state of nuclear dualism appears to confer an advantage to the ciliates by having a separate genetic store in the micronucleus, and enhances RNA synthesis by the

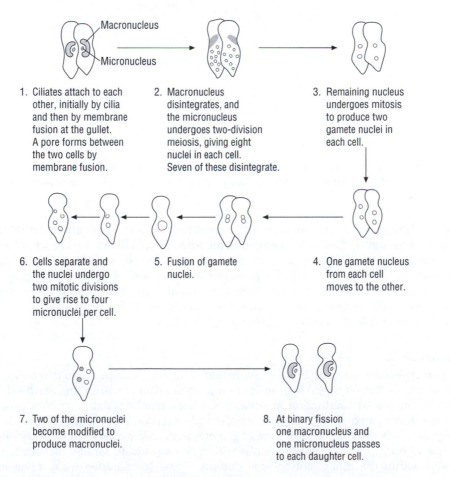

1. Ciliates attach to each other, initially by cilia and then by membrane fusion at the gullet. A pore forms between the two cells by membrane fusion.

2. Macronucleus disintegrates, and the micronucleus undergoes two-division meiosis, giving eight nuclei in each cell. Seven of these disintegrate.

3. Remaining nucleus undergoes mitosis to produce two gamete nuclei in each cell.

4. One gamete nucleus from each cell moves to the other.

5. Fusion of gamete nuclei.

6. Cells separate and the nuclei undergo two mitotic divisions to give rise to four micronuclei per cell.

7. Two of the micronuclei become modified to produce macronuclei.

8. At binary fission one macronucleus and one micronucleus passes to each daughter cell.

Figure 3. Sexual reproduction in the ciliates. From Gross T, Faull JL, Ketteridge S & Springham D (1995) *Introductory Microbiology*. With permission from Kluwer Academic.

macronucleus. This allows them to be very adaptable to changing environmental conditions.

Dinoflagellates are haploid and have unusual chromosomes, which are condensed throughout their life cycle. The chromosomes contain very little histone. Sexual reproduction commences when motile cells differentiate to become **macrogametes** or **microgametes** (Figure 4). The gametes fuse to form a zygote. Meiosis occurs followed by the degeneration of three of the four nuclear products. A haploid, motile cell then emerges from the zygote.

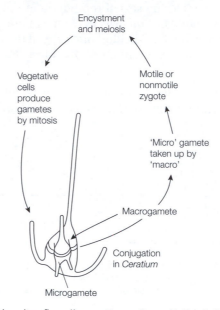

Figure 4. Reproduction in the dinoflagellates. From Gross T, Faull JL, Ketteridge S & Springham D (1995) *Introductory Microbiology*. With permission from Kluwer Academic.

Apicomplexan protists are parasites with a complex life cycle. There are both diploid and haploid phases and often two host species are infected in a life cycle. The group is characterized by a type of cell division that is called multiple division or **schizogony**. During this process, multiple divisions of haploid nuclei occur, producing many progeny. These progeny are then released into body fluids like blood where they rapidly enter new host cells and establish themselves as intracellular parasites. The malarial parasite *Plasmodium* is an example of an apicomplexan parasite (Figure 5).

Stramenopila

Diatom vegetative cells are diploid, and repeated mitotic divisions lead to a reduction in cell volume as daughter cells synthesize new frustules that fit within the inherited parent frustule. Once a 30% reduction in volume has been reached diatoms either produce a resting spore (or **auxospore**) to regain cell size or they reproduce sexually (Figure 6). Sexual reproduction begins by the formation of gametes after meiosis. In round, centric diatoms a single or multiple macrogamete forms within an **oogonium**. Motile microgametes are formed within the other mating-type diatom. These microgametes are released and fertilization of the oogonium leads to the formation of a zygote within an auxospore. The auxospore enlarges and secretes a new pair of full-size frustules. The diploid, vegetative

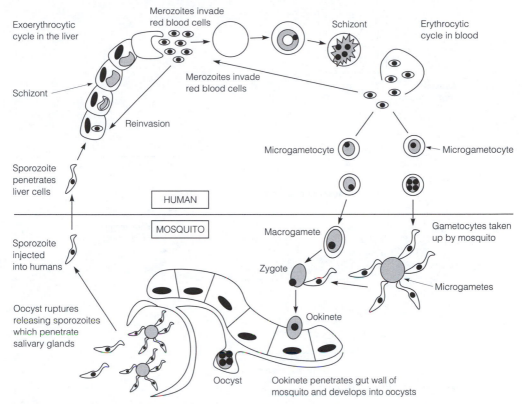

Figure 5. The life cycle of *Plasmodium* spp. indicating vertebrate and invertebrate hosts.

life cycle then continues. The long, thin diatoms, termed pennate, do not form motile gametes, but after a meiotic division fuse somatically to form a zygote.

Chrysophyte life cycles are characterized by the formation of the **statospore**. Details of life cycles vary in this phylum, but in *Dinobryon* isogamous cells seem to function as gametes. Spore formation is termed **intrinsic** and it is independent of external conditions. Around 10% of the population will encyst as a zygotic spore in any one generation, allowing populations in optimal growth conditions to maintain genetic diversity and reduce intraspecific competition.

Oomycetes are filamentous, coenocytic organisms. Asexual reproduction is characterized by the production of flagellate, motile, diploid zoospores from terminal sporangia. Sexual reproduction occurs after meiosis and gamete production in terminal or subterminal antheridia and oogonia. A few species are terrestrial and produce nonmotile gametes from sporangia and oogonia. The life cycle of the plant pathogen *Phytophthora infestans* is typical of this terrestrial group (Figure 7).

Hyphochytrids reproduce by the formation of multinucleate, unwalled protoplasts within an enlarged host cell. Cyst formation within the host cell occurs, and these are released into the soil on breakdown of the plant root where they can persist for years. The presence of a suitable host plant root breaks dormancy, and the cyst germinates by producing anteriorly biflagellate zoospores that swim to the host root and actively penetrate it.

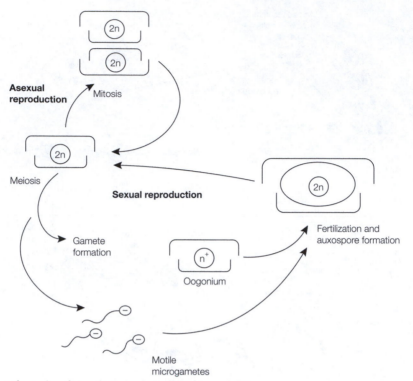

Figure 6. Life cycle of the diatoms.

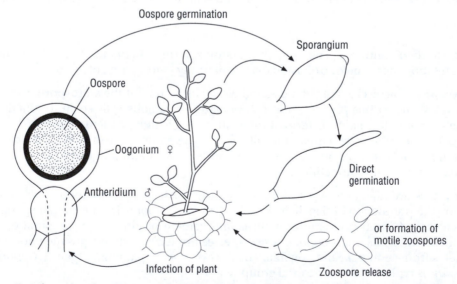

Figure 7. Life cycle of a water mold, *Phytophthora infestans*. From Deacon JW (1997) *Modern Mycology*, 3rd ed. With permission from Blackwell Science.

Life cycles in Amoebozoa

Amebae of the cellular slime molds (**Dictyostelids**) aggregate upon starvation to yield a multicellular pseudoplasmodium that differentiates into a fruiting body and generates haploid spores (Figure 8). Amebae of the acellular slime molds (**Myxogastrids**) fuse to form a diploid cell (or this is produced via haploid flagellated cells), which grows into a large multinucleate plasmodium. Sporangia arise and meiosis within them generates haploid spores that germinate to release amebae (Figure 9).

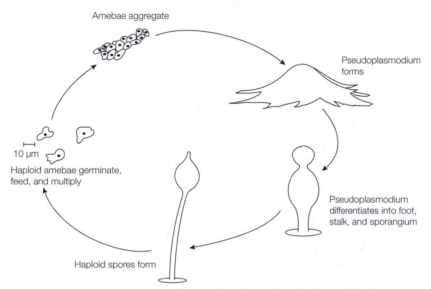

Figure 8. Life cycle of a cellular slime mold. From Sleigh M (1991) *Protozoa and Other Protists*. With permission from Cambridge University Press.

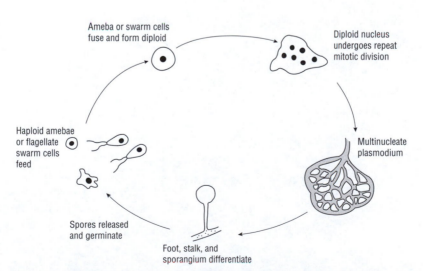

Figure 9. Life cycle of an acellular slime mold. From Sleigh M (1991) *Protozoa and Other Protists*. With permission from Cambridge University Press.

Reproduction in the pathogenic amebae is by binary fission of the cell after mitotic division of the nucleus. A single cell splits to form two identical progeny. In the pathogenic species life cycles involve an encysted stage, which can resist adverse conditions in the environment. Once the cyst enters its target host tissue, external stimuli cause excystation and the production of amebic forms that colonize the host. On resumption of adverse conditions (i.e. leaving the host in excreta or other fluids), the amebae will encyst. Figure 10 illustrates the life cycle of *Acanthamoeba*.

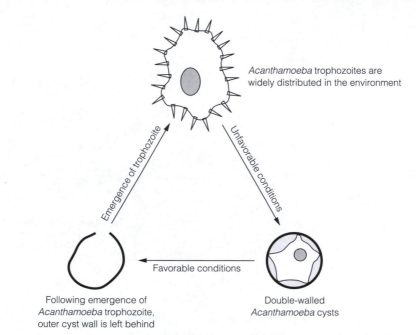

Acanthamoeba trophozoites are widely distributed in the environment

Emergence of trophozoite

Unfavorable conditions

Favorable conditions

Following emergence of
Acanthamoeba trophozoite,
outer cyst wall is left behind

Double-walled
Acanthamoeba cysts

Figure 10. Life cycle of *Acanthamoeba* spp.

J4 Beneficial effects

Key Notes

Primary productivity

Primary productivity in the oceans due to eukaryotic microbial photosynthesis is estimated to be $\sim 50 \times 10^9$ tonnes per annum. Carbon can be released as dissolved organic matter (80–90%) or as particulates (10–20%). The photosynthetic cells, associated saprophytic Bacteria, and dissolved organic matter together form the base of the aquatic food chain.

Symbiosis

Symbiotic associations can occur between species of Archaeplastida, excavates and chromalveolates, fungi and animals. In these associations, carbohydrates from photosynthesis are exchanged for nutrients from the fungus or animal partner. Excavates and chromalveolates can have internal symbionts that can be prokaryotes or other protists. Photosynthetic symbionts provide photosynthate or other factors for the heterotrophic protistan partner. Bacterial endosymbionts can provide vital cell metabolites. Ectosymbiotic prokaryotes may provide protista with motility or alternative electron acceptors. Some species are important within gut and rumen microflora of large animals, as part of a complex cellulolytic community.

Diatomaceous earth

Diatomaceous earth is formed from the silica-containing shells of diatoms. It has several commercial uses that take advantage of its chemically inert but physically abrasive qualities.

Bioluminescence

Bioluminescence is found in several of the protistan phyla and is associated with luciferin–luciferase reactions that occur within scintillons in the protistan cell. The precise function of this reaction is unknown, although it may provide defense against predators.

Related topics

(I4) The fungi and related phyla: beneficial effects
(J1) Archaeplastida, Excavata, Chromalveolata, and Amoebozoa: taxonomy and structure

(J2) Archaeplastida, Excavata, Chromalveolata, and Amoebozoa: nutrition and metabolism
(J3) Archaeplastida, Excavata, Chromalveolata, and Amoebozoa: life cycles

Primary productivity

In the ocean the total plant biomass (as **phytoplankton**) is estimated to be 4×10^9 tonnes (dry weight). Annual net **primary production** is around 50×10^9 tonnes. Productivity ranges from 25 g carbon m^{-3} in oligotrophic waters to 350 g carbon m^{-3} in eutrophic waters. This is released as **dissolved organic matter** (**DOM**), which accounts for 80–90% of organic matter present in the sea. The DOM is used as a nutrient source by heterotrophic Bacteria; populations between 10^4 and 10^6 ml^{-1} are commonly found in nutrient-poor (**oligotrophic**) waters, and much higher populations are found in nutrient-rich (**eutrophic**) waters (Section J5). The cells (both alive and as dead particulates), DOM, and prokaryotic populations form the base of the aquatic food chain.

Symbiosis

Archaeplastida, excavates, and chromalveolates can form beneficial associations with fungi, other eukaryotic microorganisms, insects, and higher animals. These symbioses can be specific and physically and physiologically intimate or they can be relatively non-specific and merely loose ecological associations.

Endosymbionts are symbionts that live inside other organisms. Photosynthetic species can occur within Alveolata and Euglenozoa (and others) and their presence allows the protistan dual organism to adopt a phototrophic habit. The photosynthetic partner is confined to a membrane-bound vacuole, but it is capable of cell division. There is a two-way exchange of materials where the products of nitrogen metabolism of the heterotrophic protistan are utilized by the photosynthetic partner and the products of photosynthesis are utilized by the heterotrophic partner.

The **zooxanthellae** are symbiotic dinoflagellates that are found as coccoid cells within animal cells. They are enclosed in intracellular double-membrane-bound vacuoles, which remain undigested. They are found in excavates and chromalveolates, hydroids, sea anemones, corals, and clams where they provide glycerol, glucose, and organic acids for the animal, and the symbiotic alga gains CO_2, inorganic nitrogen, phosphates, and some vitamins from the animal. **Reef-building corals** are only able to build reefs if they have their symbiotic photosynthetic partner, and many of the animal hosts are at least partly dependent on the algal partner for carbohydrates. **Radiolaria**, responsible for massive primary productivity in the oceans, are wholly dependent on their photosynthetic symbiont for carbohydrate.

Many aerobic protista contain bacteria as endosymbionts. *Amoeba proteus* has a symbiotic Gram-negative bacterium that is essential to the survival of the ameba, and *Paramecium aurelia* has specific symbiotic algae that are responsible for the secretion of killer factors that are important in competition within environments and mating.

There are a large number of Bacterial ectosymbionts. Spirochetes and many other species of Bacteria are often attached to the euglenozoal pellicle, arranged in very specific patterns. For example *Mixotricha paradoxa* is an inhabitant of termite guts. Its motility within this environment depends on the co-ordinated movement of adherent spirochetes.

Ecto- and endosymbiotic Bacteria are also found associated with anaerobic alveolata from sulfur-rich environments. *Kentrophoros lanceolata* has a dense mat of sulfur Bacteria on its dorsal surface. These symbionts provide alternative electron acceptors to the protistan.

Excavates and chromalveolates can be symbiotic with insects and higher animals, and are typically found within fermentative guts as mentioned above. They are part of a complex ecosystem where cellulose is broken down by fungi and prokaryotes, and their metabolic products are fermented by the protista and prokaryotes to provide fatty acids for their hosts. Some of the protistan population also predate the prokaryotic population to provide control of numbers.

Diatomaceous earth

At death, diatom cells fall through the water column to the seabed. The inert nature of the frustule silica, silicon dioxide (SiO_2), means that it does not decompose but accumulates, eventually forming a layer of diatomaceous earth. This material has many commercial uses for humans, including filtration, insulation, and fire-proofing, and as an active ingredient in abrasive polishes and reflective paints. Recently, it has been used as an insecticide, where the abrasive qualities of diatomaceous earth are used to disrupt insect cuticle waxes, causing desiccation and death.

Bioluminescence

Many of the marine dinoflagellates are capable of **bioluminescence**, where chemical energy is used to generate light. The light is in the blue–green range, 474 nm, and can be emitted as high-intensity short flashes (0.1 second) either spontaneously, after stimulation, or continuously as a soft glow.

Bioluminescence is created by the reaction between a tetrapyrrole (**luciferin**) and an oxygenase enzyme (**luciferase**). The entire reaction is held within membrane-bound vesicles called **scintillons**, where the luciferin is sequestered by **luciferin-binding protein** (**LBP**) and held at pH 8. Release of light is stimulated by either a cyclical or mechanical stimulation of the scintillon, which leads to pH changes within it. Luciferin is released from LBP, and luciferase is able to activate it. Activated luciferin exists for a very brief time before it returns to its inactivated form, releasing a photon of light energy (Figure 1).

The distribution of scintillons within the photosynthetic cell varies over 24 hours. During the night they are distributed throughout the cytoplasm, but in daylight they are tightly packed around the nucleus. The function of bioluminescence for the organism is uncertain. It appears to be useful in defense against predators. It has become important in medicine and science because the coupled system of luciferin/luciferase can be used to mark cells. Once tagged with the bioluminescent marker, marked cells can be mechanically sorted from other cells, visualized by microscopy or targeted for therapy.

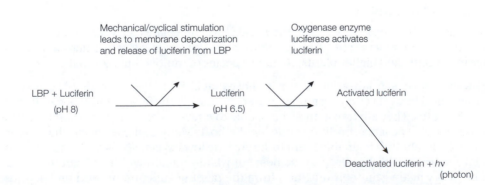

Figure 1. Luciferin–luciferase reaction within scintillon. LBP, luciferin-binding protein.

J5 Detrimental effects

Key Notes

Parasitic relationships	A wide range of organisms can be parasitized by members of the protista. They may enter via the mouth and colonize the gut where they can cause asymptomatic, mild disease. More specialized parasites cause greater effects on the gut and cause disease symptoms. Some highly specialized parasites can enter via the gut but move into other tissues to complete their life cycles. Blood parasites are transferred via vectors from previously infected hosts. They have a highly specialized life cycle.
Toxic blooms	Toxic blooms occur when nutrient-rich water supports population explosions of some photosynthetic species of protista. Some of these species can produce toxins that affect marine animals and humans that come into contact with them.
Eutrophication	The presence of excess nutrients in an aquatic ecosystem will support high levels of bacterial and eukaryotic microbial growth. The oxygen demand for this growth is great and will exceed supply, creating anoxic conditions. This is called eutrophication.
Related topics	(J1) Archaeplastida, Excavata, Chromalveolata, and Amoebozoa: taxonomy and structure (J2) Archaeplastida, Excavata, Chromalveolata, and Amoebozoa: nutrition and metabolism	(J3) Archaeplastida, Excavata, Chromalveolata, and Amoebozoa: life cycles (J4) Archaeplastida, Excavata, Chromalveolata, and Amoebozoa: beneficial effects

Parasitic relationships

Parasitism of other organisms by excavates and chromalveolates is a very common nutritional strategy. Hosts can range from other members of the eukaryotic microorganisms, including aquatic and higher plants, to many species of multicellular animals.

The lowest level of specialization is seen in those that inhabit the intestines of animals as commensals. They gain entry via food or water and exit via the feces. Many species live on gut prokaryotes. They are transient visitors, able to encyst before they are shed to insure that they survive adverse conditions outside the host. Many of these species have a fermentative metabolism to enable them to survive the low oxygen levels of the gut. A more specialized group attaches itself to the host gut wall to remain within the gut for longer periods. They again gain their nutrition from the passing gut contents and prokaryotes. These may cause some irritation to the gut epithelium.

The third group derive their nutrition from the gut epithelium and cause considerable damage in doing so. They have a number of pathogenicity factors including toxins and proteolytic enzymes called proteolysins. A more specialized form of parasitism is found in species that infect orally but then migrate into other tissues for part of their life cycle.

Parasites that have a life cycle that includes a blood-borne phase are highly specialized. They require vectors to reach their chosen niche, usually the blood-sucking insects. Part of the life cycle is completed in the vector. Trypanosomes and the malarial parasites are two of the most significant human protistan parasites. During the blood-borne phase of disease, parasite metabolism can be anaerobic and mitochondrial metabolism is switched off.

Toxic blooms

When environmental conditions for growth are optimal, photosynthetic Archaeplastida, excavates, and chromalveolates can grow exponentially, leading to high local populations (Section J1). This can occur in several species of marine dinoflagellates that contain poisons that are toxic to fish, invertebrates or mammals, depending on the species of alga and the class of compound they produce. The toxins are accumulated in the digestive glands of shellfish and when consumed by man cause **paralytic shellfish poisoning**. Many of these compounds are neurotoxic. **Saxitoxin**, accumulated in shellfish and accidentally consumed, causes numbness of the mouth, lips, and face, which reverses after a few hours.

Toxins can also accumulate in higher animals like fish. For example **ciguatoxin** from *Gambierdiscus toxicus* accumulates in muscle tissue of grouper and snapper and when eaten causes gastric problems, central nervous system damage, and respiratory failure.

Toxins can also be formed by other species. Some species of diatoms can produce **domoic acid**, which can accumulate in mussels and, when consumed by humans, causes **amnesic shellfish poisoning**, a short-term loss of memory that can occasionally cause death. Some of the golden-brown species that give rise to these types of poisoning are highly pigmented members of the Chrysophyceae, and high cell densities can be seen as so-called **red tides** in the sea. Other toxins from species of the Prymnesiophyceae can affect gill function of fish and molluscs.

Eutrophication

The presence of high populations of photosynthetic protista, their DOM products, and the high numbers of Bacteria that can be supported by large amounts of organic matter lead to a condition known as **eutrophication**. Water is in a nutrient-rich state, biomass is high, and oxygen demand exceeds supply. Anoxic conditions rapidly develop, leading to the death of aerobic organisms.

The decomposition of dead material by prokaryotes leads to further demands for oxygen, and the entire environment becomes anaerobic, allowing for the growth of anaerobic Bacteria and Archaea with the production of methane, hydrogen sulfide, hydrogen, and many other products of anaerobic metabolism (Section E4). Eutrophication is commonly seen around untreated sewage outfalls and dairy farm waste run-off, and in freshwater streams polluted by nitrate-rich agricultural field run-off.

K1 Virus structure

Key Notes

Definitions

Viruses are obligate intracellular parasites and vary from 20 to 400 nm in size. They have varied shape and chemical composition, and at their most fundamental level are composed simply of protein and a DNA or RNA genome. The intact structure is referred to as virus particle or virion, and consists of a protein capsid enclosing the genome, which may be further surrounded by a glycoprotein/lipid membrane known as the envelope. Viruses are resistant to antibiotics and all cellular life is subject to infection by one or more known viruses.

Methods of study

Virus morphology has been determined by electron microscopy (using negative staining), immunoelectron microscopy, electron cryomicroscopy, X-ray crystallography, and atomic force microscopy.

Virus symmetry

Virus capsids usually have helical or icosahedral symmetry. In many cases the capsid is engulfed by a membrane structure (the virus envelope). Helical symmetry is formed as protein subunits arranged around the virus nucleic acid in an ordered helical fashion. The icosahedron is a regular geometric (cuboid) shape, which consists of repetitions of many protein subunits assembled so as to resemble a sphere.

Virus envelopes

Virus envelopes are acquired by the capsid as it buds through membranes of the infected cell and contain viral (and in some cases, cellular) glycoproteins. Some envelopes have few glycoproteins, for example, human immunodeficiency virus (HIV), while others have many, for example, herpes simplex virus (HSV). The virus envelope contains the receptor molecule that allows the particle to attach to and infect the host cell.

Related topics

(F10) Bacteriophages

Definitions

Viruses are obligate intracellular parasites that only show activity within the cells of living organisms. All life on earth – plant, animal, bacterial, fungal – plays host to one or more viruses and in order to persist in the environment viruses must be capable of being passed from host to host and of infecting and replicating in susceptible host cells. Viruses cannot be viewed with a light microscope due to their small size; they range from approximately 20 to 400 nm, although some filoviruses can reach lengths of over 1000 nm. A virus particle can be defined as a structure that has evolved to transfer genetic information (nucleic acid) from one cell to another. The nucleic acid found in the particle is either DNA or RNA, and may be single- or double-stranded, linear or circular, intact or segmented. The simplest of virus particles consists of a protein coat (sometimes made up of only one type of protein, repeated hundreds of times) surrounding a strand of nucleic acid. The protein coat surrounding the nucleic acid is referred to as the capsid (Figures 1 and 2) and is composed of proteinaceous units or capsomers. The type of capsomer depends on the overall shape of the capsid, but in the case of icosahedral capsids the capsomers are either pentamers or hexamers. Capsomers themselves consist of subunits, often called protomers, which are a collection of one or more nonidentical protein subunits that bind together to form the building blocks for capsomer and capsid assembly (e.g. virus proteins VP1, VP2, VP3, and VP4 of picornaviruses). Within the capsid lies the nucleic acid genome of the virus, which may exist in a nucleic acid–protein complex known as a nucleocapsid, often referred to as the core of the virion. While some authors use these structural terms interchangeably, the distinction between nucleocapsid and capsid should be appreciated.

The protein coat of more complex viruses is further surrounded by a membrane, the **envelope**, usually derived from modified regions of a cellular membrane as the virus escapes the infected cell. The envelope has **peplomers** (projections) formed of virally encoded glycoproteins, that are visible as a 'fringe' around the virion when viewed with an electron microscope (Figures 1 and 2). The space between the capsid and envelope

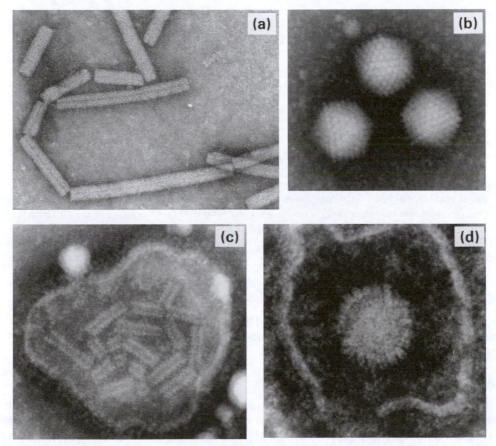

Figure 1. Examples of viruses from main groups according to 'standard' morphology: (a) unenveloped/helical (tobacco mosaic virus); (b) unenveloped/icosahedral (adenovirus); (c) enveloped/helical (paramyxovirus); (d) enveloped/icosahedral (herpesvirus). Courtesy of Professor CR Madeley and Dr IL Chrystie, St Thomas' Hospital, London.

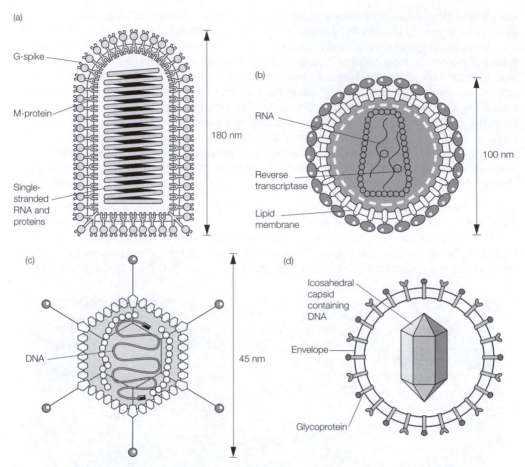

Figure 2. Diagrammatic representation of the structure of four virus particles. Rabies virus (a) and HIV (b) have tight-fitting envelopes, herpes simplex (d) has a loose-fitting envelope, whereas adenovirus (c) is nonenveloped. The diagrams are based on electron microscopic observation and molecular configuration exercises. Adapted from Strelkauskas A, Strelkauskas J & Moszyk-Strelkauskas D (2010) *Microbiology: A Clinical Approach*. Garland Science.

is not empty; it too contains virally encoded proteins and is referred to as the **matrix** or (when describing herpesviruses) the **tegument**. The complete, fully assembled virus (enveloped or naked) is termed the **virion**.

Methods of study

While the **electron microscope** (EM) had been known for many years, the invention of the **negative-staining technique** in 1959 revolutionized studies on virus structure. In negative-contrast EM, virus particles are mixed with a heavy metal solution and dried onto a support film. The metal ions do not (ideally) interact with virions but instead provide an electron-opaque background against which the virus can be visualized. Compounds such as uranyl acetate, sodium silicotungstate, and methylamine tungstate are especially good for viruses due to the high-resolution images they can produce. Observation under the EM has thus allowed the definition of **virus morphology** at the 20–50Å resolution level. In addition, negative staining of thin sections of infected cells has allowed definition of structures that appear during virus maturation and their interactions with cellular proteins. Newer developments include the use of **immunoelectron microscopy**, where the binding of antibodies labeled with gold particles allows viruses at low concentration, or which grow poorly in tissue culture, to be identified more readily. **Electron cryo-microscopy** reduces the risk of artifacts that may be unavoidable when using negative staining. High concentrations of virus are rapidly frozen in liquid ethane while on carbon grids. Electron micrographs can be digitized, allowing three-dimensional reconstructions to be predicted – resolutions of 9Å have been achieved using this method. **Atomic force microscopy** (AFM) allows direct imaging of the three-dimensional morphology of viruses, and where used to date it has confirmed the predictions from digitized EM images. **X-ray diffraction** of virus crystals is the ultimate approach in determining the ultrastructure of virion morphology. At present, only simple viruses can be crystallized. More complex viruses are analyzed by attempting to form crystals of subparticular molecules. The X-ray diffraction pattern of the virion particle allows mathematical processing, which can predict the molecular configuration of the virus particle (Figure 3).

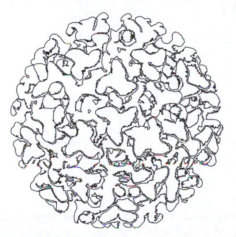

Figure 3. Three-dimensional reconstruction of an icosahedrally symmetric virus particle.

Virus symmetry

The capsids of virions tend to have one of two symmetries – **helical** or **icosahedral** (geometric). Helical symmetry can be loosely described as having a 'spiral staircase' structure, with an obvious axis down the center of the helix. The protein subunits of a helical capsid mirror the helical turns of the nucleic acid. A diagram of such a structure (tobacco mosaic virus) is shown in Figure 4, and electron micrographs of helical viruses are shown in Figures 1a and 1c. Viruses with a helical capsid structure include measles, rabies, and influenza. Some helical viruses appear to be 'open-ended' while others are enclosed structures (Figure 2a) where the protein subunits seal the capsid at one or both ends. Many viruses have geometric (sometimes called cuboid) symmetry, tending towards a spherical shape. Obtaining a true sphere is not possible although the protein subunits come together to produce a geometric structure that is very close to being spherical. These virus capsids are always fully closed and are usually based on an **icosahedron**. A regular icosahedron forms from the assembly of identical subunits arranged to give **20 equilateral triangular faces**, **30 edges and 12 vertices** (corners). The structure exhibits multiple symmetries, with **2-fold axes running along each edge**, **3-fold axes centered in each triangular face**, and **5-fold axes at each corner** (Figure 4). The minimum number of capsomers required to construct an icosahedron is 12, each composed of five identical subunits forming the adjoining corners of the capsid structure. Many viruses have more than 12; Adenoviruses display such an arrangement as illustrated in Figure 4. Adenoviruses have 12 projecting fibers, one present at each vertex, making this capsid highly distinguishable from that of other viruses. The maturation and assembly of icosahedral viruses is very complex and much of what happens is unknown. Usually icosahedral capsids form via a complex but structured array of molecular-assembly procedures that eventually give rise to the mature capsid. These may be self-assembly processes or they may involve virus non-structural proteins, acting as temporary **scaffolding** proteins, which do not form part of the mature capsid.

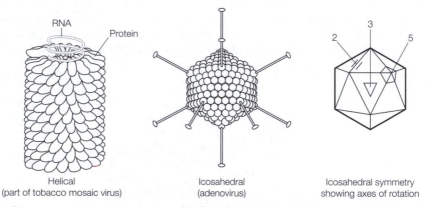

Helical
(part of tobacco mosaic virus)

Icosahedral
(adenovirus)

Icosahedral symmetry
showing axes of rotation

Figure 4. Diagrammatic representation of helical and icosahedral symmetry. Adapted from Harper DR (2011) *Viruses: Biology, Applications and Control*. Garland Science.

Virus envelopes

Many viruses enclose their capsid in a membrane **envelope**, which they gain by the process of budding; the capsid wraps itself in a coating of lipid bilayer as it passes through the membrane. Many virus envelopes are the result of budding from the plasma membrane (e.g. measles and influenza viruses) but some viruses bud through intracellular membranes (e.g. herpes simplex virus, hepatitis C virus) such as the endoplasmic reticulum (ER) or Golgi and these enveloped virions are subsequently transported out of the cell within a vesicle, hence they temporarily have a second membrane. Finally, some viruses (e.g. baculoviruses) gain an envelope without budding across any membrane, but the exact nature of this process is unknown. Indeed the process of budding itself is not fully understood, but it is known to involve specific domains within viral proteins that interact with cellular proteins, and also with the cellular ubiquitin-proteasome system. Virus envelopes can be loose and amorphous (e.g. herpes simplex virus, Figures 1 and 2) or tightly bound to the capsid (e.g. human immunodeficiency virus, HIV). Embedded within the virus envelope are a number of virus-encoded **glycoproteins**, which are involved in virus attachment to cells and serve as targets for the immune response. Most virion envelopes contain exclusively viral glycoproteins, but retroviruses (e.g. HIV) also contain host glycoproteins.

K2 Virus taxonomy

Key Notes

Virus taxonomy	The International Committee on Taxonomy of Viruses (ICTV) considers a wide range of characteristics (morphology, genome, physicochemical and physical properties, proteins, antigenic and biological properties) to classify viruses into orders, families, genera, and species, and uses these criteria to provide naming conventions for new viruses. Virus genome characteristics are grouped according to the Baltimore system, which defines seven classes of virus, alongside the taxonomic groupings.
Virus orders	These are groupings of families of viruses that share common characteristics that define membership of an order. They are designated by the suffix *-virales*. There are six orders; *Caudovirales, Herpesvirales, Mononegavirales, Nidovirales, Picornavirales*, and *Tymovirales*.
Virus families	These are groupings of genera of viruses that share common characteristics defining membership within the family, but are distinct from the members of other families within a given order. They are designated by the suffix *-viridae*. Some are further divided into subfamilies, distinguished by the suffix *-virinae*.
Virus genera	These are groupings of species of viruses that share common characteristics and that are distinct from other species within their genus. These are designated by the suffix *-virus*.
Virus species	These represent a group of viruses having several consensus characteristics in common, but that do not necessarily share a single defining property.
Related topics	(B1) Prokaryotic systematics (K3) Virus genomes

Virus taxonomy

Early attempts to classify viruses were based on their pathogenic properties, and the only feature common to many of the viruses placed together in such groupings was that of **organ tropism** or disease symptom (e.g. viruses causing hepatitis or respiratory disease). Other more fundamental aspects (e.g. **virus structure** and **composition**) led to the consensus that these initial attempts at classification were inadequate and in the late 1950s and early 1960s, hundreds of new viruses were isolated, highlighting the need for a robust classification system. In 1966 the **International Committee on Nomenclature of Viruses (ICNV)** was established at the International Congress of Microbiology held in Moscow. From this beginning, the present classification scheme has evolved, and acceptance of the characteristics to be considered and their respective weighting has become universal.

The **International Committee on Taxonomy of Viruses** or **ICTV** provides a universal taxonomy database consisting currently (2010) of 6 orders, comprising a total of 87 families, 19 subfamilies, and 348 genera. To date, 2288 species of virus have been discovered, but hundreds remain unassigned to any of the taxonomic groupings, largely because of lack of data and the continual emergence of new viruses.

As techniques in serology, microscopy, and molecular biology (particularly nucleotide sequencing) have advanced, taxonomy encompasses more and more characteristics and the ICTV classification scheme is supported by extensive verifiable data. Taxonomic factors (Table 1) now include morphological, physical, and physicochemical properties of the virion, such as genomic, protein, lipid, carbohydrate, antigenic, and biological properties. Actual virus definition requires the determination of several hundred characters, but the rapidity of genome sequencing now makes classification of newly emerging viruses much easier; for example, the virus first associated with severe acute respiratory syndrome (SARS) was identified as a Coronavirus within just a few months of its emergence in 2002.

In parallel with the ICTV taxonomy is the **Baltimore** classification scheme, which considers the composition of the virus genome and its replication strategy. This simple but elegant system, devised in 1970 by Nobel laureate David Baltimore, defines seven classes of virus, explored more fully in the next section (Section K3).

Virus orders

An order represents a grouping of families of viruses that share common characteristics that make them distinct from other orders and their families. Orders are designated by the suffix -*virales* (note the use of italics when stating virus taxonomy) and there are six orders currently approved by the ICTV. The main characteristics of viruses within each order are summarized here and in Table 2, with examples of viruses pathogenic to humans. The viruses within each order may span several host families; for example the **Mononegavirales** includes viruses that infect animal and plant species. Viruses within this order all possess enveloped capsids and single-stranded, negative-sense, nonsegmented genomes (a definition of 'sense' pertaining to virus genomes is given in the next section). The order contains four families, the *Bornaviridae*, *Paramyxoviridae*, *Rhabdoviridae*, and *Filoviridae*. The **Nidovirales** are single-stranded, positive-sense, nonsegmented enveloped RNA viruses and comprise the *Coronaviridae*, *Arteriviridae*, and *Roniviridae* families; all infect vertebrate hosts. **Caudovirales** are nonenveloped viruses with double-stranded, nonsegmented linear DNA genomes and viruses within the three families of this order (*Siphoviridae*, *Myoviridae*, and *Podoviridae*) are all **bacteriophages**; that is they are bacteria-specific. The order **Herpesvirales** contains viruses that infect a wide range of animals; enveloped, icosahedral capsids contain a double-stranded, nonsegmented linear DNA genome. There are three families within this order, the *Alloherpesviridae*, *Malacoherpesviridae*, and *Herpesviridae*, which includes the Herpesviruses that infect humans, such as herpes simplex virus. Viruses within the **Picornavirales** have single-stranded, positive-sense, nonsegmented RNA genomes within naked icosahedral capsids. Within the order, viruses infect a very broad range of hosts, ranging from plants to animals and insects. **Tymovirales** have similar genomes, but their capsids are enveloped; these viruses are confined to plant hosts. Many viruses have been successfully assigned to one of these six orders, but a greater number of known viruses currently remain unassigned.

Table 1. Some of the properties of viruses used in taxonomy

Virion properties
    Morphology
        Virion size and shape (symmetry)
        Presence or absence and nature of peplomers
        Presence or absence of an envelope
    Physicochemical and physical properties
        Virion molecular mass (M_r)
        Virion buoyant density (in CsCl, sucrose, etc.)
        Virion sedimentation coefficient
        pH and thermal stability
        Cation stability (Mg^{2+}, Mn^{2+})
        Solvent and detergent stability
        Irradiation stability
    Genome composition and organization
        Type of nucleic acid (DNA or RNA)
        Size of genome in kb/kbp
        Strandedness: ss or ds
        Linear or circular
        Sense (positive-sense, negative-sense, ambisense)
        Number and size of segments
        Nucleotide sequence
        Number and position of open reading frames
        Presence of repetitive sequence elements
        Presence of isomerization
        G+C content ratio
        Presence or absence and type of 5′ terminal cap
        Presence or absence of 5′ terminal covalently linked protein
        Presence or absence of 3′ terminal poly (A) tract
    Proteins
        Number, size, and functional activities
        Details of special functional activities, e.g. transcriptase, reverse transcriptase, hemagglutinin
        Neuraminidase and fusion activities
        Amino acid sequence or partial sequence
        Glycosylation, phosphorylation, myristylation property of proteins
        Epitope mapping
    Lipids and carbohydrates
        Content, character, etc.
    Replication
        Strategy of replication
        Transcriptional and translational characteristics
        Site of accumulation of virion proteins
        Site of virion assembly
        Site and nature of virion maturation and release
Antigenic properties
        Serologic relationships, especially as obtained in reference centers
Biologic properties
        Natural host range
        Mode of transmission in nature
        Vector relationships
        Geographic distribution
        Pathogenicity, association with disease
        Tissue tropisms, pathology, histopathology

Table 2. Taxonomic chart of selected virus families

Order	Main characteristics	Typical members	Diseases caused
Mononegavirales	Nonsegmented negative-sense ssRNA enveloped capsid	Measles virus Rabies virus Ebola virus	Measles Rabies Ebola (hemorrhagic fever)
Nidovirales	Nonsegmented positive-sense ssRNA enveloped capsid	SARS coronavirus	Severe acute respiratory syndrome
Caudovirales	Nonsegmented linear dsDNA nonenveloped capsid	(No human viruses)	
Herpesvirales	Nonsegmented linear dsDNA enveloped icosahedral capsid	Herpes simplex virus	Cold sores, genital infections
		Varicella-zoster virus	Chickenpox, shingles
		Cytomegalovirus	Mild fever; severe disease in immunocompromised patients
		Epstein–Barr virus	Glandular fever; association with certain malignancies, e.g. Burkitt's lymphoma
Picornavirales	Nonsegmented positive-sense ssRNA nonenveloped icosahedral capsid	Hepatitis A virus	Infectious hepatitis
		Human enteroviruses	Gastric infections
		Foot and mouth disease virus	
Tymovirales	Nonsegmented positive-sense ssRNA enveloped capsid	(No human viruses)	

Virus families

Virus families represent groups of virus genera that share the common characteristics of the order, but are distinct from other families within that order; families are designated by the suffix -*viridae*. Most of the families have **distinct virion morphology**, **genome structure**, and **strategies of replication**. Families as groupings have proved to be excellent models for classification and some (e.g. *Herpesviridae*) require subfamilies (suffix -*virinae*) due to the complex relationships between individual members. The *Herpesviridae* are further classified into three subfamilies, the *Alphaherpesvirinae*, *Betaherpesvirinae*, and *Gammaherpesvirinae*.

Virus genera

Virus genera group virus species that share the common characteristics of the family, but are distinct from other genera within that family; genera are designated by the suffix -*virus*. For example, the genera *Simplexvirus* and *Varicellovirus* are distinct members of the *Alphaherpesvirinae* subfamily. The criteria for designating genera vary from family to family, but include genetic, structural, and other differences.

Table 2. Taxonomic chart of selected virus families – *continued*

Unassigned	Family	Typical members	Diseases caused
	Poxviridae	Vaccinia virus	None known; rare vaccine-induced encephalitis
		Variola major	Smallpox (now eradicated)
	Papillomaviridae	Human papilloma viruses	Warts, association with some cancers (e.g. cervical cancer)
	Hepadnaviridae	Hepatitis B virus	Serum hepatitis, association with hepatocellular carcinoma
	Orthomyxoviridae	Influenza virus	Influenza
	Picornaviridae	Poliovirus	Poliomyelitis
		Rhinovirus	Common cold
	Togaviridae	Rubella virus	German measles
	Flaviviridae	Yellow fever virus	Yellow fever
		Hepatitis C virus	Serum hepatitis, association with hepatocellular carcinoma
	Bunyaviridae	Hantavirus	Hemorrhagic fever
	Retroviridae	Human T lymphotrophic virus 1 (HTLV)	Adult T-cell leukemia, lymphoma
		Human immunodeficiency virus (HIV)	Acquired immune deficiency syndrome (AIDS)

Virus species

A definition of this taxonomic level has proven troublesome, and the ICTV currently defines a virus species as 'a polythetic class of viruses that constitutes a replicating lineage and occupies a particular ecological niche'. Essentially, classification at the level of virus species cannot rely on any single property, but on a combination of shared characteristics. Assigning viruses to a final division, that of virus strain, is likewise difficult and generally reflects the sharing of distinctive characteristics that are too minor to merit distinction at the species level.

K3 Virus genomes

Key Notes

Types of viral genome

Viral genomes are diverse in size, structure, and nucleotide composition. They can be linear or circular molecules of dsDNA, ssDNA, dsRNA or ssRNA and they can be either segmented or nonsegmented. Single-stranded genomes may be wholly positive (gene-coding) or negative (anti-coding) sense, or may combine regions of both.

Techniques of study

Traditionally genomes were studied by a range of techniques that included restriction enzyme digestion, buoyant density analysis, thermal denaturation, nuclease sensitivity, and electron microscopy (EM) imaging. Advances in molecular techniques, particularly nucleotide sequencing, have allowed rapid and extensive analysis of viral genomes and their coding potential.

Baltimore classification of viral genomes

Viruses can be grouped according to the properties of their genome described above. In combination with a consideration of their gene expression strategy, these characteristics form the basis of the Baltimore classification system.

DNA viruses

Viruses containing dsDNA genomes can reach large sizes, with ssDNA viruses usually being smaller. In addition to their protein-coding sequences, many larger genomes (e.g. *Herpesvirus*, *Adenovirus*) have repetitive sequences that are involved in regulation or genome copying. Small viruses maximize their limited genome by the use of overlapping genes and regulatory sequences.

RNA viruses

Viruses with RNA genomes are very diverse; many have a single-stranded genome but some are dsRNA. Of the ssRNA viruses, some act directly as mRNAs, initiating protein synthesis very rapidly, while others require copying of the genome first. Segmented viral genomes (e.g. *Reovirus*, *Orthomyxovirus*) are divided into two or more physically distinct molecules of nucleic acid packaged into a single virion.

Related topics

(F3) DNA replication
(F4) Transcription

(F5) Messenger RNA and translation
(K2) Virus taxonomy

Types of viral genome

The viral genome carries the nucleic acid sequences responsible for the **genetic code** of the virus and, logically, larger genomes carry more genes, allowing these viruses to encode greater numbers of proteins, which may be structural (part of the virion) or non-structural. In infected cells the genome must be **transcribed** and **translated** into proteins, which requires the recognition of protein-encoding RNA molecules (messenger RNAs or **mRNA**) by cellular ribosomes. The processes of gene expression need to be appreciated in order to understand virus replication; they are well described in Section F of this book, and also more fully within the *Instant Notes Molecular Biology* volume. The virus genome, or a complementary copy of it, must also serve as a template from which new genomes can be synthesized; these ultimately assemble with viral proteins into progeny virions. Viral genomes can be of **DNA** or **RNA** in either **single-** or **double-stranded** form, although some viruses (e.g. hepatitis B virus) have strands of unequal size, so their genomes are only partially double-stranded. Genomes can be **linear** or **circular** in structure and some viruses spread their information between separate nucleic acid molecules (i.e. they can be **segmented**) that must all be present for successful infection of a host. Virus genomes vary enormously in size; the smallest virus infecting humans is hepatitis B virus, with a genome size of just 3300 nucleotides, whereas some herpesvirus genomes are over 500 000 nucleotides. The largest virus genome sequenced to date is that of the marine mimivirus, at over 1 181 000 nucleotides; larger than the genome of some bacteria. The gene sequences within a virus genome must ultimately be translated into virus proteins and this requires either direct recognition of the viral genome as a messenger RNA (mRNA) by cellular ribosomes, or the conversion of viral genetic information into new mRNA transcripts within the infected cell. There are to date seven known schemes by which viruses provide mRNAs for translating their genomes into proteins and these distinct strategies form the basis of the Baltimore classification of viruses mentioned in the previous section and outlined in more detail below (Figure 1).

Double-stranded (**ds**) genomes are generally more stable, which is reflected in the range of sizes found for viruses of each type. The largest virus genomes are of dsDNA (e.g. *Poxviridae*, *Herpesviridae*, and the mimivirus mentioned above) with single-stranded (**ss**) genomes found in smaller sizes, although the use of segmented genomes allows more information to be encoded in multiple smaller (and hence less fragile) genome segments. As indicated above, all segments must of course be present in a single virion for infectivity. Single-stranded genomes are further categorized as being **positive** (+) sense or **negative** (−) sense. This refers to the relationship of the genome sequence to that of the mRNA representing its coding capacity. Where the genome sequence is the same (excepting the presence of thymidine instead of uracil) as that of the encoded mRNA, this is known as gene-sense or positive-sense. Conversely, genome sequences that are complementary to the mRNA transcript are known as anti-sense or negative-sense. This has significance in ssRNA virus genomes; many that are (+) sense can be directly recognized as mRNAs and translated by cellular ribosomes immediately they enter a suitable cell. These genomes are said to be **infectious** as they are capable of initiating virus replication even in the absence of any virus proteins. Viruses with an infectious (+)ssRNA genome include poliovirus (*Picornaviridae*), SARS (*Coronaviridae*), *yellow fever virus*, *dengue virus* and hepatitis C virus (all *Flaviviridae*), and rubella (*Togaviridae*). A notable exception is the human immunodeficiency virus (HIV; *Retroviridae*), which has a **noninfectious** (+)ssRNA genome, and all negative-sense RNA genomes are noninfectious. The genomes of these viruses must first be transcribed into mRNAs and this requires virus enzymes (cellular enzymes cannot transcribe from an RNA template), which must be brought into the cells within the virus capsid. Such viruses include the influenza (*Orthomyxoviridae*), measles

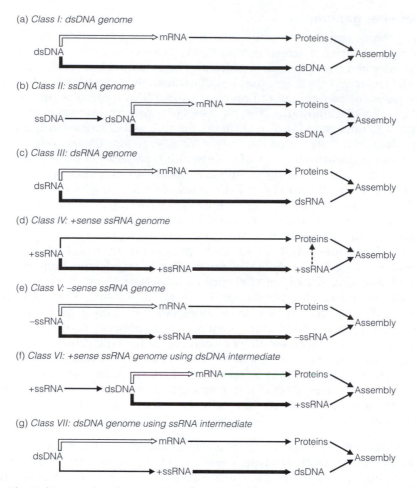

(a) *Class I: dsDNA genome*

(b) *Class II: ssDNA genome*

(c) *Class III: dsRNA genome*

(d) *Class IV: +sense ssRNA genome*

(e) *Class V: –sense ssRNA genome*

(f) *Class VI: +sense ssRNA genome using dsDNA intermediate*

(g) *Class VII: dsDNA genome using ssRNA intermediate*

Figure 1. The Baltimore classification identifies several routes by which viruses with different genome compositions achieve mRNA transcription (shown as open arrows) and genome replication (thick arrows).

and mumps (*Paramyxoviridae*) viruses and the highly virulent rabies (*Rhabdoviridae*) and Ebola (*Filoviridae*) viruses. Some ssRNA viruses have regions of both positive- and negative-sense sequence and are known as ambisense genomes; they are noninfectious genomes, and include Lassa fever virus (*Arenaviridae*).

Techniques of study

Advances in molecular biology have given rise to techniques that have superseded more traditional methods for studying nucleic acids such as melting temperature (T_m) and cesium chloride or sucrose gradient buoyant density analyses. In particular, nucleic acid sequencing has made the analysis of viral genomes simpler, more efficient, and rapid. Viral genomes from most of the families have been totally sequenced and their genes identified, allowing the prediction, and in many cases characterization, of viral gene products. Bioinformatic comparisons of virus genes with known sequences stored on computer databases have revealed fascinating similarities between viral and cellular genes that have been conserved during the co-evolution of viruses and their hosts. Virus genes have been cloned into a variety of vectors and analyzed by a number of techniques (e.g. site-directed mutagenesis) to study their role during virus replication and host infection and immunity.

A common technique still used for studying DNA virus genomes is to digest them using **restriction enzymes**, which cut DNA into small fragments at specific nucleotide sequences. The fragments are separated according to size using agarose gel electrophoresis, and this allows a characteristic restriction enzyme **map** to be derived for each virus genome. Direct digestion of RNA genomes is not possible, but with the discovery of the viral enzyme **reverse transcriptase**, it is possible to make a DNA copy of an RNA genome; this complementary DNA (**cDNA**) can then be digested with restriction enzymes to provide a genome restriction map.

Baltimore classification of viral genomes

The **Baltimore** classification of viruses considers several parameters for the grouping of related viruses, based on genomic properties including: (1) the **composition of the nucleic acid** (i.e. DNA or RNA); (2) the **number of strands** (i.e. single- or double-stranded); (3) the **sense of single-stranded RNA genomes** (i.e. positive or negative); (4) the **strategy for obtaining mRNA transcripts for gene expression**. Each of the seven classes of virus, as defined by this classification scheme, shows one or more distinctions within these criteria. The system does have its limitations (for example, ssDNA viruses are not further divided according to genome sense or segmentation, and ambisense genomes are not classed separately) but it nevertheless forms a useful consideration in the determination of virus taxonomy. The genomic characteristics of the different classes are summarized in Figure 1.

DNA viruses

Class I viruses all have dsDNA genomes, which require the transcription of mRNA from the genome molecule; the virus may use the host cell RNA polymerase II enzyme (e.g. the *Papillomaviridae*, which can give rise to warts) or may encode its own (e.g. the *Herpesviridae*). A few Class I viruses (those within the family *Polydnaviridae*) have a segmented genome but most have a nonsegmented genome.

Herpesviridae and *Poxviridae* are among the largest Class I viruses and both families contain members that infect humans. For example the herpes simplex virus genome is 152–154 kbp and that of cytomegalovirus is 235 kbp, while the now eradicated smallpox virus and its relative vaccinia virus have genomes of approximately 190 kbp. The large genomes of these viruses encode many proteins and their virions are correspondingly complex. The structure of herpesvirus genomes is unique in that several **isomers** of the same molecule can exist due to the presence of one or more regions of **repetitive** (identical) **sequences**, covalently joined to **unique sequence** regions. For example, within the genome of herpes simplex virus two unique sections (of differing length; a **unique long** (U_L) and a **unique short** (U_S) region) are each flanked by **inverted repeats**, the **terminal repeat long** (TR_L) and **terminal repeat short** (TR_S) regions. These repeats give the genome **terminal redundancy** and allow structural rearrangement of the unique regions that gives rise to four isomers, all of which are functionally equivalent (Figure 2).

The genomes of *Adenoviridae* are smaller, with sizes ranging between 30 and 38 kbp and also have terminal redundancy; each strand has 100–140 bp of **inverted repeat** at the ends, which allows the denatured single DNA strands to self-anneal to form **pan-handle** structures, an essential step during replication. Adenovirus genomes are covalently

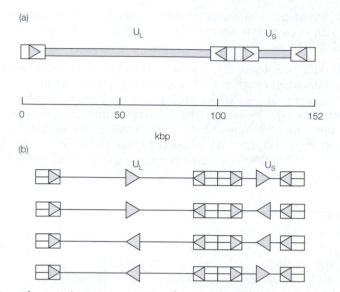

Figure 2. (a) Some herpesvirus genomes (e.g. herpes simplex virus) consist of two covalently joined sections of unique sequence, U_L and U_S, each flanked by inverted repeats, represented by arrowed boxes. (b) This organization permits the formation of four different forms of the genome. Adapted from Cann A (2005) *Principles of Molecular Virology*, 4th ed. With permission from Academic Press.

linked to a 55 kb **terminal protein** at the 5′ end of each DNA strand and this protein acts as a primer for the synthesis of new DNA strands (Figure 3).

The family *Papillomaviridae* (e.g. human papillomavirus 16 and 18, which are associated with cervical cancer) have very small genomes (about 8 kbp) that form a **supercoiled** circular structure and associate with four cellular histone proteins (H2A, H2B, H3, and H4). These viruses have between 8 and 10 genes, but even this small coding capacity is impressive for such a small DNA molecule and is achieved by several mechanisms. Firstly, papillomavirus genomes contain overlapping gene sequences (Figure 4) and use shared promoter sequences (there are only seven within the genome). Secondly, the virus uses **splicing** to generate multiple, different mRNAs from individual primary RNA transcripts, a technique used by many viruses (RNA splicing was in fact discovered as a result of studies on virus genomes). Thirdly, these viruses encode proteins with multiple

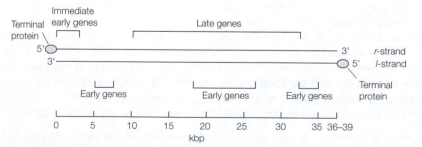

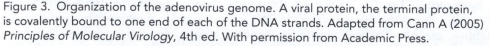

Figure 3. Organization of the adenovirus genome. A viral protein, the terminal protein, is covalently bound to one end of each of the DNA strands. Adapted from Cann A (2005) *Principles of Molecular Virology*, 4th ed. With permission from Academic Press.

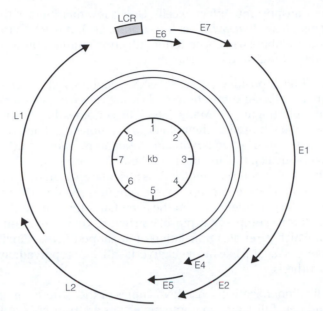

Figure 4. The organization of the small papillomavirus genome contains overlapping genes (arrowed lines) with shared promoters, including the regulatory sequences in the long control region, LCR.

functions, such as the E2 protein of human papillomavirus 16, which has functions linked with genome transcription, replication, segregation, and packaging. Many small virus genomes use these strategies and others, for example, *Polyomaviridae* use both DNA strands to encode proteins, so that a section of genome can be used in both directions. The larger virus genomes also use some or all of these strategies to an extent, but the majority of their genes are distinct and possess their own promoters.

Class II virus genomes are ssDNA and can be of positive or negative sense, and either circular or linear. All require the synthesis of a dsDNA intermediate from which transcription of their mRNA proceeds. Most ssDNA viruses are nonsegmented, but some are segmented. Furthermore, some viruses within the *Begomovirus* genus (Family *Geminiviridae*) have two genomic segments that are packaged separately into two incomplete but joined icosahedral capsids.

Parvovirus genomes are linear nonsegmented ssDNA molecules of about 5 kb and like several of the Class I virus genomes they show terminal redundancy. The ends of parvovirus genomes have **palindromic sequences** of about 115 nucleotides that self-anneal to form looped dsDNA regions, essential for the initiation of genome replication. Following genome replication, most newly assembled virions contain DNA strands of negative (–) sense, but some can contain positive (+) sense molecules. These very small genomes contain only two genes: **rep**, which encodes proteins involved in transcription, and **cap**, which encodes the coat proteins. Originally only one parvovirus was known to infect humans (parvovirus B19), but since 2005 several new viruses have been identified, associated mostly with gastrointestinal infections.

RNA viruses

All **Class III** virus genomes (dsRNA) are segmented. Transcription of mRNAs is from the genome molecules and requires a virally encoded RNA polymerase that must be packaged into the virion for transport into infected cells. Only a few members of the *Rotavirus* and *Orthoreovirus* genera (Family *Reoviridae*) infect humans, but dsRNA viruses are a wide-reaching class and members have been found infecting animals, birds, fish, and insects and also plants, algae, fungi, and bacteria.

Class IV viruses all have positive (+) sense ssRNA genomes that can serve directly as mRNA transcripts for protein translation, and as such they possess a range of features that promote this role (Figure 5). Many (+)ssRNA genomes have 7-methylguanine, a modified ribonucleotide, bound at their 5′ end. This molecule, known as the **universal cap**, protects the 5′ terminus of eukaryotic mRNAs and promotes mRNA translation by ribosomes in eukaryotic cells. In members of the *Caliciviridae* and *Picornaviridae* (e.g. poliovirus, hepatitis A virus) the cap molecule is absent, replaced by a small viral protein called VPg. Translation of these uncapped virus genomes requires the presence of additional sequences in the genome (between the 5′ end and the start of the coding region) that self-anneal to form a complex hairpin-like structure known as an **internal ribosome entry site** (IRES), which promotes translation of uncapped RNAs. Finally, the 3′ end of most Class IV genomes (again, like most eukaryotic mRNAs) is **polyadenylated** to ensure faithful protein synthesis.

During virus replication, negative-sense RNA copies of the genome are produced, from which mRNAs and new full-length genome molecules are transcribed. Transcription of all viral RNAs requires a virus-encoded RNA polymerase, synthesized rapidly in cells by translating the invading genome. Most Class IV genomes are small, ranging between 4000 and 13 000 nucleotides, although the SARS coronavirus (SARS-CoV) and its relatives are much larger, reaching lengths of 28 000–31 000 nucleotides.

Class V viruses all have (–)ssRNA genomes and many have segmented genomes. The genomes are not translated directly by ribosomes, so protein synthesis first requires transcription of mRNA from the genomic RNA; a few molecules of viral RNA polymerase are incorporated into the virion for this purpose. Interestingly, many viruses pathogenic to humans are found in this class, including the viruses responsible for mumps, measles, rabies, and influenza, and various hemorrhagic fevers such as Ebola, Marburg, and Lassa fever (the latter has an ambisense genome).

An important Class V family with segmented genomes is the Family *Orthomyxoviridae*, containing the influenza viruses, including those that infect humans. These viruses have seven or eight genome segments, each coding for one or two viral proteins (Table 1) and while a full complement of genome segments is necessary for infectivity, the source of each segment can be changed. This is known as **reassortment**, and allows the genomic segments of two different (but related) influenza viruses to swap, giving rise to new combinations. This sudden emergence of a 'new' virus genotype leads to the epidemics and pandemics of influenza that we have seen many times during our history.

On first appearance, **Class VI** virus genomes are like those of some Class IV viruses; they are of (+)ssRNA and possess both a universal 5′ cap and a 3′ polyadenylated tail (Figure 5). There are two RNA molecules within each virion, but they are identical and both carry the full coding potential of the virus, hence the genome is referred to as **diploid**, and is considered to be nonsegmented. The genomic RNA, however, is not translated by ribosomes on entry into host cells but instead undergoes a complex replication strategy whereby it is first copied into a dsDNA molecule and then permanently cloned into the host cell chromosome. These steps require several virus-encoded enzymes (including a novel DNA polymerase, **reverse transcriptase** (**RT**), and an **integrase**) that must be brought into the cell along with the genome. Hence like Class V viruses, a few copies of these

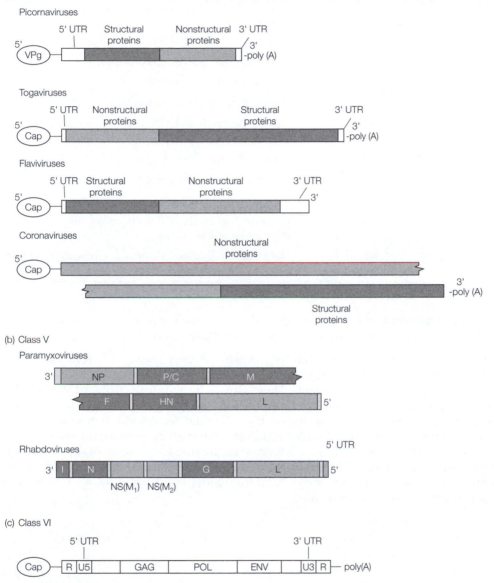

Figure 5. Diagrammatic representation of selected RNA virus genomes from (a) Class IV, (b) Class V, and (c) Class VI viruses. The position of structural and nonstructural genes is indicated, along with an indication of terminal structures or bound proteins. Untranslated regions (UTRs) do not encode viral proteins and usually contain regulatory sequences. Adapted from Cann A (2005) *Principles of Molecular Virology*, 4th ed. With permission from Academic Press.

enzymes are found within the virion itself. Following integration of the dsDNA genome into a chromosome, only cellular enzymes are needed for transcription of viral mRNAs to allow viral protein synthesis to proceed. The most well studied virus family in the class is the *Retroviridae*, which includes the human immunodeficiency virus, HIV. More recent discoveries of related viruses in invertebrate and fungal species have been assigned to the

Table 1. Segments of the influenza virus genome

Segment	Size (nt)	Polypeptide(s)	Function
1	2341	PB_2	Transcription of mRNA: host cap binding
2	2341	PB_1	Transcription of mRNA (polymerase)
3	2233	PA	Replication of genome RNA
4	1778	HA	Hemagglutinin: envelope fusion
5	1565	NP	Nucleoprotein: RNA binding; part of transcriptase complex
6	1413	NA	Neuraminidase: virus egress
7	1027	M_1	Matrix protein: major component of virion
		M_2	Integral membrane protein – ion channel
8	890	NS_1	Nonstructural (nucleus): affects cellular RNA transport; disables innate immunity
		NS_2	Nonstructural (nucleus + cytoplasm): mediates export of viral RNA from cell nucleus

families *Metaviridae* and *Pseudoviridae*. Although discovered more recently, it is likely that these viruses are the ancestors of the vertebrate retroviruses.

The most recently described **Class VII** contains viruses reassigned from Class I and comprises only (to date) the *Hepadnaviridae*, which includes hepatitis B virus, and the *Caulimoviridae*, which infect plants. These viruses have only partially dsDNA genomes (some regions are ssDNA) that are transcribed to mRNA by cellular RNA polymerase II once the genome has been made fully dsDNA. Like the *Retroviridae* they encode reverse transcriptase, which is active late during infection to synthesize new genomic DNA molecules from RNA templates. Hepatitis B virus has the smallest genome of any virus infecting humans (encoding only seven proteins from gene sequences) but it is estimated that there are 350 million people chronically infected with the virus. Clearly in the virus world, size (or complexity) isn't everything!

K4 Virus proteins

Key Notes

Overview

Viral proteins, encoded by the viral genome, are either structural (capsid, matrix, envelope) or nonstructural (e.g. enzymes, oncoproteins, inhibitors of cell macromolecular synthesis, transcription factors, etc.). They may be essential for replication of new virions, or nonessential, contributing to virus infection within cells.

Methodology

Virus proteins are studied directly using techniques such as sodium dodecyl sulfate (SDS)–polyacrylamide gel electrophoresis, Western blotting, immunoprecipitation, and pulse–chase experiments, and by the use of proteases or inhibitors of post-translational modifications. Protein structures can be derived by X-ray diffraction, or bioinformatic comparison with pre-existing databases, which can also provide information about possible functions.

Protein synthesis and complexity

Viral proteins are synthesized by the translation of viral mRNAs by cellular ribosomes. Proteins are often processed following synthesis (e.g. proteolytic cleavage, glycosylation, myristoylation, acylation, phosphorylation, and palmitoylation). In many cases, virus protein synthesis is controlled at the levels of transcription and translation, which itself is influenced primarily by the virus genome.

Structural proteins

Structural proteins are those that form the capsid, nucleocapsid, matrix or envelope components of a virion. They have a role in protecting the viral genome and delivering it into host cells by protein–protein interactions with cell surface receptors. In most viruses, structural proteins are produced in abundance late in the replication cycle.

Nonstructural proteins

These proteins do not form part of the virion architecture, although they may be carried within the virion; rather they have enzymatic activities necessary for virus infection. Most have a key role within the infected cell, such as polymerase, protease or kinase activities, DNA-binding activity or gene regulation. Many viruses have proteins that inhibit host cell nucleic acid and protein synthesis in favor of viral gene expression.

Related topics

(F5) Messenger RNA and translation
(K7) Virus replication
(K9) Viruses and the immune system

(K10) Virus vaccines
(K11) Antiviral chemotherapy

Overview

The **coding capacity** of viral genomes varies from below 5 to over 100 genes. This in turn is reflected in the different complexities of virus particles and their respective replication cycles. Viral proteins are either **structural** (part of the virion architecture) or **nonstructural** and of these some are found in the virion (usually enzymes) while others are present only within infected cells. Nonstructural proteins can have a range of functions, often mimicking those of cellular enzymes, and many are multifunctional.

Virus genes and proteins are often referred to as being **essential** or **nonessential**. The former are an absolute requirement for a virus to complete its replication cycle and to produce infectious virions. Genes or proteins that can be deleted from the virus without seriously affecting replication in cell culture are viewed as nonessential. However, replication in cell culture may not reflect the *in vivo* situation and it may be more appropriate to consider such genes and proteins as having **auxiliary** functions rather than having a redundant role. Indeed many auxiliary proteins have been characterized with important functions that contribute to virus replication, mask or inhibit immune responses, or promote transmission to new hosts. For example, several members of the insect-specific *Baculoviridae* synthesize a **chitinase** enzyme within infected cells that plays no role in the replication of new virus. However, its activity degrades the chitin exoskeleton of an infected caterpillar, liquefying the animal to promote increased transmission to new hosts by spreading the virus across the soil and foliage.

Methodology

In recent years, the range and sophistication of protein analysis tools have exploded and only a few can be mentioned here. The study of virus structural proteins traditionally required the **purification of virions** from infected cells or infected-cell supernatants, free from contaminating host (or nonstructural) proteins. This can be achieved by a series of **differential centrifugation** steps, followed by **sucrose gradient-density centrifugation** (centrifugation through a sucrose gradient usually results in a sharp band of virus at a specific location on the gradient). The most commonly used technique to study virus proteins is **SDS–polyacrylamide gel electrophoresis** (separation based on size), usually accompanied by either **Western blotting** or **immunoprecipitation** (both require the binding of virus proteins by highly specific antibodies).

The location of virus nonstructural proteins within cells is a useful indicator of their possible function, and many methods are used to detect proteins within cells, primarily **differential staining**, **immunofluorescence**, and **confocal microscopy** techniques. These techniques visualize, microscopically, the distribution within infected cells of fluorescent stains or fluorescence-labeled antibodies bound specifically to virus proteins. Other tools that can help characterize protein function include **pulse–chase** experiments (to determine when proteins are made in cells), **protease** or **glycosylation inhibitors** (to indicate how proteins are processed), and **co-immunoprecipitation** (to determine if individual proteins interact with others).

Bioinformatics (the **prediction of amino acid sequences** from gene sequencing) is commonly used now to infer virus protein structure and function, including the identification of probable functional **domains** within proteins. While not exhaustive (or all inclusive), there are recognizable amino acid sequences (**motifs**) that carry key functions such as sites for protease cleavage, motifs for spanning membranes, for the addition of modifications (e.g. glycosylation) or for targeting the protein to specific cellular locations or for secretion from the cell entirely. Bioinformatic comparison of conserved motifs in genes (and proteins) has led to an understanding of the relatedness of proteins from different viruses and even cells, often revealing a close association and co-evolution of a virus and its host. With the development of gene sequencing and cloning, and of highly efficient protein expression systems, individual proteins can now be examined in isolation by cloning of their gene sequences. Often, the cloning steps include addition of extra amino acids that serve as a **tag**, to aid in purification or detection of the protein.

Protein synthesis and complexity

As outlined earlier (Section F5), all proteins are synthesized from mRNA templates translated on the host cell ribosomes; the source of the mRNA varies between viruses (Section K3). During replication virus-specific proteins are generally synthesized in 'phases' categorized as **early** and **late**. The virus proteins translated in infected cells before the virus begins to copy its genome are known as **early proteins**, and the genes that encode them are **early genes**. Often this set of genes and proteins is further divided into **immediate early** (IE) and **delayed early** (DE). As their name implies, IE genes are transcribed (and hence IE proteins are synthesized) immediately a virus genome is released into the cell. Transcription of IE genes is usually carried out by the cellular enzyme RNA polymerase II, and no virus proteins other than those brought in by the virion are required. Delayed early proteins are synthesized a little later, because the expression of DE genes needs the activity of the newly made IE proteins. In the same way **late** (L) proteins are only synthesized after both the IE and DE proteins are available in the cell. Late protein production usually coincides with the start of the virus genome replication processes, which also need both IE and DE proteins to begin. Hence viruses generally show very strict **temporal control** of their protein synthesis, mostly regulated at the level of transcription of their genes. Some viruses regulate protein expression differently (such as poliovirus and HIV), but although the proteins of these viruses cannot be called early or late, their synthesis is nevertheless tightly controlled.

We have mentioned the use of overlapping gene sequences, shared promoters, and multifunctional proteins as mechanisms viruses use to maximize their small coding capacity. Hepatitis C virus, poliovirus, and HIV synthesize several proteins as a single large **polyprotein precursor** molecule, avoiding the need for separate transcriptional or translational sequences. The individual virus proteins must be released from the polyprotein by **proteolytic cleavage** by an enzyme that is part of the polyprotein itself. Like cellular proteins, virus proteins may be highly processed after their synthesis, such as being cleaved or gaining small molecules as various side groups. Two very common modifications include **phosphorylation** (the addition of phosphate groups) and **glycosylation** (the addition of sugar groups). Both modifications are highly specific, targeting certain amino acids, and often the modification is essential to the correct functioning of the protein. Other virus protein modifications include **myristoylation**, **acylation**, and **palmitoylation**, often associated with virion envelope glycoproteins and potentially important in the targeting of these proteins to specific sites in the cell plasma membrane before virus budding.

Structural proteins

Whatever the complexity, size or shape of a virus particle, it is formed from an assembly of the structural proteins and serves to protect the viral genome from UV light or desiccation, which would otherwise fragment the nucleic acid, rendering the virus noninfectious. In naked virions, structural proteins form the capsid and (where present) the nucleocapsid. Enveloped virions also have structural proteins within the matrix and embedded in the membrane envelope (Figure 1). The structural proteins of a virion mediate virus delivery into cells and their major role is to provide attachment to cell receptors, as a first step to infection. Immune responses to these proteins often protect against viral infection and our understanding of this interaction forms the basis of modern vaccine design (Section K10).

The icosahedral capsid of poliovirus is relatively simple with just four proteins, VP1, VP2, VP3, and VP4 (each cleaved from a polyprotein precursor), while the helical capsid of tobacco mosaic virus is simpler again, built with multiple copies of just one protein. In simple structures such as these, the capsid and nucleocapsid proteins can self-assemble to form the virion, but assembly of more complex capsids (such as that of herpes simplex virus or the double-shelled bluetongue virus) requires **scaffolding proteins**. These act as chaperones for the early association of the structural proteins, allowing the capsid to form, but they are then dismantled and play no part in the final virion structure.

Structural proteins must interact with the nucleic acid genome during the final stages of virion formation (assembly). **Packaging** of the genome (especially a large genome) into the confined space within a virion requires complex folding of the genome and intimate association through chemical bonding with one or more of the structural proteins. For example the N protein of rabies virus (*Rhabdoviridae*) and the NP protein of influenza virus (*Orthomyxoviridae*) both have positively charged amino acids that interact with the negatively charged RNA genome molecule, to encourage **condensation** (folding) of the RNA before packaging into the nucleocapsid. The viral proteins bind with **secondary** and **tertiary structures** in the genome that form as a result of specific nucleotide **packaging sequences** that anneal within themselves, folding the genome into regions of three-dimensional structure (not unlike the IRES structures of Class IV viruses). The correct association of structural proteins and packaging sequences becomes even more important when one considers the segmented genome of influenza virus, where eight different genomic RNA segments must each be packaged into a separate helical nucleocapsid, and all subsequently packaged within the final virion envelope.

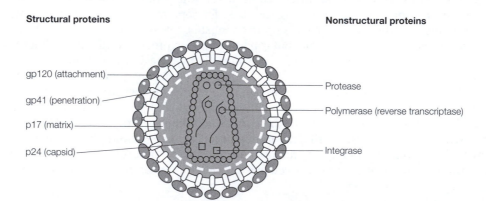

Figure 1. Virions are composed of structural proteins that enclose and protect the viral genome. The binding of structural proteins to cell receptors (attachment) initiates infection, delivering the genome into the target cell. Many viruses also carry nonstructural proteins in their virions; usually enzymes necessary for the first stages of virus replication. For example, HIV has two critical enzymes (reverse transcriptase and integrase) within its virion, and often a viral protease is present as well. Adapted from Strelkauskas A, Strelkauskas J & Moszyk-Strelkauskas D (2010) *Microbiology: A Clinical Approach*. Garland Science.

Virus envelopes contain structural glycoproteins, all of which must be targeted to the relevant cellular membrane before budding of the capsid. Many self-associate in small groups to form the tiny spikes (**peplomers**) visible on the envelope surface by electron microscopy, such as the hemagglutinin (HA, a trimer) and neuraminidase (NA, a tetramer) glycoproteins of influenza virus or the gp120 spike of HIV (a trimer; Table 1). The majority of a glycoprotein structure lies exposed on the outside of the envelope, being anchored within the envelope by a **transmembrane domain**. This part of the protein contains hydrophobic amino acids and laces back and forth across the membrane bilayer. Usually a relatively short region of the glycoprotein projects into the cell and virion interior, and is responsible for interaction with capsid or **matrix** (or **tegument**) **proteins**. The matrix proteins are not usually glycosylated but they may contain transmembrane anchor domains themselves. Hydrophobic regions can associate with the membrane or they may interact with the internal tail of the envelope glycoproteins.

Nonstructural proteins

The majority of nonstructural proteins appear only within infected cells, although a few may be carried within the virion (or help in its formation, such as the scaffolding proteins described above) but they are not part of its architecture. The number and function of nonstructural proteins varies enormously from one virus family to another, depending on the complexity of the viral genome and the replication cycle. Many are enzymes with functions critical to virus replication (e.g. the **reverse transcriptase**, **protease**, and **integrase** of HIV (Table 1), the **thymidine kinase** and **DNA polymerase** of HSV). Because they are different to cellular enzymes (even those with equivalent functions) they are key targets for antiviral drugs (Section K11). Many (e.g. HIV **Tat** (Table 1), HSV tegument proteins) have regulatory roles during the phases of virus gene transcription and others form part of nucleic acid synthesis complexes (e.g. **DNA helicase**, **DNA binding proteins**) for genome replication. The shut-off of host cell macromolecular synthesis is usually the

Table 1. Proteins of the human immunodeficiency virus (HIV) type 1

Structural proteins

gag	Polyprotein, cleaved by the viral proteinase, produces the capsid (p24) and matrix (p17) proteins
pol	Polymerase (reverse transcriptase), aspartyl proteinase, integrase, and ribonuclease
env	Polyprotein (gp160), cleaved by cellular enzymes to produce envelope proteins gp41 and gp120

Nonstructural proteins

vif	Required for production of infectious virus
vpr	Function unknown
tat	Up-regulates viral mRNA synthesis by binding to the TAR (trans-activator response element) in the transcripts from the long terminal repeat. Made from a spliced mRNA
rev	Required by mRNA transport. Made from a spliced mRNA
vpu	Virion release, receptor degradation
nef	Down-regulates CD4 and MHC, essential for virus pathogenicity

MHC, major histocompatibility complex.

function of viral nonstructural proteins and allows the cellular machinery to be devoted exclusively to virus replication; this is of course a major cause of cell death during virus infection.

There are a number of virus proteins that can act to **transform** infected cells (i.e. to effect unscheduled **mitosis**). One such example is the family *Papillomaviridae* that promotes infected cell mitosis and hence leads to the formation of warts or in some cases tumors (as do the *Polyomaviridae* and some members of the *Retroviridae*). Some of these proteins act directly to stimulate mitosis (these are **oncogenic proteins** encoded by viral **onco-genes**), others activate cellular **proto-oncogenes**, encouraging production of the cell's own mitotic stimulators, and yet others act to block cellular **tumor suppressors**, proteins that normally safeguard the cell against unwanted cell division. Nonstructural proteins have been identified with roles involving the rearrangement of the cellular cytoskeleton, with inhibition of **apoptosis** (programmed cell death, a form of resistance to infection) and the suppression or evasion of immune responses (Section K9), transmission between hosts, and with long-term (possibly lifelong) infection of a host with a virus.

K5 Cell culture and virus propagation

Key Notes

Historical perspective

The major thrust in virology research was only possible after the derivation of defined synthetic growth medium and the *in vitro* culture of eukaryotic cells. This allowed viruses to be propagated, purified, and studied outside of their hosts, and has also led to the large-scale production of purified virus particles, proteins, and vaccines.

Methods of cell culture

Cells (homogenized from tissue samples) are seeded into sterile glass or plastic vessels where the cells attach to the inner surface. Through repeated mitotic division, the number of cells expands to form a monolayer that can be subsequently developed into a long-term cell culture. Cells can be grown in a range of vessels but must be carefully maintained, free from contamination, if they are to continue dividing. Most animal cell lines grow at 37°C.

Media and buffer

Defined growth media comprise a balanced salts solution containing amino acids, vitamins, glucose, serum, a buffer, and usually a pH indicator. Such media are used routinely to support the growth of animal cells and can be made from the individual constituents but are more commonly purchased ready-made. In the majority of cases, serum is required as a supplement for cell growth.

Cell culture types

Cells *in vitro* may be considered as primary (very limited mitotic divisions), diploid (approximately 50–60 cell divisions) or continuous (potentially unlimited mitosis) cell lines. Most cell lines are adherent (requiring a surface for support) but some can grow in suspension, much like bacterial cell cultures.

Virus propagation and culture

Small aliquots of virus are added to cell cultures and, following replication, progeny virus can be harvested and assayed. Viruses may be cell-associated or released into the culture medium. Accurate quantification (titration) and proper storage (usually at −70°C) of the virus is vital.

Use of other systems

For many viruses, replication in cell culture is still not possible or is poor compared with the natural infection within a host. Influenza virus is best propagated in the allantoic cavity of embryonated chick eggs and human papillomavirus requires a complex raft culture, modeling the architecture of a stratified epithelium.

Related topics

(K6) Virus assay   (K7) Virus replication

Historical perspective

In the course of studying viruses, we have made progress in understanding the complexities of virus disease through knowledge of virus structure, replication, and host interactions. Virus research has also extended our understanding of the fundamentals of cell biology, providing insights into aspects such as mRNA splicing and protein translation, protein–protein interactions, the structure and functioning of the cytoskeleton, RNA interference, and the mechanisms of the immune system. Few of these studies would have been possible without the means of propagating viruses outside their normal hosts. This requires the ability of a virus to replicate within host cells that are themselves grown in laboratory vessels in nutrient solutions. It is worth remembering that viruses do not 'grow' in the way cells do; they replicate. For the purposes of this section, the processes that allow cells to be cultured will be referred to as cellular mitosis or division and the processes that allow the production of new virus (through replication in cultured cells) will be referred to as virus propagation or amplification. Cell culture (often called tissue culture) requires an environment in which the cells are provided with suitable nutrients and other conditions (e.g. oxygen, salts, support, space, etc.) that promote steady, regular, and continuous mitosis, achieving a doubling of cell numbers at a predictable rate. This was first achieved in the late 19th century for a number of bacteria, following the development of simple broths or **media** (Section C3) containing a few salts and sugars, but for eukaryotic cells the development of suitable media able to support continued mitosis *in vitro* did not occur until the early 1950s. Consequently early virus research was limited to bacteriophages, with an upsurge in the study of eukaryotic viruses in the last 60 years. The culture of eukaryotic cells (specifically, of animal cells) and their support of eukaryotic virus replication is the focus of this section.

There are over 3000 characterized cell lines derived from many diverse species that are kept in, for example, the American Type Culture Collection and the European Collection of Animal Cell Cultures.

Methods of cell culture

Eukaryotic cells can be cultured in the laboratory in sterile **glass** or **plastic flasks** of various sizes or in large vessels or vats (known as fermenters) for large-scale purposes. It is essential to keep the cultures free of fungal and bacterial contamination that would otherwise outgrow the cells as a result of the rich media used, destroying them. Although several antibacterial and antifungal additives are available, the primary barrier to contamination must be constant and vigilant **aseptic technique**. Reagents and equipment for use in cell culture must also be sterilized, by autoclaving (moist heat), hot-air ovens (dry heat), membrane filtration or, in the case of plastic ware, irradiation.

Once a cell line is established, it must be maintained regularly by a process known as subculturing (often called passaging, splitting or feeding the cells). Cells are harvested from the culture vessel and gently homogenized to obtain a single-cell suspension, which can then be counted, allowing a known concentration of cells to be **seeded** into a new sterile flask along with the appropriate liquid **medium** (plural = media). The flask (a plastic or glass bottle-like container) is incubated at the appropriate temperature, usually 37°C for mammalian cells. For adherent cells, the flask is incubated lying on one side and the cells attach to the inner surface of the flask wall; modern plastic culture flasks have a chemical coating to help rapid and secure adherence of cells. Once attached to the vessel wall, the cells begin to metabolize the available nutrients and soon begin to divide, forming a **monolayer** of cells over several days. The metabolic activity of the cells means that the medium will eventually be depleted of nutrients, the levels of metabolic waste products will rise, and the available surface for cell adherence will be filled (the monolayer is said to be **confluent**). These conditions are inhibitory to mitosis and, unless subcultured again, the cells will deteriorate and die. The monolayer is treated with trypsin or versene, which breaks the protein bonds that allow cell adherence to the flask and the processes described above are repeated. In addition to culturing in flasks (for long-term maintenance of the cell line and virus propagation), smaller cell monolayers can be established temporarily in the wells of specially designed multi-well culture plates, which allows for experimental infections examining virus replication, biology, and host cell interactions (Figure 1). More recently, there has been a move to encourage some cell types to grow in **suspension**, where cells do not anchor themselves to the surface of the flask or adhere to each other (e.g. hybridoma cells which secrete monoclonal antibodies).

Two natural phenomena restrict cell mitosis *in vitro*. Most cell lines exhibit **contact inhibition** whereby they suspend mitosis when in contact with surrounding cells, meaning that a monolayer remains static once it has reached **confluence**. The process of subculturing dilutes the cells into fresh vessels, giving them space to continue mitosis and maintain the cell line. The second restriction is the natural lifespan of cells, which can only undergo a defined number of mitotic divisions (known as the Hayflick limit, this is approximately 50 divisions for human cells). This limited lifespan of cells inevitably imposes a defined lifespan on a cell line, after which it can no longer continue. To preserve the cell line it is usual to store samples of cells taken from an early subculture (e.g. at a passage number between 10 and 15) by freezing them in liquid nitrogen. This allows cells to be preserved in a metabolically inert, but viable, state and to be resurrected by rapid thawing and returning to a culture flask with medium once again. Cells that have been transformed (such as those derived from tumors) do not show these limitations. Without contact inhibition cells continue dividing, piling up on top of each other, after confluence has been reached. Transformed cells also have no apparent Hayflick limit,

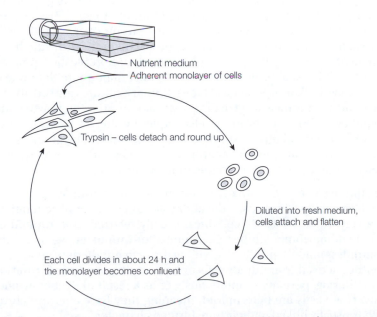

Figure 1. Cell culture. From Dimmock N & Primrose SB (1994) *Introduction to Modern Virology*, 4th ed. With permission from Blackwell Science.

and hence a greatly increased (possibly limitless) mitotic lifespan. These are often known as immortal cell lines

Media and buffers

Most media in use today are chemically defined, and are supplemented with 5–20% serum (which contains stimulants and other components necessary for cell division), although serum-free medium with added stimulants is used for an increasing number of cell types. Media contain an isotonic **balanced salts solution** with amino acids, vitamins, and glucose. In addition to serum, media can also be supplemented with antibiotics (usually penicillin and streptomycin) to help prevent bacterial contamination. Generally, mammalian cells divide well between pH 7.0 and 7.4 but cell metabolic processes affect this narrow range rapidly – exposure to oxygen causes the pH to rise and metabolic production of lactic acid causes the pH to fall. A buffer is thus incorporated into the culture medium, commonly sodium bicarbonate, in conjunction with an incubator atmosphere containing 5% CO_2. These compounds work together to absorb or release hydrogen (H^+) ions, maintaining the pH of the medium within a narrow range. It is critical to monitor the changing pH of a culture, and most media contain phenol red as a pH indicator. The compound allows for a visual measurement of pH and it is red at pH 7.4. As the cells divide over several days the production of lactic acid outstrips the buffering capacity of the bicarbonate/CO_2 system, and the indicator becomes orange at pH 7.0 and yellow at pH 6.5. A failure of cell division within the culture would mean that the medium would become more alkaline, hence the indicator would become pink at pH 7.6 and then purple at pH 7.8.

Cell culture types

Primary cell cultures are those that have undergone little or no mitotic division following the seeding of cells harvested directly from the tissue of origin. The tissue source for the majority of primary cell cultures is either laboratory animals (e.g. monkey kidney cells) or human pathology specimens (e.g. human amnion cells). Tissue samples are incubated with a proteolytic enzyme (usually trypsin) to produce a single-cell suspension. The harvested cells are then seeded into appropriate flasks with suitable medium. Cells in culture usually show epithelial (cuboid-shaped) or fibroblastic (spindle-shaped) morphology, and as it is usual for primary cultures to contain more than one cell type, they also usually show both morphological types. Primary cultures are sensitive to a wide range of viruses and are routinely used in diagnostic laboratories for growth of fresh virus isolates (from patients). However, primary cells frequently alter after only a few divisions, and many die or lose their ability to support virus replication. Hepatocytes are a good example; they are the natural target of hepatitis C virus infection in the body, but primary hepatocyte cell cultures do not support replication of this virus.

Many cells in culture will continue to divide for more than four to five generations, giving rise to a **cell line** but will eventually die after a certain number of generations, as the cells reach the Hayflick limit. These cell lines, usually referred to as **diploid cell lines**, have the normal chromosome number. **Continuous** or **immortal cell lines** are capable of potentially limitless mitosis in culture, dividing through hundreds of generations. Such cell lines are either derived from tumor cells or have undergone **transformation** during the process of culturing, perhaps spontaneously or as a result of deliberate chemical or viral treatment. These cells are **heteroploid**, meaning they have aberrant chromosome number, which is usually linked to their transformed character.

Virus propagation and culture

Most experiments in virology have required virus replication in cultured cells, although today many experiments rely entirely on the cloning of individual genes and the expression of proteins in cell cultures without full virus infection. Historically, those viruses that replicate well in cell culture have been studied in the greatest detail. Lack of *in vitro* propagation has seriously curtailed progress in research, vaccine production, and development of antiviral drugs for many viruses.

Viruses are amplified within cultures to create **virus stocks** that are then usually stored at –70°C, although some viruses remain stable at –20°C or even 4°C. It is important to record when multiple virus stocks are made from each other and to exercise caution when continuously propagating new virus stocks. The high number of genomic generations achieved during each amplification means that mutations can accumulate rapidly, potentially altering the virus over subsequent generations. It is preferable to return to an earlier stock of the virus for the production of new stocks, rather than using the next generation each time. Minimizing the level of mutation during virus stock preparation can be achieved by infecting only a low proportion (1–10%) of the cells in the culture. This is known as using a low **multiplicity of infection** (**m.o.i.**) whereby virus is added to a level equivalent to only 0.01–0.1 infectious virions per cell. The virions infect a low number of cells, leaving the remainder to continue mitosis for 24–48 hours. Following replication in the subset of infected cells, newly formed virions infect the remaining cells, resulting in significant amplification of the virus. Once replication has been achieved in the majority of cells, virus is harvested in one of two ways. Extracellular virus that has been released from cells by budding or exocytosis can be harvested directly from the culture medium. Intracellular virus that remains cell-associated must be released by cell lysis either by repeated freezing and thawing cycles or by using an ultrasonic bath. Infected cells yield various numbers of new (progeny) virus particles, ranging from 10 to 10 000 particles per cell, and this is usually quantified accurately by an infectivity assay (Section K6).

Use of other systems

For some viruses cell culture is not the chosen procedure for virus amplification. An important example is influenza virus, which instead is propagated *in vitro* using fertilized chick embryos, which have a complex array of membranes and cavities that can support the growth of viruses (Figure 2). Small aliquots of influenza virus are inoculated into the

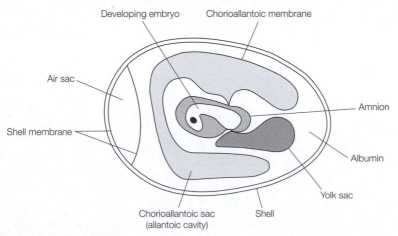

Figure 2. The embryonated egg.

allantoic cavity of the egg. The virus then attaches to and replicates in the epithelial cells lining the cavity. Virus is released into the allantoic fluid and harvested after two days growth at 37°C. Influenza vaccines are propagated in this way. The human papillomavirus (HPV) infects only stratified epithelial tissues *in vivo*, and is dependent on the maturation of cells through the epithelium to complete its replication cycle. As such the virus cannot be amplified in ordinary monolayer-type cell cultures. Rather, specialized **organotypic** cultures are required to support the complete HPV replication cycle, wherein cells must demonstrate both proliferation and differentiation. In general, these culture systems are derived from HPV-infected biopsy samples or are epithelial or keratinocyte cell cultures that are 'infected' by way of viral DNA transformation.

K6 Virus assay

Key Notes	
Virus infectivity	This is defined as the ability of a virus particle to attach, penetrate, and undergo an infectious cycle in a susceptible host cell, usually resulting in cell damage.
Virus dilution	Virus is diluted using two-, five- or ten-fold ratios in an appropriate solution (e.g. buffer or growth medium) before an infectivity assay.
Plaque assay	This is a focal assay used to detect zones of cytopathic effect (CPE), also known as plaques, in a monolayer of healthy cells. One infectious virus produces one plaque.
TCID$_{50}$	The tissue culture infective dose$_{50}$ (TCID$_{50}$) is defined as that dilution of virus that will cause CPE in 50% of replicate cell cultures.
Particle counting	The total number of virus particles (infectious and noninfectious) is determined by counting with the aid of an electron microscope. This allows determination of the particle/infectivity ratio.
Hemagglutination	The agglutination of red blood cells by some viruses (referred to as hemagglutination, HA) forms the basis of an assay that measures the number of HA units per unit volume in a given suspension.
Quantitative PCR	The processes of the polymerase chain reaction (PCR) mean that defined (user-selected) nucleic acid sequences are copied two-fold in every cycle; the final number of copies obtained can therefore be used to quantify the number of sequences (i.e. virus genomes) present in the original test sample.
Related topics	(K5) Cell culture and virus propagation (K7) Virus replication

Virus infectivity

In order to produce new virions, viruses need to be capable of replicating in a susceptible host cell. The **replication cycle** is accompanied by a number of biochemical and morphological changes within the cell (e.g. cell rounding or fusion), referred to as the **cytopathic effect** (**CPE**), which usually results in the death of the cell within a defined time **post infection** (**p.i.**). A virus will give rise to a particular CPE in a given cell type and this can be used when attempting to identify an unknown virus or to quantify a known virus. The appearance and detection of CPE regularly forms the basis of **infectivity assays** designed to determine the number of **infectious units of virus per unit volume**, and this measure of infectivity is referred to as the **titer**. An infectious unit is thought of as being the smallest amount of virus that will produce a detectable biological effect in the assay and the name given to that unit depends on the assay used, as different assays are not necessarily equivalent. Some infectivity assays are **quantal**, in that they use an 'all or nothing' approach, such as the tissue culture infective dose 50 ($TCID_{50}$) assay, while the plaque assay is **focal** in that it detects and enumerates discrete foci of infection. Both assays provide accurate quantitative information regarding the virus under test, but the focal plaque assay provides definitive quantification of the actual numbers of infective virions within a sample.

Virus dilution

Virus replication usually yields such quantities that titers are determined firstly by making accurate serial dilutions of virus suspensions, before assay. Dilutions are usually in two-, five- or ten-fold series depending on the virus, and use a suitable diluent such as cell culture medium or phosphate-buffered saline (PBS). It is important to avoid bacterial contamination (the virus assay may use cultured cells) by ensuring the use of aseptic technique and sterile diluent, tubes, pipettes or pipette tips for the transfer of volumes between each dilution. Thorough mixing of each dilution before further transfer is essential and must be done using a fresh tip to avoid carrying additional virus into the new dilution from any liquid coating the outside surface of the tip (a microliter can carry several millions of virions, causing inaccuracies in the final titer). Once diluted, virus should be assayed as soon as possible, as most viruses rapidly lose infectivity at room temperature. All the assays described here would examine several dilutions at one testing, and each dilution would be tested in duplicate (at least) to allow for a mean titer to be derived, improving accuracy and compensating for experimental inaccuracies.

Plaque assay

The plaque assay quantifies the number of infectious units (in this instance, **plaque forming units**) in a given suspension of virus. **Plaques** are localized discrete foci of infection denoted by zones of cell lysis or cytopathic effect (CPE) within a monolayer of otherwise healthy, cultured cells. Because of the great dilution at which viruses are assayed, each plaque in theory originates from a **single infectious virion** having infected a single cell, thus allowing a very precise calculation of the virus titer. The most common plaque assay is the **monolayer** assay. Here, a small volume of virus diluent (0.1 ml) is added to a previously seeded subconfluent cell culture monolayer. Following adsorption of virus to the cells, a semi-solid **overlay** (usually medium containing agar or agarose) is added to prevent the formation of **secondary** plaques, by restricting the range of movement of virions released from the infected cells, so that only adjacent cells within the monolayer are infected. Over a number of days, uninfected cells divide until the monolayer is near or at confluence, while virus replication causes small, discrete areas of infection and CPE that slowly expand as new virions infect neighboring cells, forming the plaque. Following incubation, the cell monolayers are usually fixed (a process that kills the cells but stabilizes their morphology and position) using alcohol- or formaldehyde-based solvents and stained so that infected and uninfected cells can be distinguished (Figure 1). Common stains include crystal violet (which stains infected cells, giving dark plaques) and neutral red (which leaches out of infected cells rapidly, giving pale plaques against a darker pink background of uninfected cells). The plaques are counted within each plate and the counts for plates at one of the virus dilutions are used to calculate the original virus titer. The choice of dilution depends on both the certainty of the plaque count (i.e. high enough that plaques are discrete/not overlapping) and also the need for statistical reliability (i.e. enough plaques to be considered reliable). For these reasons, choosing a dilution where each replicate of the assay holds between 20 and 100 plaques per monolayer is ideal, although the actual number that can be easily counted is often dependent on the size of the plaque and the size of the vessel used for the assay. Typical plaques are shown in Figure 1, and the assay procedure is summarized in Figure 2. The

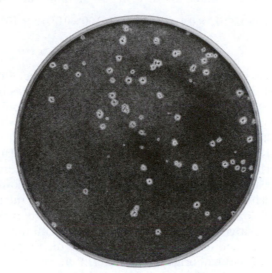

Figure 1. Herpes virus plaques on a tissue culture monolayer.

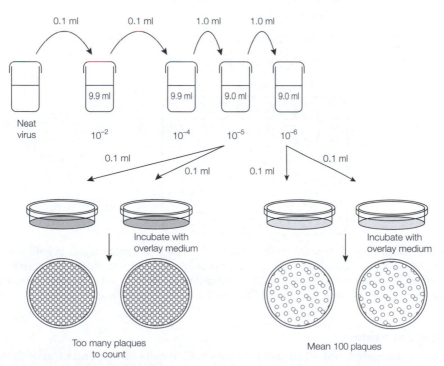

Figure 2. Diagrammatic representation of virus plaque assay (see text for calculation).

infectivity titer is expressed as the number of plaque forming units per ml (**pfu ml^{-1}**) and is obtained in the following way:

pfu ml^{-1} = (mean number of counted plaques)/(dilution × volume (ml) tested).

For example, if there is a mean number of 100 plaques from monolayers infected with 0.1 ml of virus that has been diluted to a level of 10^{-6} then the calculation is:

pfu ml^{-1} = (100)/(10^{-6} × 0.1) = 100/10^{-7} = 100 × 10^{7} = 1 × 10^{9} pfu ml^{-1}.

TCID$_{50}$

The TCID$_{50}$ is defined as **that dilution of virus required to infect 50% of a series of replicate inoculated cell cultures,** and like the plaque assay it relies on the presence and detection of CPE. Host cells are grown in confluent healthy monolayers, usually in the wells of a multi-well (24-, 48- or 96-well) tissue culture plate, to which aliquots of virus dilutions are added. The use of such plates allows between 4 and 12 replicates of each virus dilution, established in a single row or column of the plate. During incubation the virus replicates and releases progeny virions into the culture medium of each well, which in turn infect other healthy cells in the monolayer. The CPE is allowed to develop over a period of days, at which time the cell monolayers are observed microscopically, directly or following fixing and/or staining. Each well is scored for the presence or absence of CPE, and marked as positive or negative, accordingly. The numbers of positive wells at each dilution tested are used to calculate the TCID$_{50}$, which represents the dilution of virus (and hence is a measure of original virus titer) that would give CPE in 50% of the monolayers (wells) inoculated. The worked example below is derived from simulated data represented in Figure 3 and Table 1:

TCID$_{50}$ = (log$_{10}$ of highest dilution giving 100% CPE) + (½) − (total number of test units showing CPE/number of test units per dilution)

$$TCID_{50} = (-3) + (½) - (^{28}\!/_{6}) = -7.17 \quad TCID_{50} = 10^{-7.17} \quad TCID_{50} \text{ unit vol}^{-1}$$

The titer is therefore $10^{7.17}$ TCID$_{50}$ per unit vol^{-1}.

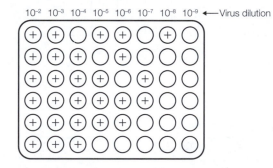

10^{-2} 10^{-3} 10^{-4} 10^{-5} 10^{-6} 10^{-7} 10^{-8} 10^{-9} ←—Virus dilution

Figure 3.   Diagrammatic representation of a TCID$_{50}$ assay. Cell monolayers in a 48-well tissue culture plate are infected with virus at increasing dilutions (i.e. six replicate wells are infected with each dilution in the range 10^{-2} to 10^{-9}). Following incubation, all wells are examined for the presence (+) or absence (blank) of CPE. See Table 1 and text for calculation.

Particle counting

As we have discussed, not all viruses replicate well in cell culture, and even of those that do, not all virus particles within a sample are infectious. Indeed, in many cases, for every

Table 1. Data from 48-well plate used to calculate TCID$_{50}$ (see Figure 3 and text for calculation)

Virus dilution	Infected test units (e.g. infected tubes)
10^{-2}	6/6
10^{-3}	6/6
10^{-4}	5/6
10^{-5}	4/6
10^{-6}	4/6
10^{-7}	2/6
10^{-8}	1/6
10^{-9}	0/6

one infectious particle up to 100 or more **noninfectious particles** may be produced from an infected cell. Loss of infectivity can be the result of many events such as incorrect virion assembly or genome packaging, the presence of a spontaneous mutation within the packaged virus genome or lack of a final virion maturation step. The total number of particles can only be determined with the aid of an **electron microscope**. The procedure relies on the use of **reference particles**, usually latex beads of uniform diameter (latex and virus particles are distinguishable when using **negative staining** with phosphotungstate), which are mixed with the virus at known concentration (i.e. a known number of reference particles per unit volume). By counting the number of virions and the number of reference particles within a grid, it is a simple matter to determine the ratio of virions to reference particles and hence calculate the virus count. The ratio of total particles to infectious particles is termed the **particle/infectivity ratio**, which is important to know when, for example, monitoring virus purification, or determining the state or age of a virus suspension.

Hemagglutination

Many viruses (including influenza virus) have the ability to agglutinate red blood cells (RBCs), a phenomenon known as **hemagglutination**, by linking cells together such that they form a mesh or network. In order for the reaction to occur, the virus needs to be present in sufficient concentration to form cross-bridges between RBCs, causing the agglutination (Figure 4a). Below this concentration, nonagglutinated RBCs remain separate and as such will sink in a solution to form a **pellet**. The hemagglutination assay is performed by mixing RBCs with diluted virus in a multi-well plate with hemispherical (round-bottomed) wells. Nonagglutinated RBCs pellet to sit as a tight red button at the base of the well, while agglutinated RBCs form a **lattice-work** structure which coats the sides of the well, remaining as a diffuse red coloration throughout the well. The assay determines the number of **hemagglutinating particles** in a given suspension of virus, i.e. it is a particle count assay, not a measure of infectivity. Nevertheless it is one of the most routinely used **indirect methods** for the determination of virus titer. The assay uses an **end-point titration** method; serial two-fold dilutions of virus are mixed with an equal volume of RBCs and the wells are observed for agglutination. The end point of the titration is the **last dilution showing complete agglutination**, which by definition is said to contain one **HA unit**. The HA titer of a virus suspension is therefore defined as the reciprocal (inverse) of the highest dilution that causes complete agglutination

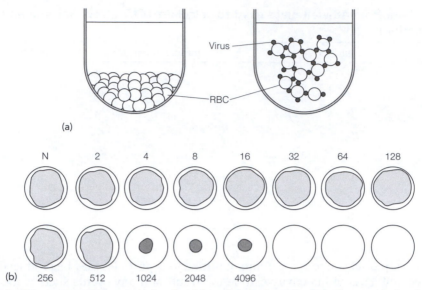

Figure 4. Schematic of hemagglutination (a) and typical results for a hemagglutination assay (b). Serial doubling dilutions of virus show the agglutination end point at 1:512.

and is expressed as the number of HA units per unit volume. An example upon which a calculation of the HA titer can be made is shown in Figure 4b. The end point in this figure is the well containing virus at a dilution of 1/512. If 0.2 ml virus dilution was added per well the HA titer would be 512 HA units per 0.2 ml or 2560 HA units ml^{-1}.

Quantitative PCR

In the past two decades, pressure to provide more rapid and selective virus identification (diagnosis) and quantification has led to the development of molecular assays that do not rely on biological or replicative properties of viruses but rather on the simple detection of their genomic sequences. The polymerase chain reaction (PCR) is a method that uses a thermostable DNA polymerase enzyme to copy a defined nucleic acid target sequence (selected by the user) in a step-wise fashion. PCR is a cyclical procedure, and for each cycle the number of copies of the target sequence theoretically doubles. **Quantitative PCR** (qPCR, sometimes called **real-time PCR**) has been used to quantify virus by determining the number of copies of a given virus sequence (the target) that are present after a known number of cycles, and subsequently estimating the number of target sequences (i.e. virus genomes) that were present in the original test sample. The assay requires accurate estimation of the number of copies of the target sequence and this is usually done by comparison with a series of diluted control sample targets where the actual numbers of molecules have been calculated. Detection of target sequences requires either a fluorescent intercalating (dsDNA-binding) dye or probe. Both systems require expensive reagents, detection equipment, and interpretation software and provide a particle (strictly speaking, genome) count, not an infectivity measurement. However with newly prepared virus the technique has been shown to be reliable and accurate (compared with infectivity assays), and is much more rapid than the plaque or TCID_{50} assays. The development, methodology, and applications of PCR are thoroughly reviewed in the *Instant Notes Molecular Biology* volume.

K7 Virus replication

Key Notes

Replication cycle

As obligate intracellular parasites, viruses must enter and replicate in living cells in order to 'reproduce' themselves. This replication cycle involves specific attachment of virus to suitable cells followed by penetration and uncoating of the genome. Nucleic acid sequences must be transcribed for protein synthesis and copied for new genomes. Maturation and assembly form new virions that are subsequently released from the cell by budding or lysis.

Attachment, penetration, and uncoating

Attachment is a very specific interaction between virus capsid or envelope proteins and a receptor on the plasma membrane of the cell. Virions are either engulfed into vacuoles by endocytosis or the virus envelope fuses with the plasma membrane to facilitate entry. Uncoating is usually achieved by cellular proteases modifying or destroying the capsid.

Genome expression and replication

Using cellular and virus-encoded enzymes and proteins, the nucleic acid genome is transcribed in a controlled fashion. The rate of transcription of mRNA (and hence its concentration within the cell) is the main regulatory mechanism. Nucleic acid is also synthesized as new genome molecules. Translated proteins may undergo post-translational modification (e.g. cleavage, glycosylation, phosphorylation).

Maturation, assembly, and release

Subunits of capsids assemble via intermediate structures, with or without the help of scaffolding proteins. Envelopes, when present, are acquired by capsids budding through one of the cellular membranes.

Related topics

(F3) DNA replication
(F4) Transcription
(F5) Messenger RNA and translation

(K3) Virus genomes
(K5) Cell culture and virus propagation

Replication cycle

The complexity and range of virus types are echoed in the various strategies they adopt in their replication cycles. Viruses, as **obligate intracellular parasites**, must **attach** to and **penetrate** suitable host cells in order to undergo a 'reproductive' cycle. This cycle is highly dependent on the metabolic machinery of the cell; all viruses lack the machinery to generate ATP or to synthesize proteins, and many other varied enzymes, proteins, and processes may be required. As a result, in most cases the virus takes over the cellular processes and orchestrates them towards its own replication. This usually results in the inhibition of host cell protein and nucleic acid synthesis, although some viruses allow cellular processes to continue for a period, and may even stimulate them (such as the human papillomavirus, which relies on continued cellular DNA synthesis in order to replicate its own genome). The cycle has a number of stages – attachment and penetration, nucleic acid synthesis and transcription, protein synthesis, assembly, maturation, and release. The outcome is the production of hundreds of progeny virions, which leave the infected cell (by **lysis** or **budding**), ultimately killing the cell and spreading to infect more host cells and tissues.

The stages of the replication cycle can be visualized indirectly using infected culture cells; following the addition of virus to the culture at a high m.o.i. (i.e. all cells are infected simultaneously), the amount of extracellular virus can be monitored by an infectivity assay, testing samples at regular intervals over time. Virus replication occurs within all cells synchronously and the changing level of extracellular virus reflects the replication cycle. This is often called the **one-step** growth curve and a typical representation is shown in Figure 1, where extracellular virus levels are shown on the y-axis, against time on the x-axis. Following virus attachment to, and penetration into, cells the level of extracellular virus falls, seen during the **eclipse phase**. During the early stages of virus replication, there is a period when no new virions are released (extracellular virus remains low), but following the start of virus assembly and release by virus, cells are lysed and increasing numbers of cell-free virions are detected (the **logarithmic** or **expansion phase**). As the cells reach the end of their capacity and the effects of virus replication take their toll, the numbers of new virions begin to plateau as the cells inevitably die (death phase). The shape of the one-step growth curve naturally varies greatly between viruses. For many bacteriophages it takes less than 60 minutes to reach the death phase, while for many animal viruses it can exceed 24 hours before maximum titers are reached.

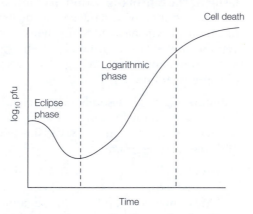

Figure 1. A typical 'one-step' growth or replicative cycle of a virus.

The replication cycles of five medically relevant human viruses have been selected to illustrate the similarities and differences of the key stages and various strategies of virus replication. Herpes simplex virus (HSV) and human papillomavirus (HPV) are both Class I viruses, poliovirus is of Class IV, influenza virus from Class V, and human immunodeficiency virus (HIV) is a Class VI virus. Their strategies are generally representative of similar viruses in each class but many variations occur.

Attachment, penetration, and uncoating

Attachment is mediated by the specific interaction between viral **attachment proteins** within the capsid or envelope and a **receptor** on the plasma membrane of the cell

(Figure 2). Importantly, the presence of the receptor determines the **cell tropism** and **species tropism** of a virus. Receptors clearly have cellular functions other than providing a binding site for viruses, but have nonetheless been paramount in affecting virus evolution. Table 1 lists examples of specific receptor sites available on host cells for attachment by viruses. HSV binds to cells through interaction of several envelope glycoproteins with various molecules including heparan sulfates, proteoglycans, and others. Two glycoproteins, gB and gC, mediate the initial receptor binding step then a third, gD, makes the critical, irreversible binding that commits the virus to infecting the cell. The poliovirus attachment protein (the capsid protein VP1) interacts with CD155, a member of the **immunoglobulin (Ig) protein superfamily** found only on many cells within humans and primates. HIV also has a limited cell tropism as a result of its attachment specificity. The attachment protein, the major envelope glycoprotein (**gp120**), attaches to the **CD4 receptor** found predominantly on human T4 lymphocyte lineages. Two further cell surface molecules, CCR5 and CXCR4, have a role in HIV attachment as **co-receptors**, and susceptible cells (macrophages and helper T lymphocytes, respectively) have one or the other in addition to CD4. The receptor bound by the HPV attachment protein, L1, is as yet unknown, but is presumably limited to cells of stratified epithelia. By contrast, the influenza virus can infect a broad range of cells within the body as its attachment protein hemagglutinin (HA) binds to N-acetylneuraminic (sialic) acid, a small carbohydrate group found on plasma membrane proteins of many cell types.

Attachment is closely followed by penetration of the virus into the cell and (usually) uncoating, where the viral genomic information is released from the capsid into the cytoplasm or nucleus (Figures 3–7). The naked capsids of poliovirus and HPV enter cells as a result of **endocytosis**, triggered by binding of the virion attachment protein and plasma membrane receptor (Figures 4,b and 5,b). The plasma membrane invaginates locally to engulf the capsid, forming a **vacuole** that transports the capsid into the cytoplasm. Fusion with a cellular lysosome lowers the pH of the vacuole, resulting in alterations of the virion structural proteins, opening a pore in the capsid that allows escape of the viral genome into the cytoplasm (Figure 2). While the poliovirus genome is immediately translated by ribosomes within the cytoplasm, the HPV genome is trafficked to the nucleus

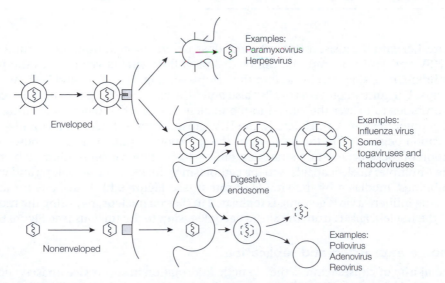

Figure 2. Methods of virus entry. Adapted from Harper DR (2011) *Viruses: Biology, Applications and Control.* Garland Science.

Table 1. Examples of receptors for viruses that infect humans

Family	Virus	Cellular receptor
Adenoviridae	Adenovirus type 2	Integrins $\alpha_v\beta_3$ and $\alpha_v\beta_5$
Coronaviridae	Human coronavirus 229E	Aminopeptidase N
	Human coronavirus OC43	N-Acetyl-9-O-acetylneuraminic acid
Hepadnaviridae	Hepatitis B virus	IgA receptor
Herpesviridae	Herpes simplex virus	Heparan sulfate proteoglycan
	Varicella zoster virus	Heparan sulfate proteoglycan
	Cytomegalovirus	Heparan sulfate proteoglycan plus second receptor
	Epstein–Barr virus	CD21 (CR2) complement receptor
	Human herpesvirus 7	CD4 (T4) T-cell marker glycoprotein
Orthomyxoviridae	Influenza A virus	Neu-5-Ac (neuraminic acid) on glycosyl group
	Influenza B virus	Neu-5-Ac (neuraminic acid) on glycosyl group
	Influenza C virus	N-Acetyl-9-O-acetylneuraminic acid
Paramyxoviridae	Measles virus	CD46 (MCP) complement regulator
Picornaviridae	Echovirus 1	Integrin VLA-2 ($\alpha_2\beta_1$)
	Poliovirus	Glycoprotein CD155
	Rhinoviruses	ICAM-1 adhesion molecule
Poxviridae	Vaccinia	Epidermal growth factor receptor
Reoviridae	Reovirus serotype 3	β-Adrenergic receptor
Retroviridae	Human immunodeficiency virus	CD4 (T4) T-cell marker glycoprotein and CCR5 or CXCR4 chemokine co-receptor
Rhabdoviridae	Rabies	Acetylcholine receptor

(Figure 4,c) with the assistance of L2, a minor structural protein from the disrupted capsid. HSV and HIV penetrate the cell by **fusion** of their viral envelope with the plasma membrane, releasing the capsid into the cytoplasm (Figures 2, 3,b, and 7,b). As well as the capsid, tegument proteins of HSV also enter the cell, and some interact with cellular microtubules to deliver the capsid to the nucleus (Figure 3,c), before uncoating of the HSV genome (Figure 3,d). Uniquely, the HIV capsid remains largely intact and genome replication begins within the framework of this structure. Influenza virus shows a combination of these penetration strategies; endocytosis of the virion is followed by release of the (multiple) nucleocapsids into the cytoplasm by fusion of the envelope and vacuole membranes, mediated by the viral HA protein (see Figure 6,b). Unusually for an RNA virus, the influenza virus genome is replicated in the cell nucleus, requiring the transport of all the nucleocapsids from the site of vacuole fusion to the nucleus (see Figure 6,c).

Genome expression and replication

The majority of viruses express their genetic information in successive phases, such that the transcription of **early** and **late genes** gives rise to a range of proteins both before (early) and after (late) genome replication has begun (Section K4). As we have described,

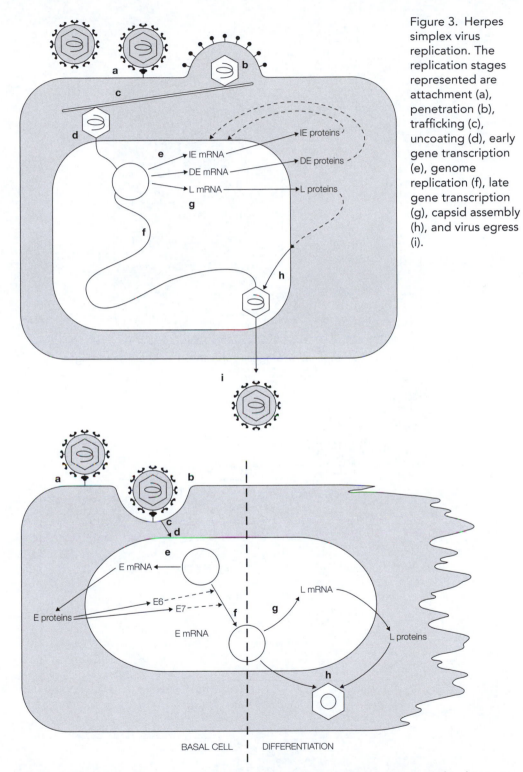

Figure 3. Herpes simplex virus replication. The replication stages represented are attachment (a), penetration (b), trafficking (c), uncoating (d), early gene transcription (e), genome replication (f), late gene transcription (g), capsid assembly (h), and virus egress (i).

Figure 4. Human papillomavirus replication. The replication stages represented (a–h) are as for Figure 3.

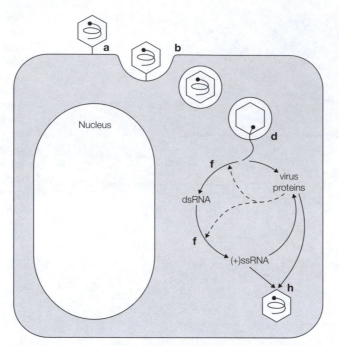

Fig 5. Poliovirus replication. The replication stages represented (a–h) are as for Figure 3.

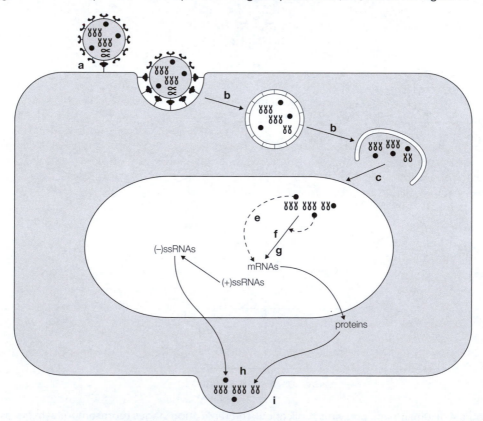

Fig 6. Influenza virus replication. The replication stages represented (a–i) are as for Figure 3.

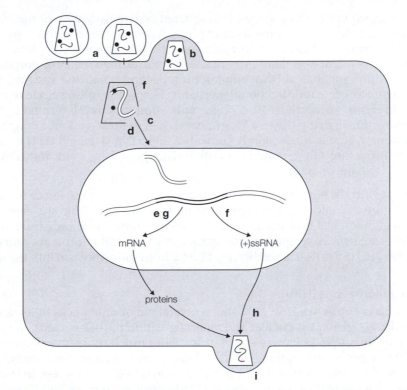

Figure 7. HIV replication. The replication stages represented (a–i) are as for Figure 3.

late virus gene expression is dependent on the prior synthesis of early proteins (both immediate early and delayed early) but it is important to grasp that for many viruses, the initiation of early gene expression itself relies on viral proteins that are enclosed within the virion and some of these are late proteins synthesized during replication within a previously infected cell. In addition, viruses also make use of cellular proteins, especially during early gene expression, before cell protein synthesis is shut off. The range and complexity of genome expression and replication strategies are very large, and to illustrate this we will consider the five viruses selected individually.

Herpes simplex virus (HSV)

Uncoating of HSV delivers the linear dsDNA genome into the nucleus of the cell (Figure 3d), along with an important protein from the tegument, VP16 (also called α-TIF). VP16 interacts with **Oct-1**, a cellular transcription factor that recognizes a specific promoter sequence (TAATGARAT, where R is a purine) known as the **octamer box**. The VP16/ Oct-1 complex preferentially transcribes five crucial HSV immediate early genes (which have promoters containing the octamer box), rather than cellular promoters with the same sequence (Figure 3e). With the translation of these five mRNAs (on cytoplasmic ribosomes) into proteins the replication cycle has begun, and subsequent expression of delayed early and late genes is inevitable. The HSV genome encodes for many proteins including several delayed early enzymes involved in the replication of the viral genome (e.g. **thymidine kinase**, **DNA polymerase**, **helicase**), which of course must return to the nucleus following their synthesis and any post-translational modification (e.g. phosphorylation, glycosylation). Replication of the dsDNA genome occurs in the nucleus and there is much evidence to suggest it is initiated by circularization of the DNA molecule. Terminal redundancy, in the form of direct repeat sequences (the a-sequences), allows the two ends to anneal together forming a circular structure. Replication of the genome is independent of cellular enzymes, and involves the viral **DNA polymerase** in association with a number of **DNA-binding proteins** and a **helicase** enzyme. Replication is thought to be by a **unidirectional** mechanism known as **rolling circle replication**, illustrated in Figure 4 of Section F10. A single **replication fork** travels around the circular genome many times, giving rise to a long linear ssDNA of multi-genomic length, known as a **concatemer**, that is subsequently converted to dsDNA (Figure 3,f). The viral DNA polymerase is the target for a number of **antiviral drugs** as it is significantly different to the host cell DNA polymerase.

Concomitant with the initiation of genome replication is the expression of the late viral proteins (Figure 3G), including the structural proteins that will form the capsid, envelope, and tegument, and nonstructural scaffolding proteins that allow capsid assembly. Late proteins are usually synthesized to higher levels than earlier proteins, and the proteins involved in HSV capsid assembly must traffic from the cytoplasm into the nucleus.

Human papillomavirus (HPV)

Although a class I virus like HSV, HPV shows several major differences in its replication cycle. The dsDNA genome is circular and replicates within the cell nucleus, but the virus does not encode a DNA polymerase and hence relies much more on cellular enzymes than HSV. Early gene transcription (Figure 4,e) produces several proteins. Two early proteins, E1 and E2, redirect cellular RNA polymerase and DNA polymerase to the expression and replication of the viral genome, while the early proteins E6 and E7 ensure that DNA polymerase is continuously available within infected cells. These proteins bind with and inhibit two cellular tumor suppressor proteins, retinoblastoma and p53, which regulate mitosis. Binding of E6 and E7 with retinoblastoma and p53 forces infected cells to enter the mitotic cycle, synthesizing DNA polymerase; virus proteins E1 and E2 direct the enzyme to replication of the virus genome as well as the cellular chromosomes. Unlike HSV, the predominant mechanism of replication is thought to be **bidirectional** (see Figures 1 and 2 in Section F3), producing circular genomic DNA molecules (Figure 4,f) but recent evidence has suggested that a rolling circle mechanism may also occur in differentiating cells of the upper epithelium. Because HPV infection causes mitosis the replicating virus genomes divide between the daughter cells, a feature that contributes to the persistence of the virus in the epithelium.

HPV encodes two late proteins (the capsid proteins L1 and L2) that are not synthesized until the infected cells move from the basal to the outer layer of the stratified epithelium and have matured (Figure 4,g). It is presumed that the promoter regulating expression of the late genes requires transcription factors not present in the basal (dividing) cells of the epithelium.

Poliovirus

As described, the poliovirus genome is translated by cellular ribosomes immediately following uncoating into the cytoplasm. Ribosomes bind to the genome at the IRES structure and produce a single large **polyprotein** of over 2000 amino acids that is subsequently cleaved by the viral protease enzyme, itself part of the full-length polyprotein. Through several cleavage steps, structural and nonstructural virus proteins are released, including the virally encoded **RNA polymerase**. This enzyme subsequently replicates the viral genome in the **cytoplasm**, a process that begins at (and is dependent on) the

3′ polyadenylated tail. Replication of genomic RNA produces both partially and fully dsRNA **replicative intermediates**. The negative-sense RNA strand serves as the template for synthesis of positive-sense RNA molecules, which either act as mRNA for increased virus protein synthesis, or become viral genomes (Figure 5,f). The viral genome molecules (and possibly the negative-sense RNA templates) are covalently associated with the viral protein VPg at their 5′ end, but it is unclear precisely when the protein becomes associated with newly synthesized RNA molecules. Levels of viral proteins and RNA continue to rise rapidly in infected cells for 1–2 days post infection.

Influenza virus

The negative-sense RNA genome molecules of influenza virus are associated with multiple copies of the **nucleoprotein** (**NP**) and an **RNA-dependent RNA polymerase** composed of three viral proteins, PA, PB1, and PB2. Primary transcription of mRNA from the genomic RNA segments (Figure 6,e) is achieved by this viral polymerase initiating from a short regulatory sequence at the 3′ end of each segment. The enzyme cannot add a 5′ cap to the viral mRNAs, but it can cleave cellular mRNAs, using their capped 5′ ends as a primer for transcription. The continuation of cellular RNA transcription is therefore important to provide the processing of viral mRNAs necessary for their translation. At first, all the genomic segments are transcribed at equivalent rates, but presently those encoding the NP and a nonstructural NS1 protein are synthesized at higher levels. At later times (Figure 6,g), transcripts representing the segments encoding the major structural proteins HA, **neuraminidase** (**NA**), and the matrix protein M1 are predominant in the infected cell. The majority of genome segments encode a single protein but two give rise to mRNAs that allow translation of two proteins as a result of splicing. Like the replication of the poliovirus genome, a complementary RNA strand (in this case positive-sense) must first be synthesized and used as a template to copy new (negative-sense) genomic RNAs (Figure 6,f). Each RNA segment is replicated separately, and neither the full-length positive-sense intermediates nor the final genomic RNA segments acquire a cap or a polyadenylated tail.

Human immunodeficiency virus (HIV)

HIV is a member of the *Retroviridae* and carries three viral enzymes within the virion capsid that initiate genomic replication within the partially opened capsid structure, following penetration of the cell cytoplasm (Figure 7,f). The enzyme **reverse transcriptase** (**RT**) copies the RNA genome into a dsDNA molecule, in partnership with a **ribonuclease** that degrades the RNA molecule as replication proceeds. Replication is thought to involve both copies of the diploid viral genome, and uses small cellular RNA molecules (transfer RNA, tRNA) contained within the capsid as primers. The final dsDNA molecule is larger than the original RNA genome, due to duplication of terminal sequences during replication. The remaining capsid structure traffics to the nucleus releasing the newly synthesized dsDNA (Figure 7,c, d), whereupon it is permanently inserted into a chromosome by the third enzyme found within the capsid, **integrase**. The integrated genome is referred to as a **provirus** and its replication and expression now use cellular enzymes. As part of the chromosome, transcription of viral mRNA (Figure 7,e–g) is by cellular RNA polymerase II, regulated by promoter sequences found within the duplicated terminal regions of the provirus, the **long terminal repeats** (**LTRs**). Many mRNAs are synthesized from the genome, some being translated into polyproteins that, like that of poliovirus, are **proteolytically cleaved** by the virus **protease** into smaller functional proteins. The envelope glycoproteins (gp120 and gp41) are cleaved from a large precursor protein before post-translational modification in the endoplasmic reticulum and are subsequently inserted into the plasma membrane of the cell. Transcription of the provirus also provides full-length RNA molecules (with 5′ cap and 3′ polyadenylated tail) that are packaged into new viral capsids as genomes.

Maturation, assembly, and release

As proteins and new genomes are synthesized in the infected cell they are channeled to various locations for **virion assembly** and **egress** (release). Depending on the virus, capsids assemble in the nucleus or the cytoplasm, either building around the genome molecule, or packaging it as a final **maturation** step following capsid assembly. Genome and capsid are released from cells by **budding** through a cellular membrane, or by causing cell lysis. Enveloped viruses may bud through the **nuclear membrane**, **plasma membrane** or even the **endoplasmic reticulum**; prior to budding the virus-specific envelope glycoproteins have been laid down in the membrane.

The steps of capsid assembly vary according to the complexity of the mature capsid. Simple capsids such as those of poliovirus and HPV assemble through the association of structural proteins into units known as capsomers. The structural proteins of poliovirus (VP0, VP1, and VP3) first bind together in trimers that then associate as pentamers; 12 pentamers form a complete but immature icosahedral capsid. Using an unknown mechanism that may in part depend on the covalently attached VPg protein, genomic RNA is sequestered into the capsid and the final maturation step, cleavage of VP0 into VP2 and VP4, seals the virion. The HPV capsid is simpler again, comprising only two proteins. Pentamers of the major structural protein L1 form the major capsid structure, along with a few molecules of the minor protein L2, thought to locate at the vertices of the icosahedron. Packaging of the viral genome appears to involve the L2 and E2 proteins, although E2 is not found in the virion. Both poliovirus and HPV exit infected cells by lysis, releasing several hundreds of progeny virions.

Mature HSV capsids have several structural proteins and are assembled in the nucleus, assisted by two **scaffolding proteins**, UL26 and UL26.5, which associate with the major capsid protein (VP5) in a double-shelled procapsid structure. UL26 is a protease that cleaves the scaffolding proteins that exit the structure leaving an empty, immature icosahedron formed predominantly of VP5. Genome packaging is an elegant process whereby the long concatemers of replicated viral genomic DNA must be cleaved by an endonuclease at a precise point, to release individual genome molecules that can be packaged. The genome enters the immature capsid at one open corner, which is then sealed by a cluster of UL6 proteins. Several alternative paths for egress of the mature capsid have been proposed and may not be mutually exclusive. The most likely process suggests an initial budding of the capsid through the inner nuclear membrane, acquiring an envelope that is then lost as it fuses with the outer nuclear membrane, releasing the naked capsid into the cytoplasm. The association of the capsid with tegument proteins is thought to occur as the capsid buds into the ER, and release of virus from the cell is due to subsequent egress via the Golgi and exocytic vesicles to the plasma membrane. In addition to exocytosis, many herpesviruses invade adjacent cells by the process of **cell–cell fusion**. The plasma membrane of an infected cell fuses with that of an adjacent uninfected cell, facilitating the entry of progeny virions that undergo a further replication cycle. This phenomenon can be seen in cell cultures, visible as large areas of **multinucleate fused cells** (**syncytia**).

Influenza virus and HIV exit the cell by budding at the plasma membrane. Packaging of HIV genomic RNA molecules involves recognition of specific sequences (known as the packaging or epsilon signal). In a similar manner, accumulation of influenza virus nucleocapsids (RNA segments associated with multiple copies of the viral NP protein and one copy each of the PA, PB1, and PB2 proteins) at the plasma membrane leads to bud-ding. Small sequences in the untranslated regions of each RNA molecule – unique to each segment – may be responsible for selective packaging, insuring one copy of each segment is packaged during budding. Electron microscopy often reveals a distinct pattern of orga-nization of the genomic segments during virus budding, supporting the idea of selective packaging, but it is likely that random packaging also occurs. This theory is supported by the routine isolation of virions containing more than the normal complement of genome molecules (eight for influenza A virus) and also by the high particle to infectivity ratio (Section K6) found in laboratory preparations of virus, suggesting that many virions fail to package a complete 'set.'

K8 Virus infections

Key Notes

Virus transmission

Viruses can gain access to the host through the skin and mucous membranes, via the respiratory or gastrointestinal tracts or through sexual contact. Understanding virus transmission between hosts, and the routes of entry into and exit from hosts, allows the study of the epidemiology of infections, which in turn helps to control virus spread.

Clinical outcomes of infection

The outcome of virus infection depends on a number of factors such as the age and immune status of the host and the virus strain. Infections may be localized at or near the site of entry or they can be systemic, where the virus spreads from its point of entry to involve one or more target organs. The outcome of a viral infection follows one of several recognizable patterns, leading to development of clinical disease symptoms, inapparent infection or even cancer. Viral infections may be eradicated by the immune system (or by the use of antiviral drugs) or they may result in long-term infections, either with or without symptoms.

Related topics

(K7) Virus replication

(K9) Viruses and the immune system

Virus transmission

Virus infection in a host is the result of virus replication within individual cells, as described in the previous section. To initiate an infection, viruses must gain access to susceptible cells within a host and in order to persist and evolve viruses need to be able to infect a large population of susceptible hosts. Entry into and **transmission** (spread) between hosts can occur by a number of routes, most of which are classed as **horizontal transmission** routes, i.e. transmission between individuals of a family, group or population.

Many viruses gain entry via the **respiratory route** (Figure 1), including influenza viruses and the rhinoviruses (many of which cause the common cold), coronaviruses (e.g. SARS), and respiratory syncytial virus (RSV). Following inhalation within droplets or dust, viruses can infect and replicate in the epithelial cells of the upper or lower respiratory tract mucosal membranes. The host responds (after several days) with the characteristic symptoms of sneezing and coughing, which of course transmit virus to new hosts. An important respiratory pathogen of infants, RSV infection remains localized at the site of entry, causing severe necrosis of the bronchiolar epithelium. The infected cells slough off, blocking the small airways and leading to obstruction of air flow and respiratory disease. A number of viruses that enter and exit the host via this route are not respiratory viruses and spread beyond the respiratory tract to set up infection (and symptoms) in the cells of the blood or other target organs (Figure 2). These may be viewed as **systemic** infections and examples include the measles, rubella (German measles) and varicella-zoster

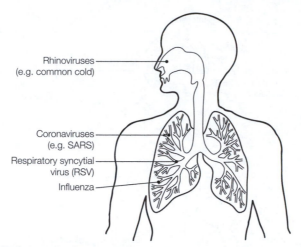

Figure 1. Routes of infection in the respiratory tract. Virus infections can produce a variety of respiratory disorders, depending on the area of respiratory tract infected. Adapted from Strelkauskas A, Strelkauskas J & Moszyk-Strelkauskas D (2010) *Microbiology: A Clinical Approach*. Garland Science.

(chickenpox) viruses. The measles virus replicates in local lymph nodes that drain from the infected respiratory tissue. The virus enters the blood (**primary viremia**) where it replicates in the cells of epithelial surfaces before entering the blood again (**secondary viremia**). At this point the patient is highly infectious but does not have the distinctive measles rash, which appears about 14 days post infection.

The **oral–gastrointestinal** mucosa is a port of entry used mainly by those viruses responsible for gut infections, such as rotavirus, Norwalk virus, and the Enteroviruses, which include poliovirus. Vast numbers of virus particles can be excreted in fecal material (e.g. in the order of 10^{12} particles per gram), facilitating the easy spread of these viruses in conditions of poor sanitation. The drinking of fecally contaminated water and consumption of contaminated shellfish or other food prepared by unhygienic food handlers are ways in which these viruses are spread. The **skin** normally provides an impenetrable barrier to pathogen invasion, but a range of viruses can enter following **trauma** to the skin. Often this is a wound such as that from an animal bite (rabies virus) or insect bite (yellow fever and dengue viruses). Several viruses are carried by an insect vector such as the mosquito or tsetse fly, and are commonly referred to as arboviruses (arthropod-borne viruses). Other viruses penetrate the skin via contaminated needles during injections, needle-stick injuries or intravenous drug abuse, and these include HIV and the hepatitis B and hepatitis C viruses. These viruses can also be transmitted by **iatrogenic** routes; infection as a result of medical interventions such as the transfusion of contaminated blood or blood products or the use of unsterilized surgical equipment. In addition to these routes, **sexual transmission** of viruses is an important mechanism of spread, and viruses such as HSV, HPV, hepatitis B, and HIV can infect hosts in this manner too.

Viruses may also be transmitted **vertically** – that is, from mother to offspring via the placenta, during childbirth, or in breast milk. Rubella virus (German measles) and cytomegalovirus (CMV) acquired by the mother during pregnancy are easily transmitted to the developing embryo and often lead to severe congenital abnormalities and/or spontaneous abortion. Some viruses, such as HSV, can infect the developing fetus *in utero* or during birth and subsequently present as an acute disease syndrome in the neonate. In

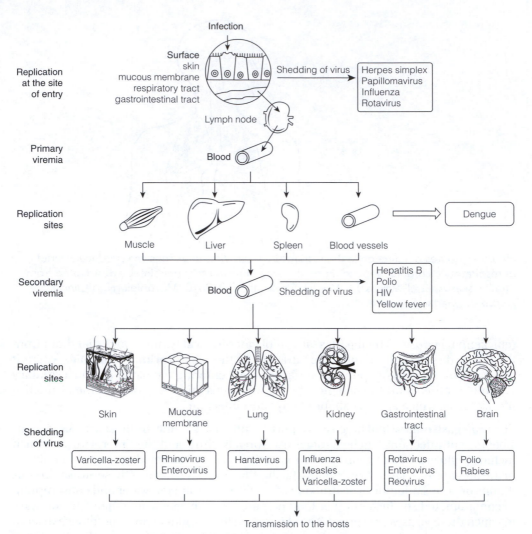

Figure 2. Virus spread within the host. Different viruses have different modes of spread within the host. Some viruses remain localized at the site of entry, whereas others may spread to involve other tissues. Routes of infection are shown, together with examples of possible clinical outcomes. Possible sites of replication are shown and some are also sites of shedding and transmission. Adapted from Strelkauskas A, Strelkauskas J & Moszyk-Strelkauskas D (2010) *Microbiology: A Clinical Approach*. Garland Science.

the case of HIV and hepatitis B, virus transmission in this way means that the neonate may be born with an asymptomatic infection with the virus persisting in a **carrier state** and developing into disease much later. HIV has also been shown to infect the offspring of infected mothers by way of breast milk.

Clinical outcomes of infection

The outcome of a viral infection is dependent on a number of factors including **age**, **immune status**, and **physiological well-being** of the host. For example, HSV infection is often fatal in the neonate but when it occurs in the older child it is not so, and also often without overt symptoms. Epstein–Barr virus (EBV) causes a very mild febrile illness in young children but the more severe infectious mononucleosis (glandular fever) in teenagers and young adults. CMV in a healthy individual may cause a mild febrile illness, but in immunosuppressed individuals it can lead to fatal pneumonia. Measles rarely causes severe complications in healthy, well-nourished children, but is fatal in many children living in developing countries where malnutrition is a compounding problem (the WHO estimate for 2008 was 164 000 deaths). For the purposes of study and comparison, the outcome of a virus infection can be categorized (Table 1) by considering the extent of virus synthesis, cellular damage, and clinical symptoms. It will become clear from Table 1, and the descriptions below, that some viruses can give rise to more than one type of infectious outcome, depending on the clinical situation. Before we describe viral pathogenic outcomes, it is important to remember that many infections are **abortive**. Entry into nonsusceptible hosts (that lack suitable cells and/or cell receptors), or attachment (and even penetration) of virus to nonsusceptible cells in such a host will not result in replication and the process of invasion is a dead end.

Inapparent (asymptomatic) infection

Many virus infections are **inapparent**, **asymptomatic** or **subclinical**, there being no apparent outward symptoms of disease. This is not due to a lack of virus replication or of cell damage and/or death within the host, but cellular damage is limited and hence not translated into obvious clinical symptoms. This is generally the case in the immune host, where there has been recovery and virus eradication by the immune system following a previous infection or vaccination. In this case, the immune system rapidly recognizes and curtails virus replication in the host following reinfection and hence the level of cell damage is limited to below detection levels. However, there are also several viruses (e.g. respiratory viruses and enteroviruses) that do not produce clinical symptoms in non-immune individuals. In the majority of individuals infected with poliovirus, the virus replicates in the epithelial cells of the gastrointestinal tract and is excreted in the feces, but causes no symptoms.

Acute infection

This is the pattern observed during many viral infections in otherwise healthy individuals. Viral replication leads to cellular damage and clinical symptoms of varying severity followed by virus eradication by the immune system and recovery, accompanied by a

Table 1. Outcomes of virus infections

Outcome	Replication	CPE	Symptoms	Examples
Inapparent	Yes	Yes	No	Poliovirus (in gut)
Acute	Yes	Yes	Yes	Poliovirus (in CNS)
Latent	No, but can reactivate	No	No, but shows recurrence	HSV, VZV, CMV, EBV
Persistent	Yes, low	Yes	Yes, minor	HPV-1, HPV-11
Chronic	Yes, high	Yes	Yes or No	HBV, HCV
Progressive	Yes	Yes, eventually	Yes, eventually	HIV, Measles virus (rare)
Neoplastic	Yes	Yes – transformation	Yes	HPV-16, HPV-18, HBV

VZV, varicella-zoster virus; HBV, hepatitis B virus; HCV, hepatitis C virus; CNS, central nervous system.

level of immunity to the virus. This is true of most viral infections encountered during childhood such as measles, mumps, chickenpox, and German measles, and most respiratory viral diseases for which a number of viruses are responsible. Indeed a broad spectrum of other viruses also follows this pattern, including hepatitis A virus (infectious hepatitis), rotavirus (gut infections), and Coxsackie virus (myocarditis, pericarditis, conjunctivitis). Poliovirus, which we have seen leads to an inapparent infection within the gut, may also replicate within the central nervous system. Here, virus replication and cell damage cause clinical symptoms; an acute infection. Until the development of excellent vaccines within the 1950s, polio was a severe, worldwide, and often fatal disease due to the muscular paralysis that resulted from viral infection of motor neurons.

Many viruses are fatal in distinct circumstances or in a percentage of victims, but some are always fatal, including cerebral rabies and HIV infections. Following the bite of a rabid animal, the rabies virus replicates within the peripheral nerves and during an incubation period of between 30 and 90 days it moves to the spinal cord and brain. During this time it is possible to treat the patient to prevent the virus reaching the central nervous system, but if it does so then replication here results in release of more viruses that then spread to virtually all the tissues of the body including the salivary glands where it is transmitted in the saliva. In only a short time (7–12 days) from this point, the patient develops a variety of abnormalities including **hydrophobia** (aversion to water), **rigidity**, **photophobia** (aversion to light), focal or generalized convulsions, and a variety of **autonomic disturbances** that precede a flaccid paralysis, coma, and death.

Acute and inapparent infections are characterized by their limited duration, as a result of the successful intervention of the host immune response. However, some viruses have ways of avoiding immune clearance, such that they are not totally eliminated from the body. These viruses persist in the host in one form or another, and give rise to long-term infections.

Latent infection

A restricted range of viruses, most being in the *Herpesviridae* family (HSV, varicella-zoster virus (VZV), EBV, and CMV), give rise to **latent infections** within the host, as a result of evading immune clearance during the **primary infection** (which can be acute or inapparent). During latent infections virus replication is curtailed as a result of viral regulatory mechanisms, significantly suppressing virus production; in fact for HSV and VZV no new virus is produced during **latency**. Intermittently, virus replication may be **reactivated** and virus production is resumed (**recurrence**), possibly leading to clinical symptoms (**recrudescence**) that may be similar or different to those observed during primary infection. Both HSV and VZV produce a primary infection in skin and mucosal cells, but a latent infection within neurons and both can reactivate from a latent state many times during the life of an individual. HSV is found latent within the neurons of the trigeminal or sacral ganglia between periods of reactivation, when recurrence can produce the typical painful cold sore lesions on the mouth or genitals. It is not fully understood what happens to reactivate the virus at the cellular level, although a number of stimuli including menstruation, exposure to UV light, and stress are responsible for initiating it. A very different clinical syndrome may result from the reactivation of VZV. The primary infection produces chickenpox, whereas reactivation is associated with the development of shingles, a localized area of extremely painful vesicles that can in some cases remain for many weeks. Additionally, some patients suffer from very severe **post-herpetic neuralgia** that can persist for months or years.

Persistent infection

HPVs establish long-term **persistent infections** with low levels of virus replication as a result of immune evasion. HPV infections occur within stratified epithelia, and the mechanism of infection permits limited immune detection. The basal cells of stratified epithelia support only early virus protein synthesis, which is poorly immunogenic, while the late proteins (L1 and L2) that trigger strong immunity, are only found in high levels in the upper layers of the epithelium, at a site removed from immune surveillance. Hence the virus evades immune clearance, and the segregation of newly replicated genomes to daughter cells within the basal layer (recall that HPV stimulates mitosis) insures the prolonged presence of the virus in the tissue.

Chronic infection

The hepatitis B and C viruses (HBV, HCV) are classic examples of viruses that give rise to a long-term infection with high levels of virus replication, known as a **chronic infection**. Between 5 and 10% of individuals infected with HBV and 80% of those infected with HCV fail to eliminate virus after acute infection, and will continue to harbor the virus within their hepatocytes for months or years. In chronic infections the very high rate of virus replication means that high levels of HBV and HCV are found in the blood of these individuals. Estimates suggest that over 300 million people worldwide carry HBV or HCV in their blood and body fluids that are continuously shed from the body, making carriers a serious transmission risk. HBV is transmitted in blood and body fluids, but is also passed vertically from mother to offspring. The exact details of HCV transmission, other than in blood and body fluids, are not known. Despite high levels of virus replication, some chronic carriers are **asymptomatic** but others suffer long-term CPE of infected hepatocytes that can result in a greatly increased risk of **cirrhosis of the liver** and **hepatocellular carcinoma**. At present, in the UK, HCV infection is the predominant condition that leads to the need for liver transplantation.

Slow, progressive infection

A very few viruses persist within the infected host for extremely long periods of time before they develop an inevitable symptomatic infection. The most well-known virus giving rise to this type of **slow, progressive infection** is HIV. The virus initiates an acute, primary infection in **T4 lymphocytes** and other cells (macrophages, monocytes, and dendritic cells), leading to the development of anti-HIV antibodies and the transmission of virus in a range of body fluids. Following this primary infection and incomplete clearance, the virus can show a period of latency in infected T4 lymphocytes that reactivates as the cells are activated some time later. If untreated, in most individuals the virus will eventually progress to more rapid replication and will ultimately cause the clinical syndrome known as **acquired immune deficiency syndrome**, **AIDS**. A small subset of individuals infected with HIV has not, after more than 20 years without antiviral treatment, shown progression of virus infection to AIDS. These individuals have a genetic mutation that results in a truncated CCR5 molecule on the surface of their macrophages. The incomplete CCR5 molecules do not function as normal co-receptors for HIV attachment, blocking penetration of the virus into cells and thereby preventing infection. This is an elegant and extremely important example of how molecular knowledge of a virus replication cycle can inform us about the progression and outcome of a serious disease, which may shed light on future treatments.

A less well-known example of a slow, progressive infection is that which develops in very rare cases of individuals infected by measles virus. In the majority of cases measles is a short-lived acute infection, but in a very small minority of cases (1 in 100 000) the virus is not completely cleared by the immune system, and establishes a low-level infection within the brain. The infection proceeds very slowly, and over time several mutations accumulate in the newly synthesized viral genomes. These mutations result in a failure to synthesize all the viral structural proteins, which means that the virus remains undetected by the immune system. Eventually (8–10 years later) infected cells die within the brain, leading to a disease known as subacute sclerosing panencephalitis (SSPE), which is untreatable and fatal.

Neoplastic growth

Infection – the introduction of viral genetic material – results in significant cellular changes designed to promote virus replication. In the majority of the examples we have looked at so far, the CPE ultimately results in cell death. However, as we have seen with HPV, infection with some viruses instead leads to sustained and often increased mitosis. This can contribute to the development of **neoplasia** (new cell growth), which may lead to oncogenic events. Viruses contribute to cancer formation by two main routes; those DNA viruses that stimulate cell mitosis do so as a result of their own, unique oncogenic proteins, while the oncogenic RNA viruses alter the expression pattern of cellular oncogenes.

Several of the DNA viruses we have mentioned have been implicated as being co-factors in the development of a range of malignancies. All have one or more proteins, essential for their normal replication cycle, which promote the cell cycle. We have described the role of the E6 and E7 proteins in HPV that block tumor suppressor proteins in infected basal cells and a range of transforming proteins (the T-antigens) in Polyomaviruses also activate mitosis in infected cells. The smallest HBV protein, the X-protein, is thought to block the action of p53 (like the HPV E7 protein) but it does so by an unknown mechanism, and only a proportion of long-term chronic HBV infections progress to hepatocellular carcinoma. The *Herpesviridae* can also, in some conditions, contribute to cancer formation, the most common member being EBV, which is associated with a number of cancers such as Burkitt's lymphoma and nasopharyngeal carcinoma.

The RNA viruses that stimulate mitosis and lead to cancerous outcomes are all, with the exception of the hepatitis C virus, members of the family *Retroviridae*. They have been assigned into three subgroups, according to their mechanism and speed of activity. The fast-acting or **transducing** retroviruses cause mitosis by expressing an oncogene that they have acquired (into their genome) from a previous infected cell. The gene is expressed in addition to their essential genes and is often mutated as a result of frequent (and uncorrected) genome copying during replication. Infection with such a virus therefore usually leads to high-level expression of an altered oncogenic protein, stimulating mitosis and forming tumors in a short period, perhaps within just a few weeks of infection. The slower *cis*-acting retroviruses induce mitosis by increasing the expression of cellular oncogenes. Integration of the viral (dsDNA) genome into a chromosome allows nearby cellular genes to be influenced by the strong promoters within the virus genome. The result is that infected cells give rise to tumors within a few months of the primary infection. The third group is the *trans*-acting retroviruses, which also increase the expression of cellular oncogenes. They do so through the action of viral proteins (transcription factors) and cancers can take several years to appear.

SECTION K – THE VIRUSES

K9 Viruses and the immune system

Key Notes

The immune system	Virus infections are countered by the host immune system. The immune system consists of a range of molecules and cells with different functional responses to viral invasion. Simplistically, the immune system may be viewed as having two arms; the rapidly engaged nonspecific innate and the slower, antigen-specific adaptive systems.
Virus antigens and recognition	Viral proteins (antigens) can stimulate the activity of immune cells, a property known as antigenicity. The smallest part of an antigen is known as an epitope. Antigens trigger activation of immune lymphocytes by binding with a receptor in their plasma membrane. Lymphocytes recognize antigens on the surface of cells, and antibodies (the products of activated B cells) can also recognize free virus.
Immune responses to viruses	The innate response acts in a nonspecific manner while the adaptive responses react specifically towards the individual virus. The innate immune system comprises lytic and phagocytic cells; the natural killer (NK) cells, polymorphonuclear leukocytes (PMNLs), and macrophages. Soluble immune proteins such as interferon and cytokines enhance the action of other immune cells. The adaptive response consists of T and B lymphocytes that ultimately effect cell lysis or virus neutralization.
Virus-induced immunopathology	The immune response towards an invading virus may itself contribute to (or in some cases be entirely responsible for) the cellular damage and disease that follow infection.
Viruses and the immuno-compromised	Individuals who are immunosuppressed suffer more severe clinical outcomes following virus infection than immunocompetent hosts. HIV infection, medication, and many genetic defects cause immunosuppression.
Evasion of the immune system	Many viruses have adapted to evade, inhibit or outpace the immune system.
Related topics	(K4) Virus proteins (K10) Virus vaccines (K8) Virus infections

The immune system

The host response to viral infection involves the immune system; a range of lymphocytes and soluble proteins that act to curtail virus replication, destroy infected cells, and block infection of new cells. The presence of virus proteins (**antigens**) stimulates nonspecific responses such as the action of phagocytic and lytic cells that attack infected host cells. Viral antigens also react specifically with lymphocytes such as B cells and T cells that respond to provide lytic cells and antibodies directed in a precise manner targeting the virus in question. A successful immune response eliminates virus and infected cells from the host and establishes a 'memory' of the virus to promote rapid response in the face of a future infection by the same (or a closely related) virus. The importance of immunological memory is discussed in Section K10 and the *Instant Notes Immunology* volume explains the full range and activities of the immune system.

Virus antigens and recognition

Viral proteins, both structural and nonstructural, are foreign to the host and many induce the activity of immune cells. The ability to stimulate an immune response is referred to as **antigenicity** or **immunogenicity** and the stimulating molecule (the **antigen** or **immunogen**) is usually a protein, although carbohydrates and lipids can also act as antigens. The smallest part of an antigen is referred to as an **epitope**. For many viruses the structural proteins are highly immunogenic and the main triggers of the immune response, particularly envelope proteins that are inserted into the plasma membrane of infected cells. Antigen recognition is the trigger for activation of immune lymphocytes and is determined by binding of the unique molecular structure of the antigen with a protein receptor in the plasma membrane of lymphocytes.

Immune responses to viruses

The immune system has two types of effectors; those of the **innate** response, which act early during infection in a general, nonspecific manner, and the cells of the **adaptive** response, which are activated a few days later and react specifically with the stimulating antigen. The innate immune system responds first in combating virus infections and comprises lytic (**natural killer**, NK) cells and phagocytic cells that are found either within the blood (**polymorphonuclear leukocytes**, PMNLs) or that can trawl the tissues (**macrophages**). NK cells have an important role in the control of some virus infections. In addition, infected cells themselves can respond to virus by the synthesis of **interferon**, a secreted protein that can induce an **antiviral state** in other, as yet uninfected cells, protecting them from viral invasion. Interferon proteins can also enhance the activity of immune lymphocytes.

The adaptive response requires several days to activate and consists of the **T lymphocytes** and, a few days later, the **B lymphocytes**. Before activation, the numbers of these cells are low but recognition of viral antigens stimulates expansion of activated cell lines that give rise to either undifferentiated memory cells or differentiated effectors. T lymphocytes recognize viral antigens (via the T-cell receptor, TCR) that have been processed and exposed on the surface of cells in conjunction with host major histocompatibility complex (MHC) proteins. There are two types of T lymphocytes important in viral infections and they recognize antigens in association with different MHC proteins. Cytotoxic T lymphocytes (CTLs, also known as T8 or CD8 cells) lyse infected cells following the binding of viral antigens in association with MHC type I proteins that are found on all cells except neurons and red blood cells. Helper T lymphocytes (T4 or CD4 cells) bind viral antigens in association with MHC type II proteins that are found only on certain antigen-presenting cells (APCs) including dendrites and macrophages. These lymphocytes are crucial for the action of both CTLs and B lymphocytes through the production of cytokines, soluble proteins that enhance or suppress immune cell functions. B cells similarly recognize viral antigens via a cell surface receptor (the B-cell receptor, BCR) and differentiated lymphocytes (plasma cells) synthesize antibody that binds with the same specificity (i.e. to the stimulating antigen). Antibodies may bind with antigen present on free virions or on infected cell surfaces. The binding of antibody to free virus can neutralize its infectivity by blocking virus attachment, penetration or uncoating. Binding of antibody to infected cells triggers the action of the Complement cascade, a series of nine soluble proteins that assemble into a pore structure (the membrane attack complex) that leads to infected cell lysis. The presence of antigen-bound antibody or Complement molecules on the surface of an infected cell can also recruit phagocytic cells (a process known as opsonization) to engulf the cell.

For acute infections, initial (primary) T-cell responses usually peak between 7 and 10 days post infection and decline within 3 weeks. The primary antibody response usually peaks later and is often barely detectable during the acute stage of infection but serum antibody levels increase dramatically 2–3 weeks post infection and may linger for several months. Primary antigen recognition and immune responses usually result in the formation of **memory B and T cells** that persist, making antigenic recognition much faster during reinfection. This **secondary** response is virtually immediate in terms of CTL and antibody response and often at a higher level. The mechanisms that combat viral infections are outlined in Table 1.

Table 1. Mechanisms to combat virus infection

Type of immunity	Effectors	Mechanism of action
Innate	Natural killer cells	Lyse infected cells when MHC levels are low
	Polymorphonuclear leukocytes	Engulf infected cells in blood
	Macrophages	Engulf infected cells in tissues
	Interferons	Induce antiviral state in uninfected cells, stimulate MHC levels
Adaptive	Cytotoxic T lymphocytes	Lyse infected cells
	Helper T lymphocytes	Secrete cytokines, assist lymphocyte maturation
	B lymphocytes	Produce antibodies
	Antibodies	Neutralize virus, activate Complement cascade
	Complement proteins	Lyse cells, recruit phagocytes

Virus-induced immunopathology

The immune response mounted against an invading virus may itself be responsible for the damage and disease state that follow infection. This is the case for the majority of the common cold symptoms seen with rhinovirus infections and the skin rash that is associated with measles. We have described symptomatic and asymptomatic chronic HBV infections; the major contributor to the symptomatic outcome is immune lysis of infected hepatocytes by CTLs. The recently emerged SARS virus itself causes little cellular damage, but it stimulates a 'cytokine storm' in patients, a heightened release of immune proteins that lead to an inappropriate and dangerous immune response. Dengue virus has four known serotypes and infection with any single serotype gives rise to a febrile illness with an accompanying rash and joint and muscle pain. However, subsequent infection with a different strain of Dengue virus, especially in children, leads to a much more severe pathology, dengue hemorrhagic shock syndrome (DHSS), thought to be due to the enhanced immune response to the second strain. Viruses may also evoke autoimmunity, probably via **molecular mimicry**, which is the production of an antigen that shares conserved sequences with a host cell protein. As a result, antibodies or T cells are produced with specificity against the viral antigen but that also react with host proteins.

Viruses and the immunocompromised

Individuals with natural or artificially induced immunosuppression are at risk of more severe clinical disease from a range of viral infections. For example, infants with the genetic disorder **severe combined immunodeficiency** (**SCID**) develop recurrent infections early in life (e.g. rotavirus in the gut, which induces prolonged diarrhea). In particular the *Herpesviridae* (such as HSV, CMV, and VZV), which are responsible for latent infections, show far more frequent **reactivation** and **recurrence** and severe recrudescence in immunocompromised individuals. **Immune suppression** is induced by medication following transplant surgery and may lead to generalized shingles (recurrence of VZV), CMV pneumonia or genital warts (HPV). Individuals with little or no antibody production (**hypogammaglobulinemia**) often harbor and excrete viruses for many years (e.g. poliovirus from the gut).

Some viruses induce immunosuppression in various ways, and to various levels. For example, HBV suppresses the induction of interferon in some hosts (which in turn reduces the CTL response), while the measles virus transiently suppresses T-cell responses during replication. Severe immunosuppression results from HIV infection of CD4 T-helper cells and macrophages. The virus persists in the host for months or years, eventually progressing to high levels of replication within CD4 cells, seriously depleting the T-helper cell population. This results in serious immunosuppression (**acquired immunodeficiency syndrome**, **AIDS**) and susceptibility to a wide range of pathogens (that normally are readily eliminated by the immunocompetent host), which are usually the cause of death in such patients.

Evasion of the immune system

Many viruses have mechanisms to inhibit, suppress or evade the immune response in order to continue replication and transmission. These include inhibition of antigen processing, suppression of interferon production, inhibition of phagocytes, and reduced immunogenicity, alteration of lymphocyte traffic, effects on cellular modulators, depression of Complement activity, resistance to the immune effectors, and rapidly appearing antigenic variation. Many viruses use several different mechanisms together.

Most of the *Herpesviridae* of course escape the immune system while latent in their host and several have other mechanisms that act during acute replication. Epstein–Barr virus (EBV) inhibits antigen presentation to immune cells and interferon production through expression of a protein similar to interleukin 10, an inhibitory cytokine. HSV can interfere with NK cell function or even kill these cells and CMV suppresses NK cell lysis by mimicking MHC proteins. Both HSV and EBV structural proteins interfere with the Complement cascade and an HSV structural protein on infected cell surfaces prevents Complement fixation or opsonization by phagocytes. The *Adenoviridae* and *Poxviridae* (such as vaccinia virus) also interfere with antigen presentation by down-regulating the expression of MHC type I protein and are capable of inhibiting the antiviral mechanism of interferon in cells.

The high mutation rate of many RNA viruses gives rise to alterations in antigenic molecules such that they are no longer recognized by the circulating immune cells and antibodies. This is evident during the long-term infection of HIV within a host and is also seen at the population level in the influenza virus. As this virus undergoes constant replication within different hosts across a population, gradual antigenic changes result from mutations, causing **antigenic drift**. This accounts for the yearly recurrence of influenza across the globe, and the need for new vaccine formulations each year, as the mutations change the viral antigens sufficiently that immunological memory does not recognize them. Occasionally, larger genetic changes occur, known as **antigenic shift**, where a complete genomic segment is altered as a result of a process known as **reassortment**. The influenza A virus that infects humans can also infect pigs and birds, where it may invade cells that also harbor porcine or avian influenza viruses. One or more of the genomic segments from these viruses may be packaged within a virion that otherwise comprises influenza A virus proteins and genomic segments. When these segments encode for the strongly immunogenic proteins such as HA or NA, the resulting new virus may be capable of replication in previously immune hosts, as the immune memory to influenza virus HA and NA molecules cannot recognize the new antigens. On several occasions during history, antigenic shift has led to severe influenza pandemics that have spread devastating illness across the globe.

K10 Virus vaccines

Key Notes

Vaccination	Vaccination is the use of a specific antigen (vaccine) to stimulate the immune response to protect against infection by the virus that possesses the antigen naturally. A good vaccine mimics the primary host response against infection with virus and protects against disease with minimal side effects. Vaccines can be infectious virus (usually modified (attenuated) to be less virulent) or inactivated (killed) virus. Subunit vaccines (viral proteins, capsids or DNA) are being developed although most are at an experimental stage.
Live (attenuated) vaccines	Attenuation is achieved by repeated passage of the virus in cell cultures or by deliberate genetic manipulation. Attenuated vaccines are effective in stimulating the full range of immune responses but they aim to produce little or no infectious symptoms. Examples are polio and the separate and combined measles, mumps, and rubella (MMR) vaccines.
Inactivated (killed) vaccines	Inactivation of viruses by chemical treatment produces (generally) less effective vaccines than attenuation, as they do not replicate within the host. However, they can still give significant protection against infection and include influenza and hepatitis A vaccines.
Subunit vaccines	These are mostly at the experimental stage and are composed of virion subunits, either individual proteins or assembled, empty capsids. They are used when attenuated or inactivated vaccines are ineffective or technologically not possible to manufacture. Those in clinical use at present are the HBV and HPV vaccines.
DNA vaccines	At present these are wholly experimental and are designed to induce immunity by the expression of viral proteins as a result of injecting the host with a plasmid containing viral DNA encoding the antigenic regions of the virus proteins.
Related topics	(F11) Plasmids (K9) Viruses and the immune system (K4) Virus proteins (K8) Virus infections

Vaccination

For many centuries it was observed that individuals surviving the common and often fatal disease **smallpox** were subsequently protected from reinfection with the disease. Indeed smallpox was so feared that the practice of deliberate infection with a view to providing such protection became widespread. The procedure (named **variolation** from variola major, the smallpox virus) applied pus from a lesion (pock) of an individual infected with what appeared to be a nonlethal strain of the virus onto a scratch in the skin. Variolation was itself often fatal if the **virulence** of the original infection was misjudged, and was succeeded in the late 18th century following the early experiments of **Edward Jenner** (and others) with **cowpox**. Jenner himself was the victim of near-fatal variolation as a boy; as a result he and his contemporaries pioneered a less dangerous procedure based on their observations that milkmaids who contracted the milder disease of cowpox, did not succumb to smallpox. Jenner introduced the cowpox virus (which gives mild, acute infections in humans) into some small boys whom he later challenged with virulent smallpox virus! Fortunately, Jenner's theory was correct and the boys were protected against the disease. The success of this procedure (named **vaccination** from the Latin *vacca* (cow) as a tribute to Jenner's work) was due to the fact that cowpox virus shares a number of antigens with smallpox virus but does not cause severe disease in humans. Cowpox replication had induced an immune response which, upon subsequent challenge with smallpox, was sufficient to prevent the closely related smallpox virus undergoing any significant infection. A hybrid cowpox virus of unknown origin, vaccinia virus (variola minor), was used by the World Health Organization (WHO) in a vaccination program that in 1977 succeeded in eradicating smallpox from the world. The WHO has an ambitious vaccination program that seeks to deliver vaccines to most children of the world and to eradicate many viruses in the near future. Since the launching of the polio vaccination program in 1998, the WHO has reported that the number of cases has fallen by 99% and the disease is now endemic in only four countries (WHO, 2008).

The majority of vaccines in clinical use today (see Table 1 for examples) are of two types, live (**attenuated**) or inactivated (**killed**) vaccines. More recent **subunit vaccines** contain selected virus proteins and many are being developed, along with several experimental **DNA vaccines**. Vaccines are administered by a number of routes – **intramuscular**, **intradermal**, **subcutaneous**, **intranasal** or **oral**.

Live (attenuated) vaccines

It is generally accepted that live vaccines are the preferred vaccines and examples are shown in Table 1. Initially these vaccines were derived by the continued passage of the **virulent** wild-type (natural) virus in cultured cells and selecting for **avirulent viruses** that would replicate *in vivo* but not cause disease. The replication of avirulent virus triggers the immune system (Section K9), which not only eliminates this virus from the host but also establishes immune memory that serves to resist the virulent wild-type virus. A vaccine created in this way during the 1950s, the Sabin poliovirus vaccine, is still in use today. Attenuation is the result of genetic mutation and of course viruses can therefore mutate again in such a way that they revert back to the wild-type genotype and phenotype (i.e. revert back to full virulence). Recent experiments have shown that, for poliovirus type 3, reversion to a virulent form involves only two amino acid changes. The problem of reversion has meant that this rather empirical approach of virus attenuation has now been replaced by defined genetic manipulation of viruses either deleting or mutating individual genes. The risk of reversion with these vaccines is minimal.

Live vaccines have distinct advantages over killed. They require only small amounts of input virus (which must replicate), they can induce local immunity (i.e. mucosal antibody), they may be given by the natural route of infection (for example, the Sabin vaccine is given orally), and they are usually less expensive. Drawbacks to their use include the problem of reversion to virulence as described above, a loss of effectiveness if not kept live (usually requiring refrigeration or freeze-drying), and poor effectiveness in individuals with other infections – for example, infections of the gut inhibit replication of

Table 1. Currently available viral vaccines

Vaccine	Vaccine type	Uses
Sabin polio	Attenuated trivalent	Routine childhood immunization
Salk polio	Killed whole virus	Immunization of immunocompromised people; universal childhood immunization in some developed countries
Measles	Attenuated (Schwarz, Moraten, others)	Routine childhood immunization
Rubella	Attenuated (RA 27/3)	Routine childhood immunization; adolescent girls; susceptible women of child-bearing age
Mumps	Attenuated (Urabe or Jeryl Lynn)	Routine childhood immunization
Measles, mumps, rubella (MMR)	Attenuated triple vaccine	Routine childhood immunization (1 or 2 doses)
Varicella-zoster	Attenuated (Oka)	Routine childhood immunization (USA); vaccination of susceptible people
Yellow fever	Attenuated (17D)	Routine or mass vaccination in endemic areas; travelers to endemic areas
Influenza	Killed whole virus or subunit (HA or NA)	Vaccination of high-risk individuals or the elderly
Rabies	Killed whole virus or recombinant G protein	Post-exposure vaccination; pre-exposure; veterinarians or travelers; subunit only used for animal vaccination
Hepatitis A	Killed whole virus	Pre-exposure vaccination: high-risk people and travelers
Japanese B encephalitis	Killed whole virus	Pre-exposure vaccination of travelers to endemic areas; routine or mass vaccination in endemic areas
Tick-borne encephalitis	Killed whole virus	Pre-exposure vaccination of travelers to endemic areas; routine or mass vaccination in endemic areas
Hepatitis B	Purified HBsAg from manipulated yeast, mammalian cells or from plasma	Routine childhood or adolescent immunization: immunization of high-risk adults; post-exposure immunization
Human papilloma	Subunit, VLPs	Immunization of young adult females

the poliovirus vaccine. Additionally, contamination with other adventitious agents can cause serious problems and live vaccines are of course limited to use in only immuno-competent individuals.

Inactivated (killed) vaccines

In addition to the Sabin attenuated poliovirus vaccine, an inactivated vaccine (the **Salk vaccine**) pioneered at the same time is still used in many countries today. Inactivation of the Salk vaccine typifies the approach to producing killed vaccines. High titers of wild-type virus are grown in cell culture and inactivated by the use of chemicals such as β-propiolactone or formaldehyde. The Salk vaccine is administered subcutaneously or intradermally to stimulate an immune response. With no subsequent virus replication (as seen with live vaccines) the immune response follows that of a primary response to an inert antigen but the necessary high levels of antibody are therefore stimulated only by multiple injections.

Killed vaccines have the advantage that they show no reversion to virulence, and are also stable and easy to store, often not requiring refrigeration. They can be administered to immune-suppressed patients and are usually not affected by interference from other pathogens, which is important in many developed countries. Killed vaccines, however, can be expensive and difficult to inactivate (no single infectious particle can remain in the preparation) and considered dangerous to manufacture (before inactivation). They do not stimulate the full scope of the immune system and require multiple (booster) doses. The recently derived **hepatitis A vaccine** is an inactivated vaccine, primarily produced because it was technically impossible to develop an attenuated hepatitis A virus capable of replication within the liver without disease. It is considered that the use of killed vaccines alone, which often allow some wild-type virus replication at local sites (e.g. poliovirus), will not be sufficient to totally eradicate viruses from the world.

Subunit vaccines

Attempts to design and produce **subunit** vaccines have been prompted by a number of factors. Many viruses do not replicate easily – or at all – in cell culture, while others are considered to be too dangerous for use as live or killed vaccines, HIV being a prime example. There is also a risk that some vaccines may become latent in the body and subsequently reactivate (e.g. HSV) and of course some may simply be poorly effective or have unacceptable side effects (e.g. influenza virus vaccines).

The basis of the approach is to develop a vaccine product that represents an incomplete composite of viral antigens that, when injected into the host, will induce protective immunity against the virulent virus. The proteins of choice for such a vaccine are usually capsid or envelope proteins; often the protein that binds to the cell receptor such as gp120 of HIV or hemagglutinin (HA) of influenza. These proteins can be extracted from the virion by chemical treatment (influenza HA), concentrated from the plasma of infected patients and inactivated (hepatitis B surface antigen, HBsAg), engineered by recombinant DNA technology (HBsAg) or synthesized as a peptide such as the experimental peptide vaccine for foot and mouth disease virus, comprising part of the VP1 capsid protein of the virion.

These vaccines may suffer from some of the drawbacks of inactivated whole virus vaccines in that they do not replicate in the host. This limits their efficacy of immune stimulation and there is much research directed to the improved presentation and immunogenicity of virus subunit vaccines. Antigens may be injected with **adjuvants**, immune-stimulating agents such as aluminium oxide or the cholera toxin B subunit protein. Another approach is to conjugate the vaccine as a chimeric protein linked to host cytokines or even antibodies that enhance the immune response. Another approach under study at present is to introduce the selected viral gene into a virus vector such as the vaccinia virus. Immunization of the host with the replicating **recombinant** virus expresses a range of antigens including the protein encoded by the cloned gene. The goal is the stimulation of immunity to both the viral vector and the virus represented by the cloned gene. No such vaccine is licensed for human use, but a recombinant vaccinia virus expressing the rabies virus G protein is used for veterinary purposes.

Few subunit vaccines are in general use at present. The HBsAg induces strong protective immune responses and for many years was prepared by purifying the protein from the serum of infected individuals. The risk of infection with this preparation is now eliminated as the current vaccine involves purification of the protein from genetically engineered yeast cell cultures. The vaccine is used routinely to induce protection against HBV in members of the medical and scientific professions. A recently approved subunit vaccine (Gardasil, manufactured by Merck) is formed by the assembly of the two HPV structural proteins, L1 and L2, into capsid-like structures (known as **virus-like particles**, VLPs). The proteins are expressed in isolation, meaning that the VLPs contain no genome and hence are not infectious. Gardasil combines proteins from four strains of HPV, providing a broad protection vaccine that is aimed at young women to immunize against infection with high-risk HPV-6, -11, -16, and -18, which are strongly associated with cervical cancers.

DNA vaccines

This is an innovative approach to vaccination that has grown out of the concept of gene therapy; the expression of novel proteins *within* a host, from extraneous gene sequences. DNA sequences encoding antigenic portions of viruses (protective antigens) are inserted into a plasmid (Section F11), which can be injected into the host as naked DNA free of protein or nucleoprotein complexes. The host cells can take up the DNA and may express the virally encoded proteins through their own transcription and translation procedures. Such vaccines theoretically stimulate strong cytotoxic T-cell responses as DNA derived from bacteria (i.e. the plasmid itself) promotes helper T-cell cytokine production. Experimental influenza vaccines in animals have proved effective in challenge experiments and clinical trials are underway with DNA vaccines for HIV, but as yet these are some way from approved clinical use.

K11 Antiviral chemotherapy

Key Notes

Historical perspective	Viruses utilize the biochemical machinery of their host cell for replication but many viruses also rely on their own proteins, distinct in form or function from host proteins, which allow selective drugs to be used as antiviral agents.
Modes of action	A large proportion of antiviral compounds are nucleoside analogs while others inhibit viral enzymes (e.g. proteases, polymerases, integrases) or processes such as penetration or egress of virus from the cell. Aciclovir, one of the most successful antivirals to date, is specifically phosphorylated by viral enzymes prior to selective incorporation into the growing virus DNA, where it acts as a chain terminator.
Effective dose, therapeutic index, and drug toxicity	Determining the antiviral effectiveness of a compound is achieved usually in tissue culture by assaying its ability to reduce viral infectivity. Ideally, strong activity will be achieved at a concentration low enough to cause minimal toxicity to uninfected cells.
Drug targeting, design, and clinical trials	Advancing knowledge of virus replication at a molecular level promotes a targeted approach to drug design. Determining the three-dimensional structure of viral proteins (particularly enzymes) has allowed drugs to be designed that react specifically with viral molecules. Clinical trials have several phases that assess the pharmacokinetics, pharmacology, metabolism, and clinical efficacy of the compound.
Drug resistance	Viruses can mutate to develop drug-resistant strains, a common problem affecting the long-term effectiveness of antiviral chemotherapy. Combinatorial approaches and strict adherence to drug-taking regimes can help to prevent this.
Related topics	(K7) Virus replication (K8) Virus infections

Historical perspective

Viruses are resistant to the action of antibiotics that target bacteria and other microbes. As obligate intracellular parasites with very restricted genetic coding capacity, viruses rely heavily on utilizing the metabolic machinery of the cell for their replication. However, replication involves (and is reliant on) one or more specific proteins encoded by viruses and this has led to the development of a number of successful antiviral agents that can, at some level, selectively inhibit viral protein functions and therefore curtail virus infection. Chemotherapeutic agents fall into three broad groups. **Virucides** include detergents and solvents that directly inactivate viruses. **Antivirals** (the main focus of this section) inhibit virus replication and aim to achieve this with little or no effect on host cell metabolism. Finally, **immunomodulating agents** (e.g. therapeutic interleukins and interferons) attempt to enhance the immune response against viruses to promote virus clearance.

Modes of action

Virus replication involves multiple stages, described in Section K7, and each stage (**attachment and penetration**, **uncoating of the nucleic acid**, **transcription and translation**, **genome replication**, and **release of mature progeny**) can be a target for antiviral drug intervention, as illustrated for HIV in Figure 1. To date, the most common target is viral nucleic acid metabolism by compounds known as **nucleoside analogs** or **non-nucleoside inhibitors**. Many of the first antivirals targeted against HIV acted against the viral polymerase (reverse transcriptase) such as azidothymidine (AZT) and lamivudine (also effective against HBV). A range of antivirals are effective towards different stages of the influenza virus replication cycle such as amantidine, which blocks uncoating by blocking the ion channel that forms by the viral M2 protein; ribavarin, which inhibits the viral RNA polymerase; and zanamivir, which prevents the final egress of the virus by inhibiting neuraminidase activity. Many viruses produce a virus-specific protease to process their proteins and this has been a valuable target for drug inhibition against HIV (e.g. ritonavir, saquinavir). A series of inhibitors of *Picornaviridae* act by directly binding to the virion capsid, blocking the interaction between the virion and the receptor on the cell surface that facilitates penetration and uncoating.

Perhaps the most successful antiviral compound is the nucleoside analog **acyclovir** (ACV – now named **aciclovir** and sold under the trade name **Zovirax**), with more than 40 million patients treated. The drug inhibits the replication of HSV and has been administered prophylactically with no ill effects for over 20 years to individuals to suppress recurrences of genital herpes. ACV is an analog of the natural nucleoside guanosine (Figure 2) and in this form is inactive and harmless to cells. Activation requires three enzymatic phosphorylation steps, to the active drug molecule ACV triphosphate. The **HSV thymidine kinase** enzyme can convert ACV to ACV monophosphate (ACV-MP), but ACV is a poor substrate for cellular enzymes, which do not perform this step. Hence ACV remains inactive in uninfected cells. By contrast, the conversion of ACV to ACV-MP in HSV-infected cells is rapidly followed by conversion to the triphosphate (ACV-TP) form through the action of cellular enzymes and it then enters the nucleotide pool within these cells. ACV-TP competes with guanosine triphosphate as a substrate for **HSV DNA polymerase** (cellular DNA polymerases are much less sensitive) and as the ACV-TP is linked into the growing chain of replicating viral DNA it forms a **chain terminator**. There is no 3′-OH group on the ACV sugar moiety to link with the next nucleotide residue and hence the growing chain of virus DNA can extend no further, terminating virus replication. Table 1 highlights the modes of action of a range of antiviral compounds.

Interferon is a natural human product (a cytokine) that acts on the surface of normal cells to render them immune to virus replication. The compound is used to treat hepatitis B and C infections (usually accompanied by ribavirin) but is toxic, the side effects mimicking the symptoms of influenza. A newer version, polyethylene glycol interferon (Peginterferon; Roche), is less toxic and more stable, allowing a reduced dosing regime of one subcutaneous injection per week, rather than three.

Effective dose, therapeutic index, and drug toxicity

Because of the intimate replication of viruses within cells, and their absolute reliance on cellular protein and energy metabolisms, the development of antiviral compounds faces the difficult challenge of **selective toxicity** – interference with viral replication needs to

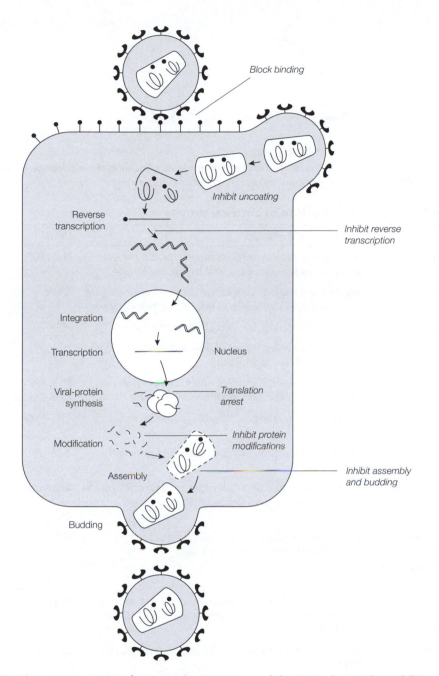

Figure 1. The various stages of virus replication. Antiviral drugs can be used to inhibit any of these steps. From Yarchoan R, Mitsuya H & Broder S (1988) *Sci. Am.* 259, 110–119.

be achieved without unacceptable damage to the host (uninfected cells). The activity of antiviral compounds is assessed at an early stage in their development, quantifying their ability to interfere with viral growth in tissue culture. The replication of the target virus is assayed by, for example, $TCID_{50}$ or plaque assay, comparing levels of infection within cells in the presence of various concentrations of the drug candidate (with drug-free

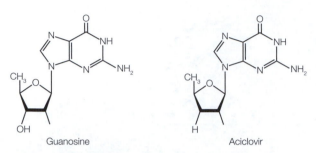

Guanosine   Aciclovir

Figure 2. Chemical structures of guanosine and aciclovir, revealing the exchange of a single hydroxyl (OH) group for hydrogen (H) within the ribose sugar ring.

Table 1. Mechanism of action of antiviral drugs

Drug	Mechanism(s) of action	Virus
Aciclovir	Nucleoside analog, inhibits nucleic acid synthesis: active form produced by viral thymidine kinase	HSV, VZV
Ganciclovir	Nucleoside analog, blocks nucleic acid synthesis by inhibiting viral thymidine kinase and other enzymes	CMV, VZV
Idoxuridine	Nucleoside analog, inhibits DNA synthesis	HSV
Vidarabine	Nucleoside analog, blocks DNA synthesis by inhibiting DNA polymerases	HSV
Ribavirin	Nucleoside analog, inhibits nucleic acid synthesis, possibly by inhibiting viral RNA polymerase	RSV, HCV, influenza, Lassa fever virus
Amantadine Rimantadine	Inhibits virus coating, blocks M2 H$^+$ ion channel	Influenza
Lamivudine	Reverse transcriptase inhibitor, inhibits genome replication	HIV, HBV
Tamiflu Zanamivir Oseltamivir	Inhibits neuraminidase cleavage of sialic acid, blocks release of virus from infected cells	Influenza
Azidothymidine Neviparine Dideoxyinosine Dideoxycytidine	Nucleoside analogs, inhibit nucleic acid synthesis by inhibition of reverse transcriptase	HIV
Indinavir Ritonavir Saquinavir	Anti-protease inhibitor, prevents cleavage of virus proteins, blocks infection of new cells and virion maturation	HIV
Foscarnet	Blocks protein synthesis by inhibition of RNA and DNA	CMV, HSV
Interferon	Activates uninfected cell enzyme pathway that interferes with new virus transcription	HBV, HCV

control infections alongside). The resulting level of virus infectivity is plotted against the drug concentration and the concentration of compound that reduces virus titer by 50% is expressed as the **effective dose 50** concentration (ED_{50}) (Figure 3); if achievable, the ED_{90} concentration is also determined. Tissue culture can also be used to assess the toxicity

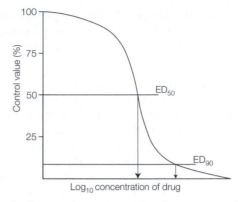

Figure 3. Determination of effective dose (ED_{50}, ED_{90}) for an antiviral drug in tissue culture.

of compounds, although this is often tested in animals and humans. Animals play a vital role in the study of toxicity and a number of statutory tests are carried out to determine the risks and side effects before new compounds enter clinical trials. Drug toxicity is expressed as the **selective index**, which determines the ratio between the concentration at which the drug inhibits cell proliferation or DNA synthesis (i.e. it is toxic to uninfected cells), and the concentration at which the drug inhibits virus replication. A high value indicates a selective drug with low toxicity, while a ratio approaching equivalence indicates that the compound is toxic. For compounds with a less than ideal selective index, a consideration of **benefit/risk** ratio can be appropriate, as side effects may be tolerated more if the risk from the disease is high (e.g. AIDS). Some antiviral agents have a level of toxicity which, while acceptable for some diseases (e.g. AIDS), would not be tolerated for less severe diseases.

Drug targeting, design, and clinical trials

Initially antiviral compounds were discovered in an empirical fashion; chemical synthesis of a wide range of compounds followed by testing of antiviral activity in cell culture. The increasing knowledge of the molecular basis of virus replication processes, the availability of complete **nucleotide sequences** of virus genomes and **three-dimensional protein structure** information means that compounds can now be more precisely designed to interact with specific targets and active sites on essential viral proteins. In practice the most useful targets are viral enzymes that have properties or activities different to those of the counterpart enzymes in host cells (e.g. **thymidine kinase**, **DNA polymerase**, **reverse transcriptase**, and **protease**). Through gene cloning, it is possible to express, purify, crystallize, and examine the three-dimensional structure of individual virus proteins and use this information to determine molecular structures predicted to interact with particular sites in the viral protein. In addition, features such as low toxicity and production costs are highly important, given the enormous cost of developing a potential candidate to clinical application and market. The majority of costs are split between the scientific research necessary to identify and test the compounds, and clinical trials that establish its effectiveness and safety *in vivo*. Clinical trials must pass, in order, through several tightly regulated phases.

Phase I involves administering the drug candidate to healthy human volunteers where studies on the pharmacokinetics, pharmacology, and metabolism of the compound are monitored.

In **Phase II** the compound is administered to disease patients, with a view to assessing data similar to those of Phase I, as the metabolism may be different compared with a healthy individual. Usually in excess of 100 patients are needed for these trials to give reliable data.

In **Phase III** the drug is usually tested for its clinical efficacy by comparing with **placebos** or existing drugs. The main aim is to determine the **benefit/risk ratio** for the therapeutic course, and this phase requires 100–1000 patients. In many, but not all Phase III trials, the drug and placebo are administered randomly, such that neither the test subjects nor the trial administrators know which individual has received which treatment. This is known as a double-blind trial and is important to safeguard the unbiased observation, recording, and interpretation of clinical outcomes. In certain circumstances Phase III trials are conducted without placebos (i.e. all patients receive the drug), as withholding of the drug and its potential therapeutic value would be unethical.

Phase IV studies are usually conducted following marketing approval and increased experience of treating patients, providing more information on safety and efficacy.

Sadly most potential antiviral compounds, although good inhibitors of virus protein activity in biochemical tests, or even of viral replication in cell culture, fail to pass successfully through all of the phases of clinical studies and are ultimately rejected.

Drug resistance

Probably the greatest efforts in recent years have been devoted to the development of antiretroviral agents to combat the replication of HIV. Effective treatments for this virus must involve combinations of drugs to attempt to overcome the problem of virus **resistance** to antivirals. The high rate of virus replication in a host means that there is a high rate at which mutations occur in the virus genome – compounded in RNA viruses by the lack of any proofreading activity in RNA polymerases. As a result, changes occur in the amino acid sequences of virus proteins, including those that serve as targets of antivirals (in a similar manner to the antigenic drift described for influenza virus in Section K9). Drug binding and activity are diminished or abolished and hence the virus adapts to become resistant to drugs that had previously been effective. The development of resistance to antivirals is a persistent problem with HIV, where the integration of the HIV genome into the chromosome of host cells makes for very long-term replication over months and years. HIV mutations that give rise to resistance to AZT occur with a very high frequency, and treatment of HIV infection requires combining drugs that target different virus enzymes, e.g. two nucleoside analogs combined with an anti-protease inhibitor. Such regimes have been highly effective in reducing virus loads and raising CD4 cell counts in HIV-infected individuals and in reducing the risk of viral resistance. Most important are measures that insure that patients adhere to the regime of drug taking, which in some cases is highly complex and involves a number of different drugs routinely. Due to genomic integration, antiviral agents are unlikely to eliminate HIV from the patient, but they are effective in reducing symptoms, delaying the progression of infection to AIDS by several years.

SECTION K – THE VIRUSES

K12 Plant viruses

Key Notes

Historical aspects Plant viruses were discovered over a century ago and have featured greatly in contributing to our knowledge of virus structure (e.g. tobacco mosaic virus, turnip yellow mosaic virus, and tomato bushy stunt virus).

Plant viruses Plant viruses are diverse in size, shape, and biochemistry and are present in many virus families, which also include animal viruses (e.g. *Rhabdoviridae, Reoviridae*).

Disease and pathology Plant viruses are responsible for extensive economic loss estimated at over $70 billion worldwide annually. They cause necrosis, wilting, mosaic formation, and other damage, which reduces yields and value of crops, etc.

Transmission, infection, and systemic spread Plant viruses are transmitted mainly by invertebrate animals (e.g. aphids, leafhoppers) but also 'manually' by contaminated implements or vertically through infected seeds. The viruses gain entry by penetrating cuticles of plant cells and need to spread systemically to cause disease (via plasmodesmata) or by cell division.

Control of plant virus disease Infected plants are virtually impossible to 'cure.' Control is by use of naturally resistant plant varieties or more recently genetically manufactured resistant varieties and by eradication of the transmission vector.

Viroids and satellites These are 'virus-like' infectious agents composed of small RNA genomes folded into a complex tertiary structure. Some encode a capsid protein, others a nonstructural protein, and yet others appear to have no coding capacity at all. Nevertheless, their potential for agricultural damage remains significant.

Related topics
(K1) Virus structure
(K2) Virus taxonomy
(K3) Virus genomes
(K4) Virus proteins
(K7) Virus replication

Historical aspects

The early studies on infectious diseases in the late 19th century quickly demonstrated that microscopic organisms were the cause of many illnesses, and a number of bacteria were isolated by filtering the lysate of diseased cells and examining them by cultivation and microscopy. However, several diseases failed to reveal an agent that could be grown or seen with the microscopes of the day, and yet diseased cell lysates could faithfully transmit the symptoms to new hosts. Many of these first steps into the discovery of viruses (the word 'virus' comes from the Latin for 'poison') were done using plant viruses and our modern understanding of virus structure, replication, assembly, and even the nature of virus genetics was derived from studies of the tobacco mosaic virus (TMV). The virus, which causes mosaic patterning of the leaves of the tobacco plant (see Figure 2), was first described in 1892. TMV was the first virus to be observed under the electron microscope, the first to demonstrate the intrinsic infectivity of a naked viral genome, and the first virus to be assembled *in vitro* from purified preparations of viral RNA and coat protein molecules. Studies with turnip yellow mosaic virus and tomato bushy stunt virus revealed the morphological details of icosahedral viruses. Ironically, in more recent times the molecular study of plant viruses has trailed behind that of animal viruses, as plant cells are more difficult to culture than animal cells.

Plant viruses

Plant viruses have traditionally been named with a combination of the host plant and the type of disease produced (e.g. tobacco mosaic, turnip yellow mosaic, tomato bushy stunt, cauliflower mosaic, tomato spotted wilt). Plant viruses are diverse in their morphology, nucleic acid composition, and replication. It is beyond the scope of this book to detail all viruses but Figure 1 highlights the various morphologies and groupings of plant viruses. They may be **dsDNA** (*Badnaviridae*, e.g. rice tungrobacilliform virus), **ssDNA** (*Geminiviridae*, e.g. maize streak virus), **dsRNA** (*Reoviridae*, e.g. clover wound tumor virus) or **ssRNA** viruses of positive (*Comoviridae*, e.g. tobacco ringspot virus) or negative (*Rhabdoviridae*, e.g. potato yellow dwarf virus) sense. Like animal viruses, plant virus genomes can be segmented or nonsegmented, and enclosed in helical or icosahedral capsids, although several have amorphous capsids with no defined shape. Envelopes are found only within the *Bunyaviridae* and *Rhabdoviridae* families and there are no Class VI plant viruses.

Disease and pathology

Plant virus infections can affect photosynthesis, respiration, nutrient availability, and hormonal regulation of growth, with morphological outcomes that include mosaic formation, yellowing, molting or other color disfiguration, stunting, wilting, and necrosis. These conditions, resulting from cell and tissue damage, can significantly reduce the yield and commercial value of crops and the impact of plant viruses globally is estimated to cost $70 billion annually. Plants and crops of all types can be affected; for example, swollen shoot disease devastates cocoa trees and is a serious and ongoing threat to the cocoa bean industry across large parts of western Africa. Rice tungro disease is the most important rice crop disease in Southeast Asia and is caused by a co-infection of two viruses. Disease symptoms are due to infection by rice tungro bacilliform virus (Family *Caulomoviridae*), while transmission between host plants relies on rice tungro spherical virus (Family *Sequaviridae*) and an insect vector, the leafhopper. Despite such obvious signs of infection, the way in which plant viruses cause cell damage has been difficult to understand. Plant viruses code for very few proteins, but as is the case with many viral proteins, they are usually multifunctional. Many of the nonessential (and as yet unknown) functions of plant virus proteins may interact detrimentally with host proteins, resulting in the observed pathologies.

Transmission, infection, and systemic spread

The stem and leaf surfaces of plants can in many ways be compared to the human skin; they do not have receptors for virus attachment and have a protective function. For plant viruses to be infectious they must penetrate this barrier and gain access to the metabolic machinery of the plant cell; hence it is often through wounds that viruses enter plant tissues. Many plant viruses depend on a vector for their transmission, commonly

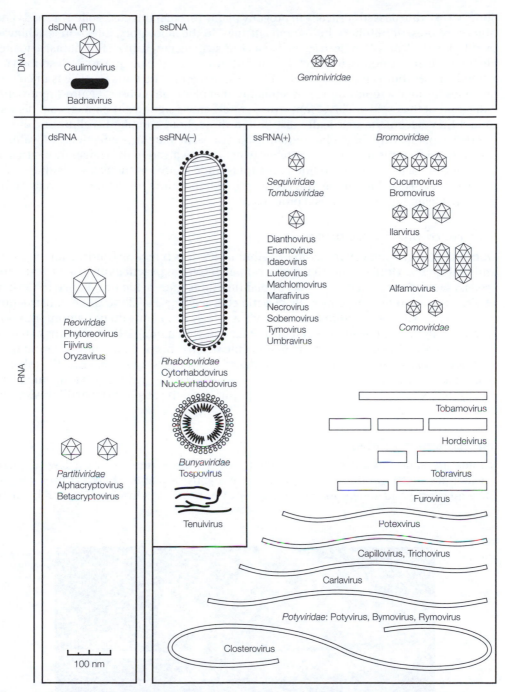

Figure 1. Families and genera of plant viruses. Reproduced from *Eighth Report of the International Committee on Taxonomy of Viruses*. Springer-Verlag.

invertebrates such as **aphids**, **leafhoppers**, **mealy bugs**, **whiteflies**, **thrips**, **mites**, and **soil nematodes**. A few viruses can be transmitted as a result of contact with **contaminated implements** (e.g. spades, hoes) or by direct contact between neighboring plants.

Some viruses are passed vertically to progeny plants through infected pollen or seeds and others are present in tubers, bulbs, and cuttings. In the laboratory, infection is achieved by rubbing the leaf with a cloth soaked in a virus suspension, using a fine abrasive to create the necessary wounds. In effect this procedure mimics the plaque assay described for animal viruses, but is not as accurate (Figure 2). On entry the virus particle is uncoated and goes through a replication cycle similar to that of animal viruses. Figure 3 represents a diagram of the cycle of TMV. Plant viruses rarely cause significant damage and disease unless they become **systemically** distributed throughout the plant. Failure for this to happen explains why some plants are resistant to particular virus infections. Movement is facilitated by virus movement proteins, which are involved in the transport of virus or virus nucleic acid through the fine pores (**plasmodesmata**) in the cell walls that interconnect plant cells. Movement also occurs through the companion and sieve cells of the phloem, facilitated by viral capsid proteins.

Control of plant virus disease

Once infected, it is almost impossible to eliminate a virus infection from a plant by use of antiviral agents. Until recently a reliance on **horticultural practice** (the use of virus-free seeds, eradication of vectors, choice of planting time, etc.) helped to reduce the extent of virus infection but more recently **genetic engineering** has allowed the construction of plants that show **natural resistance** to virus infections. Plants do not mount immune responses but there is evidence that prior infection with a nonpathogenic virus can protect plants from infection by a more pathogenic strain. This phenomenon, referred to as **pathogen-derived resistance**, has been induced deliberately by engineering the tobacco plant genome to contain the coat protein gene of TMV. The resulting **transgenic** plant was protected from challenge by the wild type but as yet the mechanism of this resistance is not clearly understood.

Viroids and satellites

Viroids are small, naked circular infectious **ssRNA molecules** of just a few hundred nucleotides that form the smallest known pathogens of plants. Unlike viruses, viroids have no capsid and appear to encode no mRNA or proteins. Viroids are spread by plant

Figure 2. Focal assay of tobacco mosaic virus on the leaf of a plant. From Dimmock N & Primrose SB (1994) *Introduction to Modern Virology*, 4th ed. With permission from Blackwell Science.

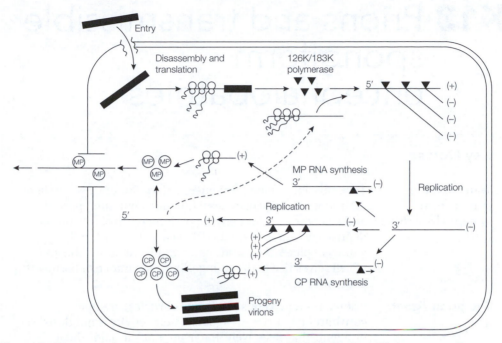

Figure 3. Diagram of stages of tobacco mosaic virus (TMV) infection. All the events shown are presumed to occur in the cytoplasm of infected cells. MP, movement protein; CP, coat protein.

propagation (e.g. cuttings and tubers), through seeds, and by manual mishandling with contaminated implements. Approximately 25 viroids have been identified to date and they can cause serious plant diseases. The first to be examined in detail was potato spindle tuber viroid (PSTVd), which is responsible for significant annual loss to the potato industry. The RNA is a covalently closed circle and ranges in size from 246 to 357 nucleotides in length. The RNA has a complex secondary and tertiary structure, which gives it a rod-like shape and resistance to nucleases. The RNA does not have a characteristic open reading frame and so does not act as mRNA. How this piece of RNA causes disease is largely unknown, but it is thought that it may interfere with either the action or selectivity of the host cell RNA polymerase II enzyme, or with RNA splicing.

Satellites are dependent pathogens that require host cell or helper virus proteins for full replication. They are of two types: **satellite viruses** encode their own capsid protein while **satellite nucleic acids** are enclosed in the capsid of a helper virus infecting the same host cell. The hepatitis D virus (HDV) is a satellite nucleic acid comprising a circular RNA molecule of 1700 nucleotides, and is the smallest known pathogen of humans. HDV only replicates in cells infected with HBV and as a result its transmission and treatment are linked to that of its helper virus, HBV. The HDV genome encodes two proteins, known as delta antigens, and their presence appears to contribute to the replication of both satellite and helper, often leading to severe clinical outcome.

K13 Prions and transmissible spongiform encephalopathies

Key Notes

Transmissible spongiform encephalopathies	Transmissible spongiform encephalopathies (TSEs) are fatal neurodegenerative diseases with long incubation periods and a very specific underlying pathology. Insoluble protein deposits (plaques or amyloids) accumulate within the kidneys, spleen, liver, and significantly the brain. The loss of cells gives rise to the sponge-like appearance of tissues at post mortem.
The prion agent	Prions (proteinaceous infectious particles) are not conventional infectious agents. Their activity is not destroyed by agents that selectively inactivate nucleic acids (heat, ultraviolet light, ionizing radiation, nuclease enzymes) but is sensitive to urea, SDS, phenol, and other protein-denaturing chemicals. Prions are therefore thought to possess no nucleic acid but consist solely of protein and the pathological agent appears to be a modified cellular protein that folds incorrectly.
Animal TSEs	Scrapie, the most extensively studied TSE, causes a neurological disease in sheep and goats and is probably transmitted orally and vertically. Bovine spongiform encephalopathy (BSE) appeared in 1986 and it has been suggested that this was due to extensive feeding of cattle with contaminated bone meal. TSEs show strain variation and animals show different genetic susceptibilities.
Human TSEs	Creutzfeldt-Jakob disease (CJD) and Gerstmann-Straussler-Scheinker disease (GSS) are sporadic and familial diseases that affect approximately 1 in 10^6 individuals annually worldwide. CJD can also be acquired iatrogenically. A TSE known as Kuru resulted from the cannibalistic ritual of the Fore people of New Guinea. The disease, transmitted orally, had an incubation period of up to 30 years but is now extinct. BSE has been identified as the prion agent associated with a new variant CJD (vCJD) that has claimed 118 known victims since 1995.

Transmissible spongiform encephalopathies

The **transmissible spongiform encephalopathies** (TSEs) are a unique suite of progressive neurodegenerative diseases that show both infectious and hereditable transmission characteristics. TSEs have been observed in various mammals including humans, and as yet no treatment is available. The pathogenic agent replicates in a number of tissues including spleen and liver, before appearing in high concentrations in the brain and central nervous system (CNS) towards the terminal stages of the disease. Ultimately, severe degeneration of the brain and spinal cord lead to clinical signs and the condition is then invariably fatal within several months. Pathologically the disease is characterized by the appearance of abnormal protein deposits (**amyloids**) in the kidney, spleen, liver, and brain. Amyloids arise from the accumulation of various proteins, which take the form of **plaques** or **fibrils**. Amyloidosis is also a characteristic feature of Alzheimer's disease, and although this condition is not transmissible, a prion-based pathology has not been totally eliminated. The protein deposits are insoluble and hence cytotoxic, leading to cell death and a sponge-like appearance of the brain, which gave the spongiform encephalopathies their name; microscopic holes can be visualized in thin sections of brain tissue taken at post mortem. There is no conventional immune response to the agent, although the immune system plays an important part in the development of the disease before the agent gets into the CNS. Diagnosis of TSEs is difficult and for some time could only be made by immunohistochemical staining of the prion protein in tissue taken at post mortem, although there have been some examples of diagnosis based on detection of the prion protein in cerebrospinal fluid (CSF) or tonsillar tissue.

The prion agent

There has been much speculation as to the molecular nature of the infectious agents responsible for these disease conditions. In 1972 Stanley Prusiner proposed that these agents were not viruses but a unique class of infectious agent totally free of nucleic acid and consisting solely of protein. Prusiner coined the term **prion** (from **proteinaceous infectious particle**) to describe the agent and was awarded the Nobel Prize in 1997 for his work investigating prions and TSEs. Evidence for the proteinaceous, nucleic acid-free nature of the agent has come from a number of experiments in which the chemical and physical nature of the infectious agents has been examined. They are resistant to heat inactivation at temperatures that would readily destroy DNA or RNA and are resistant to both ultraviolet light and ionizing radiation, treatments that normally damage microbial genomes. No susceptibility to DNAse, RNAse or enzymatic hydrolysis is detectable but the agents are sensitive to treatment with urea, SDS, phenol, and other protein-denaturing chemicals. These characteristics indicate that the prion has, as proposed, a proteinaceous character lacking an infectious or transmissible nucleic acid.

Of course all proteins are encoded by nucleic acids and the prion protein (**PrP**) is no exception, being encoded by a cellular gene, the *Prnp* gene. Found in all mammalian and bird genomes examined to date the protein is a glycoprotein of 208 amino acids that cycles between the endosomal and plasma membranes of various cell types, including neuronal cells. The biological function of the cellular protein, **PrP^c**, is unknown although evidence suggests a role in transmembrane transport or signaling. Mice lacking this protein develop normally, suggesting redundancy in its function and critically such mice are also resistant to infection by disease-forming prion proteins (designated **PrP^{sc}** taken from the disease scrapie), indicating the requirement for the cellular protein in disease. How do PrP^c and PrP^{sc} differ and how is the disease state achieved? The answer lies in the different conformations of the two proteins; the PrP^c has a high proportion of alpha-helical structure but can undergo a conformational change to the pathogenic PrP^{sc} form, which has reduced helical structure as regions refold into beta-sheets (Figure 1). This altered form of the protein shows increased resistance to proteases and reduced plasma membrane expression within the cell and, more importantly, increased self-aggregation, forming long fibrils of

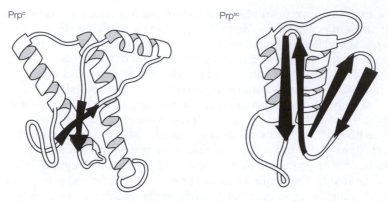

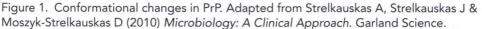

Figure 1. Conformational changes in PrP. Adapted from Strelkauskas A, Strelkauskas J & Moszyk-Strelkauskas D (2010) *Microbiology: A Clinical Approach*. Garland Science.

protein that eventually become insoluble, causing the cellular damage. How PrP^c converts to PrP^{sc} and whether the process is reversible is unclear, but there seems strong evidence that the presence of PrP^{sc} ultimately converts more and more PrP^c until the cell is destroyed. The concept of prion protein conformational conversion supports the three disease routes observed with TSEs. Inheritance of a mutated *Prnp* gene can lead to familial (hereditary) disease, a spontaneous transcription error during gene expression would give sporadic disease and acquisition of the PrP^{sc} isoform initiates transmissible disease.

As with conventional microbes there are different strains of prion that appear to 'breed true' in terms of the characteristic symptoms they produce following infection of new hosts. For example, scrapie can be transmitted to both sheep and goats where, despite variations in the amino acid sequence of their PrP proteins, it displays identical disease characteristics. Likewise some strains of BSE can be propagated by several animal species (each with their own different prion protein) and the three-dimensional conformation of the refolded PrP^{sc} is identical regardless of the host species. These findings suggest that the final structure of PrP^{sc} and its disease characteristics are independent of the host. By contrast, transmission of scrapie (via contaminated feed) into cattle gives distinct characteristics, and confirms that the contribution of infecting and host prion proteins is not yet understood.

In humans there are two common versions (polymorphisms) of the prion protein that differ in a single amino acid (**valine or methionine at coding position 129**). There are also rare mutant forms, many of which are associated with inherited susceptibility to prion disease. The polymorphic variation at position 129 does not itself lead to disease, but influences the clinical characteristics of a disease-promoting mutation that can occur at position 178. Mutation of the normal aspartate to asparagine at this position leads to CJD if there is valine at position 129, but to fatal familial insomnia (FFI) if the individual possesses methionine at amino acid 129. Sheep likewise have several different forms of the prion protein linked to susceptibility, whereas cattle have two forms that do not appear to be associated with susceptibility to BSE.

Animal TSEs

Examples of TSEs in animals include **scrapie** in sheep and goats, **transmissible mink encephalopathy** (TME), **feline spongiform encephalopathy** (FSE), and bovine spongiform encephalopathy (BSE, labeled 'mad cow disease' by the media). **Scrapie** has been recognized as a distinct infection in sheep for over 250 years. A major investigation into its etiology followed the vaccination of sheep for louping-ill virus with formalin-treated extracts of ovine lymphoid tissue, unknowingly contaminated with scrapie prions. Two years later, more than 1500 sheep developed scrapie from this vaccine. The scrapie agent has been extensively studied and experimentally transmitted to a range of laboratory animals, e.g. **mice and hamsters**. Infected sheep show severe and progressive neurological symptoms, such as abnormal gait. The name derives from the fact that sheep with the disorder repeatedly scrape themselves against fences and posts. The natural mode of transmission between sheep is unclear although it is readily communicable and the placenta has been implicated as a source of prions, which could account for horizontal spread within flocks. In Iceland scrapie-infected flocks of sheep were destroyed and the pastures left vacant for several years. However, reintroduction of sheep from flocks known to be free of scrapie for many years eventually resulted in scrapie, suggesting that it was able to retain infectivity in the soil for several years. Sheep have also been infected by feedstuff contaminated with BSE, to which they are susceptible.

Bovine spongiform encephalopathy (**BSE**) appeared in Great Britain in 1986 as a previously unknown disease. Affected cattle showed altered behavior and a staggering gait and post-mortem investigation revealed protease-resistant PrP^{sc} in the brains of the cattle and the typical spongiform pathology. The origins of BSE suggest that cattle were infected with meat and bone meal (MBM) given as a nutritional supplement, contaminated with an unknown bovine TSE (research has shown that the BSE and scrapie prions are different, hence scrapie is not the ancestor of BSE). MBM was initially prepared by rendering the offal of sheep and cattle using a process that involved steam treatment and hydrocarbon solvent extraction but in the late 1970s the solvent procedure was eliminated from the process, resulting in high concentrations of fat in the MBM. It is postulated that this high fat content protected prions in the sheep or beef offal from being completely inactivated by the steam. Since 1988 the practice of using dietary protein supplements for domestic animals derived from rendered sheep or cattle offal has been forbidden in the UK. Statistics argue that this food ban has been effective in controlling the epidemic (Figure 2) and British beef has been reported as being free of BSE (June 2006).

Human TSEs

Prior to the mid 1990s there were four forms of human TSE identified; **Creutzfeldt-Jakob disease** (CJD), the rare, inherited **fatal familial insomnia** (FFI), **Gerstmann-Straussler-Scheinker disease** (GSS), and the now extinct **Kuru**, which was the first human TSE to undergo extensive investigation. The first cases were recorded in the 1950s and occurred in male and female adolescents and adult women members of the **Fore people** in the highlands of New Guinea and at one point was the leading cause of death of women of the tribes. The Fore people practiced **ritual cannibalism** as a rite of mourning for their dead, the women and children but not the adult men taking part in the ceremony. The disease demonstrated progressive loss of voluntary neuronal control followed by death within 1 year after the onset of symptoms. As the cannibalistic ritual ceased so did transmission of the disease and although some individuals developed symptoms after an incubation period of up to 30 years (usually it was much shorter), the disease is now eradicated. Kuru demonstrated clearly that a human TSE could be acquired via the oral route.

Patients with CJD present with a progressive subacute or chronic decline in cognitive or motor function associated with recurrent seizures. CJD is rare, with an annual worldwide incidence of approximately **one per million population** and can be **familial** (inherited;

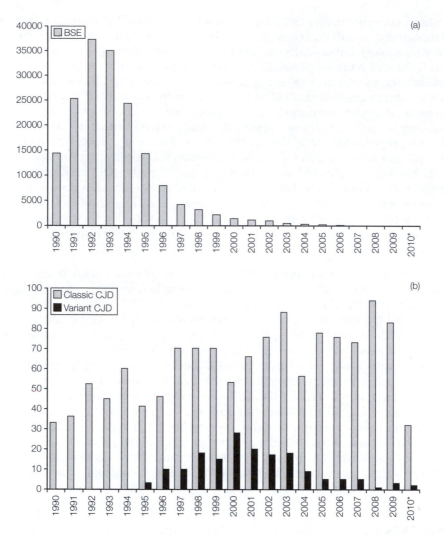

Figure 2. Reported incidence of (a) BSE and (b) vCJD in the UK between 1990 and September 2010 (i.e. partial data for 2010* only). Data obtained from the World Organization for Animal Health and the National CJD Surveillance Unit, University of Edinburgh. For comparison purposes, classic CJD includes the combined annual figures for familial, iatrogenic, and sporadic CJD shown in chart b.

10% of cases), or acquired **sporadically** or **iatrogenically** (cases have been recorded following corneal grafts, the administration of pituitary hormones or the use of electrodes contaminated with PrP^{sc}).

In April 1996 a new variant of CJD (**vCJD**) was described in the UK. The disease has features that distinguish it from other forms of CJD including an **early age of onset** (average 27 years as opposed to 65 for CJD), suggesting a shorter incubation period, a **prolonged period of illness** (average 13 months as opposed to 3 months for CJD), and a **psychiatric presentation** as opposed to **neurological symptoms**, with the **absence of the typical electroencephalography (EEG) appearances of CJD**. In terms of pathology vCJD resembles Kuru, particularly the type of amyloid plaque formation. Evidence has accumulated that the BSE prion has been transferred (most likely via the oral route) to humans and is responsible for vCJD. The source of the infected material is unknown, although BSE-contaminated cattle products are a likely candidate. In 1989 the human consumption of specified bovine offal (brain, spleen, thymus, tonsil, and gut) was prohibited in the UK and in 1996 the ban was extended to sheep offal. Furthermore, the brain and CNS are now removed from all cattle at abattoirs before the distribution of meat.

Up to September 2010, 118 cases of deaths as a result of vCJD have been confirmed, with a further 51 deaths likely to be due to this TSE (post-mortem confirmation pending). There is strong epidemiological evidence that the epidemic has peaked (Figure 2) but it is worth considering that all but one case to date have occurred in individuals homozygous for valine at the position 129 polymorphism described above. Since all forms of the polymorphism support BSE in a mouse system, it is possible that there is an as yet symptom-free cohort of infected individuals that are still incubating vCJD. Although it is considered highly unlikely, the possibility of human-to-human transmission remains.

Further reading

General reading:

Madigan, M.T., Martinko, J.M., Stahl D.A. and Clark D.P. (2011) *Brock Biology of Microorganisms*, 13th Edn. Pearson Education, Upper Saddle River, NJ.

Willey, J., Sherwood, L. and Woolverton, C. (2011) *Prescott's Microbiology*, 8th Edn. McGraw-Hill Higher Education, Columbus, OH.

Singleton, P. and Sainsbury, D. (2006) *Dictionary of Microbiology and Molecular Biology*, 3rd Edn. John Wiley & Sons, New York.

Tortora, G.J., Funke, B.R. and Case, C.L. (2009) *Microbiology: An Introduction*, 10th Edn. Pearson Education, Upper Saddle River, NJ.

Advanced Reading:

Section B Systematics

Feselstein, J. (2003) *Inferring Phylogenies*. Sinauer Associates, New York.

Priest, F. and Goodfellow, M. (2000) *Applied Microbial Systematics*. Springer Verlag, Berlin.

Stackebrandt, E. (2006) *Molecular Identification, Systematics and Population Structure of Prokaryotes*. Springer Verlag, Berlin.

Section C Microbiology

Alberts, B., Johnson, A., Lewis, J., Raff, M., Roberts, K. and Walter, P. (2008) *Molecular Biology of the Cell*, 4th Edn. Garland Science, New York.

Atlas, R.M. and Bartha, R. (1997) *Microbial Ecology: Fundamentals and Applications*, 9th Edn. Benjamin-Cummings Publishing Co., Redwood City, CA.

Cappucino, T.G. and Sherman, N. (2010) *Microbiology: A Laboratory Manual*, 4th Edn. Pearson Education, Upper Saddle River, NJ.

Hames, B.D. and Cooper N. (2011) *Instant Notes in Biochemistry*, 4th Edn. Taylor & Francis, Oxford.

Irving, W., Boswell, T. and Ala'Aldeen D. (2005) *Instant Notes in Medical Microbiology*, Taylor & Francis, Oxford.

Isaac, S. and Jennings, D. (1995) *Microbial Culture*. Garland Science, New York.

Maier, R.M., Pepper I.L. and Gerba, C.P. (2008) *Environmental Microbiology*. 2nd Edn. Academic Press, Amsterdam

Section D Microbial Growth

Panikov, N.S. (1995) *Microbial Growth Kinetics*. Springer Verlag, Berlin.

Smith, H.L. and Waltman, P. (1995) *The Theory of the Chemostat*. Cambridge University Press, Cambridge.

Section E Microbial Metabolism

Hames, B.D. and Cooper N. (2011) *Instant Notes in Biochemistry*, 4th Edn. Taylor and Francis, Oxford.

Nelson, D.L. and Cox, M.M. (2008) *Lehninger Principles of Biochemistry*, 4th Edn. Palgrave Macmillan, Basingstoke.

Nicholls, D.G. and Ferguson, S.J. (2002) *Bioenergetics 3*, 3rd Edn. Academic Press, Amsterdam.

Section F Prokaryotic DNA and RNA Metabolism

Abedon, S.T. and Lane-Calender, R. (2005) *The Bacteriophages*. Oxford University Press, Oxford.

Brown, T. (2005) *Genomes 3*, 3rd Edn. Taylor & Francis, Oxford.

Latchman, D. (2006) *Gene Regulation*. Taylor & Francis, Oxford.

Lindahl, T.R. and West S.C. (1995) *DNA Repair and Recombination*. Royal Society, London.

Turner, P.C., McLennan, A.G., Bates, A.D. and White, M.R.H. (2005) *Instant Notes in Molecular Biology*, 3rd Edn. Taylor & Francis, Oxford.

Section G Industrial Microbiology

Garbutt, J. (1997) *Essentials of Food Microbiology*. 2nd Edn. Hodder Arnold, London.

Glazer, A.N. and Nikaido, H. (2007) *Microbial Biotechnology: Fundamentals of Applied Microbiology*. Cambridge University Press, Cambridge.

Waites, M.J., Morgan, N.L., Rockey, J.S. and Higton, G. (2001) *Industrial Microbiology: An Introduction*. Blackwell Scientific Publishing, Oxford.

Section H Eukaryotic Microbes: An Overview

http://www.tolweb.org

Section I The Fungi and Related Phyla

Hibbert D.S. *et al.* (2007) A higher level phylogenetic classification of the fungi. *Mycological Research* 111: 509–547.

Alexopoulos C.J., Mims C.W. and Blackwell M. (1996) *Introductory Mycology*, 4th Edn. John Wiley & Sons, New York.

Deacon, J.W. (2006) *Fungal Biology*, 4th Edn. John Wiley & Sons, New York.

Section J Archaeplastida, Excavata, Chromalveolata, and Amoebozoa

Edward Lee, R. (2008) *Phycology*, 4th Edn. Cambridge: Cambridge University Press.

Lynn, D.H. (2008) *The Ciliated Protozoa: Characterization, Classification, and Guide to the Literature*, 3rd Edn. Springer Verlag, Berlin.

http://www.biani.inige.ch/msg/Amoeboids/Amobozoa.html

Section K The Viruses

Cann, A.J. (2005) *Principles of Molecular Virology*. Academic Press, London.

Collier, L. and Oxford, J. (2006) *Human Virology*, 3rd Edn. Oxford University Press, Oxford.

Digard, P., Nash, A. and Kandall, R. (2005) *Molecular Pathogenesis of Viral Infections, 64th Symposium of the Society for General Microbiology*. Cambridge University Press, Cambridge.

Dimmock, N., Easton, A. and Leppard, K. (2001) *Introduction to Modern Virology*. Blackwell Scientific Publishing, Oxford.

Strauss, J. and Strauss, E. (2002) *Viruses and Human Disease*. Academic Press, London.

Wagner, E. and Hewlett, M.J. (2004) *Basic Virology*, 2nd Edn. Blackwell Scientific Publishing, Oxford.

Abbreviations

A	adenine	FAD	flavin adenine dinucleotide (oxidized)
ABC	ATP-binding cassette		
ACP	acyl carrier protein	FADH2	flavin adenine dinucleotide (reduced)
ADP	adenosine 5′-diphosphate		
Ala	alanine	FAMEs	fatty acid methyl ester analysis
AMP	adenosine 5′-monophosphate	FMN	flavin mononucleolides
A-site	amino-acyl site (ribosome)	G	guanine
ATP	adenosine 5′-triphosphate	GC	gas chromatography
ATPase	ATP synthase	G-phase	gap phase (bacterial cell cycle)
BHK	baby hamster kidney	GTP	guanosine 5′-triphosphate
Bp	base pair	HA	hemagglutination
C	cytosine	Hfr	high frequency recombination
cAMP	cyclic adenosine 5′-monophosphate	HIV	human immunodeficiency virus
		HMP	hexose monophosphate pathway
CAP	catabolite activator protein	HSV	herpes simplex virus
CAT	chloramphenicol acetyl transferase	I	inosine
		ICNV	International Committee on Nomenclature of Viruses
cfu	colony-forming unit		
CMV	cytomegalovirus	Ig	immunoglobulin
CNS	central nervous system	IHF	integration host factor
CoA	coenzyme A	Inc group	incompatible group (of plasmids)
CPE	cytopathic effect	IS	insertion sequence
C-phase	Chromosome replication phase (bacterial cell cycle)	Kb	kilobase
		KDO	2-keto-2-deoxyoctonate
CRP	cAMP receptor protein	KDPE	2-keto-2-deoxy-6-phosphogluconate
CTL	cytotoxic T lymphocyte		
Da	Dalton	Lac	lactose
d-Ala	D-alanine	LBP	luciferin-binding protein
DAP	meso-diaminopimelic acid	LPS	lipopolysaccharide
DGGE	denaturing gradient gel electrophoresis	m.o.i.	multiplicity of infection
		MAC	membrane-attack complex
D-Glu	D-glutamic acid	MCP	methyl-accepting chemotaxis protein
DHA	dihydroxyacetone		
DMSO	dimethylsulfoxide	MEM	minimal essential medium
DNA	deoxyribonucleic acid	MHC	major histocompatibility complex
dNTP	deoxyribonucleoside triphosphate		
		MPN	most probable number
DOM	dissolved organic matter	mRNA	messenger ribonucleic acid
D-phase	division phase (bacterial cell cycle)	MRSA	methicillin-resistant *Staphylococcus aureus*
Ds	double-stranded	MTOC	microtubule organizing centre
EF	elongation factor	NAD+	nicotinamide adenine dinucleotide (oxidized form)
EM	electron microscopy		
ER	endoplasmic reticulum	NADH	nicotinamide adenine dinucleotide (reduced form)
FACS	fluorescence activated cell sorting		

NADP+	nicotinamide adenine dinucleotide phosphate (oxidized form)	PSI and II	photosystems I and II
		P-site	peptidyl site (ribosome)
NADPH	nicotinamide adenine dinucleotide phosphate (reduced form)	qPCR	quantitative PCR
		R	resistance (plasmid)
		RBC	red blood cell
NAG	N-acetyl glucosamine	redox	reduction-oxidation
NAM	N-acetyl muramic acid	RER	rough endoplasmic reticulum
NB	nutrient broth	RNA	ribonucleic acid
NTP	ribonucleoside triphosphate	rRNA	ribosomal RNA
O	operator	RT-qPCR	reverse transcriptase linked quantitative PCR
OD	optical density		
Omp	outer membrane protein	RuBisCo	ribulose bisphosphate carboxylase
P	promoter		
PCBs	polychlorinated biphenyls	S	Svedberg coefficient
PCR	polymerase chain reaction	snRNA	small nuclear ribonucleic acid
PEP	phosphoenol pyruvate	SPB	spindle pole bodies
Pfu	plaque-forming unit	ss	single-stranded
Pfu	*Pyrococcus furiosus* polymerase	SSCP	single-stranded conformation polymorphism
PHB	poly-β-hydroxybutyrate		
Phe	phenylalanine	T	thymine
Pi	inorganic phosphate	TCA	tricarboxylic acid
PLFA	phospholipid-linked fatty acid analysis	TCID	tissue culture infective dose
		tRNA	transfer RNA
PMF	proton motive force	Trp	tryptophan
PMN	polymorphonucleocyte	TSB	tryptone soya broth
PPi	inorganic pyrophosphate	U	uracil
PPP	pentose phosphate pathway	UDP	uridine diphosphate
PS	photosystem	UDPG	uridine disphosphate glucose
		UV	ultraviolet light

Index